COMPLEX PLANETARY SYSTEMS

IAU SYMPOSIUM No. 310

COVER ILLUSTRATION:

IAU Symposium 310 takes a broad look at the complexity of planetary systems, in terms of the formation and dynamical evolution of planets, their satellites, minor bodies and space debris, as well as to the habitability of exoplanets, in order to understand and model their physical processes. The main topics covered are diverse, including: studies of the rotation of planets and satellites, including their internal structures; the long term evolution of space debris and satellites; planetary and satellite migration mechanisms; and the role of the Yarkovsky effect on the evolution of the rotating small bodies. Intended for researchers and advanced students studying complex planetary systems, IAU S310 appeals to non-specialists interested in problems such as the habitability of exoplanets, planetary migration in the early Solar System, or the determination of chaotic orbits. This volume provides a valuable insight into the state-of-the-art research in this exciting interdisciplinary field.

IAU SYMPOSIUM PROCEEDINGS SERIES

INTERNATIONAL ASTRONOMICAL UNION

UNION ASTRONOMIQUE INTERNATIONALE

COMPLEX PLANETARY SYSTEMS

PROCEEDINGS OF THE 310th SYMPOSIUM OF THE INTERNATIONAL ASTRONOMICAL UNION HELD IN NAMUR, BELGIUM JULY 7–11, 2014

Edited by

ZORAN KNEŽEVIĆ
Astronomical Observatory, Belgrade, Serbia

and

ANNE LEMAITRE
Centre naXys, University of Namur, Belgium

Shaftesbury Road, Cambridge CB2 8EA, United Kingdom

One Liberty Plaza, 20th Floor, New York, NY 10006, USA

477 Williamstown Road, Port Melbourne, VIC 3207, Australia

314–321, 3rd Floor, Plot 3, Splendor Forum, Jasola District Centre, New Delhi – 110025, India

103 Penang Road, #05–06/07, Visioncrest Commercial, Singapore 238467

Cambridge University Press is part of Cambridge University Press & Assessment, a department of the University of Cambridge.

We share the University's mission to contribute to society through the pursuit of education, learning and research at the highest international levels of excellence.

www.cambridge.org
Information on this title: www.cambridge.org/9781107078680

First published 2014

A catalogue record for this publication is available from the British Library

ISBN 978-1-107-07868-0 Hardback

Table of Contents

Small bodies, asteroids and space debris

Preface

The huge number of available observations (from ground and space) and their high precision, as well as the computational power and speed of the present-day computers have spectacularly changed the nature and the accuracy of the dynamical models, especially for planetary evolution studies. Indeed, any phenomenon, any planetary system, and any n-body problem appears now much more complex than it was originally thought. Consequently, it cannot be described by a simple dynamical model. To provide a complete understanding of the global behavior of the real system, several levels of modeling should be taken into account at the same time.

To unravel the complexity of such systems, to understand the underlying phenomena and to build a model able to reproduce the observations in a realistic way, requires collaboration and interdisciplinarity of scholars from very different fields.

Celestial mechanics has decisively evolved in that direction during the last decade. Our symposium aimed to bring forward these efforts. Thanks to this highly complex analysis, taking into account as many aspects as possible in the models, realistic results are obtained, giving answers to several astronomical questions, especially but not only in planetary sciences.

The results obtained and discussed at the symposium, with this perspective and level of complexity and interdisciplinarity are numerous. Let us mention

- the concept of habitability of a planet, mixing biological and dynamical criteria,
- the studies of the rotation of planets and satellites, requiring data and models of their internal structure and thus interactions with geologists and physicists,
- the understanding of the long term evolution of space debris and satellites, which requires the introduction of drags, frictions or shadowing effects,
- the modeling of migration mechanisms, including the interaction between the planets and the gas and planetesimal disk,
- the discovery of new mechanisms to explain the size, the location and the composition of natural satellites,
- the cascade of resonances and sub resonances necessary to explain a specific observation,
- the role of the Yarkovsky thermal effect on the evolution of the rotating small bodies,
- the NEO chaotic orbit determination used in impact monitoring,
- the development of sophisticated symplectic integrators and of increasingly precise tools for detection and measure of chaos.

All these results have shown the disappearing of the formal historical border between analytical and numerical approaches.

This symposium focused on the main dynamical aspects of planetary science, pointing in each topic to the recent results obtained thanks to the synergy of different disciplines, and to the treatment of the problem in its full complexity.

The speakers (and especially the key speakers) were invited to show results of this interdisciplinary approach, describing the complexity of the system and emphasizing the outcomes of its new description. The SOC gave priority to talks and posters resulting from collaborations between teams and disciplines. The meeting was a success, with 129 participants, coming from 30 different countries. The program was organized in 6 sessions:

- Session 1 : Introduction and motivation
- Session 2 : Rotation, N-Body and algorithms
- Session 3 : Exoplanets

- Session 4 : Small bodies (asteroids and space debris)
- Session 5 : Solar System and natural satellites
- Session 6: Formation of planetary systems

The opening session was reserved for the welcoming messages, from the University, from the research center naXys and from the organizers. The formal part was followed by J. Laskar's (IMCCE, Observatory of Paris) plenary talk on *Chaotic diffusion in the Solar System and the astronomical calibration of geological timescale.* This topic was very representative of the spirit of the symposium: it is not possible to solve a real astronomical problem without considering its full complexity and interdisciplinarity.

The choice of the key speakers was especially successful; the SOC wanted to choose the new generation of scientists, and they all answered with enthusiasm, giving excellent, lively and interesting talks (35-40 minutes), well documented and at the top of the present research in the field. T. Van Hoolst for the rotation, C. Beaug and A. Correia for the exoplanets, A. Morbidelli and A. Crida for the Solar system, S. Jacobson and A. Rossi for the small bodies and debris and S. Raymond for the formation of the planetary systems, proved their expertise and communication skills.

The SOC selected 76 contributed talks of 18-20 minutes; the questions that followed the talks were numerous and very interesting, giving suggestions for future work or comparisons. The majority of the presentations referred to very recent research, published this year or even only submitted. The organization of the meeting, with two coffee breaks, allowed to have discussions, give comments and rise questions just after the talks.

58 posters were exhibited during the whole week; the researchers had the possibility to present their results during the related session, in the form *one slide, one minute.* The exercise was interesting, very well used by most of them. Three special prizes were given to the best posters: the first one for the master students (won by D. Skoulidou, from Greece), the second one for the PhD students (won by G. Tsirvoulis from Greece) and the third one for confirmed researchers (won by H. Jang-Condell from USA). The jury was composed of D. Scheeres, D. Hestroffer and M. Yseboodt.

The symposium was held at the premises of the University of Namur, to which the organizers owe a gratitude for the warm hospitality.

Last but not the least, we gratefully acknowledge the support of the sponsors listed on page *xxi* which made this conference possible.

Anne Lemaitre, chair SOC,
Anne-Sophie Libert, chair LOC
Namur, October 10, 2014

THE ORGANIZING COMMITTEE

Scientific

A. Lemaitre (Chair), Belgium
T. Carletti, Belgium
V. Dehant, Belgium
J. Laskar, France
A. Morbidelli, France
D. Scheeres, USA
C. Beaugé, Argentina
A. Celletti, Italy
Z. Knežević, Serbia
A. Milani, Italy
D. Nesvorny, USA
K. Tsiganis, Greece

Local

A.S. Libert (Chair)
A. Lemaitre
B. Noyelles
M. Sansottera
A. Vienne
D. Navarro Ortega
Ch. Lhotka
A. Petit
S. Sotiriadis

Acknowledgements

The symposium is coordinated by the IAU Division I (Fundamental Astronomy), and supported by Division III (Planetary System Sciences).

The Local Organizing Committee operated under the auspices of the University of Namur.

Funding by the
International Astronomical Union,
Fonds National de la Recherche Scientifique,
Service Public de Wallonie,
Namur - Europe - Wallonie,
Namur Complex Systems Research Center,
and
University of Namur,
is gratefully acknowledged.

CONFERENCE PHOTOGRAPH

Participants

Ramiro **Alvarez**, Universidad de Guanajuato, Mexico ramiro@astro.ugto.mx

Kyriaki **Antoniadou**, Aristotle University of Thessaloniki, Greece kyant@auth.gr

Pierre **Auclair-Desrotour**, IMCCE, Observatoire de Paris, France pauclair-desrotour@imcce.fr

Rose-Marie **Baland**, UCL / ORB, Belgium balandrm ¡balandrm@oma.be¿

Roman **Baluev**, Pulkovo Observatory, Saint Petersburg, Russia r.baluev@spbu.ru

David **Bancelin**, Institute for Astrophysics, University of Vienna, Austria david.bancelin@univie.ac.at

Giulio **Ba**, University of Pisa, Italy bagiugio@gmail.com

Cristian **Beaug**, Universidad Nacional de Cordoba Argentina beauge@oac.uncor.edu

Herv **Beust**, IPAG, Grenoble, France Herve.Beust@obs.ujf-grenoble.fr

Bertram **Bitsch**, University of Lund, Sweden bert@astro.lu.se

Emeline **Bolmont**, LAB, Bordeaux, France emeline.bolmont@obs.u-bordeaux1.fr

Borislav **Borisov**, Shumen University, Shumen Bulgaria b.st.borisov@abv.bg

Gwenal **Bou**, Universit Pierre et Marie Curie, Paris France boue@imcce.fr

Ramon **Brasser**, Academia Sinica, Tawan brasser@gate.sinica.edu.tw

Timoteo **Carletti**, University of Namur, Belgium timoteo.carletti@unamur.be

Daniel **Casanova**, University of Namur, Belgium daniel.casanova@unamur.be

Sebastien **Charnoz**, University Paris Diderot, France charnoz@cea.fr

Sourav **Chatterjee**, University of Florida, USA chatterjee.sourav2010@gmail.com

Steve **Chesley**, Jet Propulsion Laboratory, USA steve.chesley@jpl.nasa.gov

Alexandre **Correia**, University of Aveiro, Portugal correia@ua.pt

Alexis **Coyette**, UCL - ORB, Belgium alexis.coyette@observatoire.be

Aurlien **Crida**, Observatoire de la. Cte d'Azur, France crida@oca.eu

Frabrizio **De Marchi**, University of Pisa, Italy fabrizio.demarchi@for.unipi.it

Katherine **Deck**, MIT, USA kdeck@mit.edu

Vronique **Dehant**, Royal observatory of Belgium, Belgium v.dehant@oma.be

Russell **Deitrick**, University of Washington USA deitrr@astro.washington.edu

Florent **Deleflie**, IMCCE/GRGS France Florent.Deleflie@imcce.fr

Jean-Baptiste **Delisle**, IMCCE - Observatoire de Paris, France delisle@imcce.fr

Sara **Di Ruzza**, SpaceDyS, Pisa, Italy saradiruzza@gmail.com

Alex **Dias de Oliveira**, Observatrio Nacional Rio de Janeiro, Brazil oliveira.astro@gmail.com

Joanna **Drazkowska**, Heidelberg University, Germany drazkowska@uni-heidelberg.de

Rudolf **Dvorak**, University of Vienna, Austria dvorak@astro.univie.ac.at

Christos **Efthymiopoulos**, Academy of Athens, Greece cefthim@academyofathens.gr

Siegfried **Eggl**, IMCCE - Observatoire de Paris, France siegfried.eggl@imcce.fr

Vacheslav **Emel'yanenko**, Institute of Astronomy, RAS, Moscow, Russia vvemel@inasan.ru

Davide **Farnocchia**, Jet Propulsion Laboratory, USA Davide.Farnocchia@jpl.nasa.gov

Kohei **Fujimoto**, Texas A&M University, USA kfujimot@tamu.edu

Mattia **Galiazzo**, University of Vienna, Austria mattia.galiazzo@univie.ac.at

Nikolaos **Georgakarakos**, University of Central Macedonia, Greece georgakarakos@hotmail.com

Antonio **Giorgilli**, Universita degli Studi di Milano, Italy antonio.giorgilli@unimi.it

Ioannis **Gkolias**, University of Rome "Tor Vergata" , Italy gkolias@mat.uniroma2.it

Mikael **Granvik**, University of Helsinki & Finnish Geodetic Institute, Finland mgranvik@iki.fi

Octavio Miguel **Guilera**, Instituto de Astrofsica de La Plata, Argentina oguilera@fcaglp.unlp.edu.ar

Rustam **Guliyev**, Shamakhy Astrophysical Observatory Turkey rustamdb@gmail.com

Nader **Haghighipour**, University of Hawaii, USA nader@ifa.hawaii.edu

Tom **Hands**, University of Leicester, United Kingdom tom.hands@le.ac.uk

Daniel **Hestroffer**, IMCCE - Observatoire de Paris, France hestro@imcce.fr

Subhon **Ibadov**, Institute of Astrophysics, Academy of Sciences, Tajikistan ibadovsu@yandex.ru

Tamara **Ivanova**, Institute of applied astronomy of Academy of Sciences, Russia itv@ipa.nw.ru

Anatoliy **Ivantsov**, ASRI-Technion, Isral ivantsov@imcce.fr

Seth **Jacobson**, Observatoire de la Cte d'Azur, France seth.jacobson@oca.eu

Hannah **Jang-Condell**, University of Wyoming, USA hjangcon@uwyo.edu

Zoran **Knezevic**, Astronomical Observatory of Belgrade, Serbia zoran@aob.rs

Alexey M. **Koksin**, Tomsk State University, Russia shefer@niipmm.tsu.ru

Eiichiro **Kokubo**, National Astronomical Observatory of Japan, Tokyo, Japan kokubo@th.nao.ac.jp

Tamas **Kovacs**, Observatory of Budapest, Hungary tkovacs@konkoly.hu

Irina **Kovalenko**, IMCCE - Observatoire de Paris ,France ikovalenko@imcce.fr

Katherine **Kretke**, Southwest Research Insitute, USA kretke@boulder.swri.edu

Eduard **Kuznetsov**, Ural Federal University, Russia eduard.kuznetsov@urfu.ru

Jacques **Laskar**, IMCCE - Observatoire de Paris, France laskar@imcce.fr

Man Hoi **Lee**, The University of Hong Kong, Hong-Kong mhlee@hku.hk

Adrien **Leleu**, IMCCE - Observatoire de Paris, France leleu.adrien@gmail.com

Anne **Lemaitre**, University of Namur, Belgium anne.lemaitre@unamur.be

Christoph **Lhotka**, University of Vienna, Austria christoph.lhotka@univie.ac.at

Jian **Li**, Nanjing University, China ljian@nju.edu.cn

Anne-Sophie **Libert**, University of Namur, Belgium anne-sophie.libert@unamur.be

Helene **Ma**, University of Pisa, Italy helenema@mail.dm.unipi.it

Thomas I. **Maindl**, University of Vienna, Austria thomas.maindl@univie.ac.at

Christian **Marchal**, ONERA, France clbmarchal@wanadoo.fr
Jean-Luc **Margot**, University of California, Los Angeles, USA jlm@astro.ucla.edu
Stefano **Mar**, University of Pisa, Italy maro@mail.dm.unipi.it
Andrea **Milani**, Department of Mathematics, University of Pisa, Italy milani@dm.unipi.it
Helena **Morais**, University of Aveiro, Portugal helena.morais@gmail.com
Alessandro **Morbidelli**, Observatoire de la Cte d'Azur, France morby@oca.eu
Magda **Murawiecka**, Adam Mickiewicz University, Poland magda.murawiecka@gmail.com
Alexander **Mustill**, Lund Observatory, Sweden alex@astro.lu.se
Juan F. **Navarro**, University of Alicante, Spain jf.navarro@ua.es
Benjamin **Nelson**, Pennsylvania State University, USA benelson@psu.edu
Vasily **Nikonov**, Lomonosov Moscow State University, Russia nikon_v@list.ru
Bojan **Novakovic**, University of Belgrade, Serbia bojan@matf.bg.ac.rs
Benot **Noyelles**, University of Namur, Belgium benoit.noyelles@unamur.be
Rocio Isabel **Paez**, University of Rome "Tor Vergata", Italy paez@axp.mat.uniroma2.it
Alexis **Petit**, University of Namur, Belgium alexis.petit@unamur.be
Cristobal **Petrovich**, Princeton University, USA cpetrovi@princeton.edu
Arnaud **Pierens**, Observatoire de Paris, France arnaud.pierens@obs.u-bordeaux1.fr
Elke **Pilat-Lohinger**, University of Graz, Austria elke.pilat-lohinger@univie.ac.at
Eva **Plvalov**, Astronomical Institute Academy of Sciences, Slovak Republic plavala@slovanet.sk
Elena **Popova**, Pulkovo Observatory, Saint Petersburg, Russia m02pea@gmail.com
Alexandre **Pousse**, IMCCE - Observatoire de Paris, France apousse@imcce.fr
Billy **Quarles**, NASA Ames Research Center, USA billy.l.quarles@nasa.gov
Viktor **Radovic**, Department of Astronomy, Belgrade, Serbia rviktor@matf.bg.ac.rs
Darin **Ragozzine**, Florida Institute of Technology USA dragozzine@fit.edu
Nicolas **Rambaux**, Observatoire de Paris, France Nicolas.Rambaux@imcce.fr
Sean **Raymond**, Laboratoire d'Astrophysique de Bordeaux, France rayray.sean@gmail.com
Francoise **Remus**, Observatoire de Paris, France francoise.remus@obspm.fr
Davide **Ricci**, Instituto de Astronomia of Mexico, Mexico indy@astrosen.unam.mx
Andy **Richard**, IMCCE - Observatoire de Paris, France arichard@imcce.fr
Philippe **Robutel**, IMCCE - Observatoire de Paris, France robutel@imcce.fr
Mara Paula **Ronco**, Instituto de Astrofsica de La Plata, Argentina mpronco@fcaglp.unlp.edu.ar
Alexey **Rosaev**, NPC NEDRA, Jaroslavl, Russia hegem@mail.ru
Aaron **Rosengren**, IFAC - CNR, Italy a.rosengren@ifac.cnr.it
Alessandro **Rossi**, IFAC - CNR, Italy a.rossi@ifac.cnr.it
Galina **Ryabova**, Tomsk State University, Russia rgo@rambler.ru
Tatiana **Salnikova**, Lomonocov Moscow State University, Russia tatiana.salnikova@gmail.com
Marco **Sansottera**, Universit degli Studi di Milano, Italy marco.sansottera@unimi.it
Daniel **Scheeres**, University of Colorado, USA scheeres@colorado.edu
Giulia **Schettino**, University of Pisa, Italy giulia.schettino@gmail.com
Daniele **Serra**, University of Pisa, Italy dserra@mail.dm.unipi.it
Natalia **Shakht**, Pulkovo Observatory, Saint Petersburg, Russia shakht@gao.spb.ru
Takashi **Shibata**, University of Tokyo, Japan shibata.takashi@nao.ac.jp
Bruno **Sicardy**, Observatoire de Paris, France Bruno.Sicardy@obspm.fr
Despoina **Skoulidou**, Aristotle University of Thessaloniki, Greece dskoulid@physics.auth.gr
Valeriy **Snytnikov**, Novosibirsk State University Russia snyt@catalysis.ru
Sotiris **Sotiriadis**, University of Namur, Belgium sotiris.sotiriadis@unamur.be
Federica **Spoto**, University of Pisa, Italy spoto@mail.dm.unipi.it
Bonnie **Steves**, Glasgow Caledonian University, United Kingdom B.Steves@gcu.ac.uk
Olga **Stoyanovskaya**, Novosibirsk State University, Russia o.p.sklyar@gmail.com
Winston **Sweatman**, Massey University, New Zeland w.sweatman@massey.ac.nz
Georgios **Tsirvoulis**, Astronomical Observatory of Belgrade, Serbia gtsirvoulis@aob.rs
Timothe **Vaillant**, IMCCE - Observatoire de Paris, France timothee.vaillant@ens-lyon.fr
Giovanni **Valsecchi**, IAPS-INAF, Italy giovanni@iaps.inaf.it
Tim **Van Hoolst**, Royal Observatory of Belgium, Belgium tim.vanhoolst@oma.be
Stefaan **Van wal**, University of Colorado at Boulder, USA stefaanvanwal@msn.com
Dimitri **Veras**, University of Warwick, United Kingdom d.veras@warwick.ac.uk
Roberto **Vieira-Martins**, Observatorio Nacional, Rio de Janeiro, Brazil rvm@on.br
Alain **Vienne**, IMCCE - Lille1, France alain.vienne@univ-lille1.fr
Tamara **Vinogradova**, Institute of Applied Astronomy, Saint Petersburg, Russia vta@ipa.nw.ru
Marie **Yseboodt**, Royal Observatory of Belgium, Belgium m.yseboodt@oma.be
Li-Yong **Zhou**, Astronomy Department, Nanjing University, China zhouly@nju.edu.cn

Complex Planetary Systems
Proceedings IAU Symposium No. 310, 2014
Z. Knežević & A. Lemaitre, eds.

doi:10.1017/S1743921314007698

The libration and interior structure of large icy satellites and Mercury

Tim Van Hoolst[1,2]

[1] Royal Observatory of Belgium
Ringlaan 3, B-1180 Brussels, Belgium
email: tim.vanhoolst@oma.be

[2] Institute of Astronomy, KU Leuven
Celestijnenlaan 200D, B-3001 Leuven, Belgium

Abstract. Longitudinal librations are periodic changes in the rotation angle of a planet or satellite. Their observation and subsequent interpretation have profoundly increased our understanding of the interior structure of Mercury. Likewise, libration is thought to provide important constraints on the interior structure of icy satellites. Here we study the libration of Mercury and large icy satellites rotating synchronously with their orbital motion and explain how it depends on the interior structure.

Keywords. planets and satellites: individual (Mercury), geodesy, rotation, icy satellites

1. Introduction

Most large and medium-sized (radius larger than 200 km) satellites of the solar system are in a synchronous spin-orbit resonance in which their rotation period is equal to their orbital period. The situation arises as a result of dissipation in the satellite associated with tides raised by the central planet. In this stable state of synchronous rotation the longest axis of the satellite points towards the central planet at pericenter. Mercury is in a 3:2 spin-orbit resonance as has been postulated for the first time by Giuseppe Colombo based on observations of Pettengill and Dyce in 1965. Before that time, it was thought that Mercury's spin was in a 1:1 resonance with its orbital motion, just like the Moon and other satellites in their orbit around their central planet. The resonant rotation of Mercury is stable like the 1:1 resonance of the rotation of the satellites.

Stability, however, does not imply that the rotation rate is constant. Because of the difference in rotation period and orbit period, the long axis of Mercury rotates with respect to the direction to the sun. Also the long axis of satellites is generally not oriented in the direction to the central planet due to the eccentricity of the satellite orbits and the resulting variable orbital speed. The central body (the sun or the central planet) therefore exerts a time-variable gravitational torque on Mercury and on synchronously orbiting satellites. Consequently, Mercury and the satellites vary their rotation rate about the mean rate and the direction of the long axis librates with respect to its orientation at constant mean rotation rate (forced longitudinal libration).

Here we focus on the libration at the orbital period of Mercury (often referred to as the annual libration) and of the satellites (mostly referred to as the diurnal libration). We describe the expected libration behavior and show that observations of librations provide information on the interior structure. For example, a global liquid layer such as a subsurface ocean strongly affects the rotational dynamics and leads to a transfer of angular momentum between solid and liquid layers. Rotation observations have already been used to prove that Mercury's core is at least partially liquid and can demonstrate the existence of an internal ocean in icy satellites. Librations can also be used to further characterize different layers. As a prime example of this, the size and mean density of Mercury's core have recently been well constrained by libration observations (Hauck *et al.* 2013, Rivoldini & Van Hoolst 2013).

2. Mercury

From the determination of the epoch and time delay that maximize the cross correlation of back-scattered radar signals at two antennas separated by a long baseline, Margot *et al.* (2007, 2012) have estimated the annual libration amplitude of Mercury to be (38.5 ± 1.6) arcsec, which corresponds to a maximum displacement at the equator with respect to the equilibrium rotation of (455 ± 19) m. In 1976 Peale had already realized that the observation of the annual libration of Mercury at about 88 days (the period of revolution of Mercury around the sun) is central to improve our understanding of the interior of Mercury. In particular it allows estimating the sizes of the core and the silicate shell (mantle plus crust). The reason for this lies in how Mercury reacts to the gravitational forcing of the sun. This can most easily be described by the equation for the conservation of angular momentum.

We will neglect in this analysis Mercury's small obliquity (the observed value is (2.04 ± 0.08) arcmin, Margot *et al.* 2012) and assume principal axis rotation at a rate Ω in this study of Mercury's rotation rate. Only the polar or z-component of the angular momentum must then be considered. For an entirely solid planet, we have

$$C\dot{\Omega} = L, \tag{2.1}$$

where

$$L = \frac{3}{2}(B - A)\frac{GM_{\odot}}{r^3}\sin 2\xi \tag{2.2}$$

is the solar torque on Mercury's permanent figure. We neglect the tidal torques as their effect is well below the observational precision (Van Hoolst *et al.* 2012). The distance r is the distance between the mass centers of the sun and Mercury, and ξ is the angle between the direction of Mercury's axis of smallest moment of inertia A (also called the long axis) and the direction to the sun:

$$\xi = f - \phi, \tag{2.3}$$

where f is the true anomaly and ϕ the rotation angle between the long axis of Mercury and the major axis of Mercury's orbit considered fixed. The torque is seen to depend on the orbital and rotational motion of Mercury through its dependence on the angle ξ and the distance r. It also depends on the equatorial shape of Mercury as given by the equatorial moment of inertia difference $B - A$. An accurate value for this quantity has only recently become available by the determination, by means of radio tracking the MESSENGER spacecraft, of the Hermean gravity field, in particular the degree-two and order-two coefficient $C_{22} = (B - A)/(4MR^2) = (0.809 \pm 0.01) \times 10^{-5}$ (Smith *et al.* 2012). Prior to this accurate determination, interpretations of observed librations were seriously hampered (e.g. Margot *et al.* 2007).

Since the rotation is close to the 3:2 spin-orbit resonance, we introduce the small libration angle $\gamma = \phi - \frac{3}{2}M$, where M is the mean anomaly. By rewriting the torque in terms of the libration angle and the mean and true anomaly and expanding the time-dependent factor $\sin(2\gamma + 3M - 2f)/r^3$ of the torque in terms of the mean anomaly, we

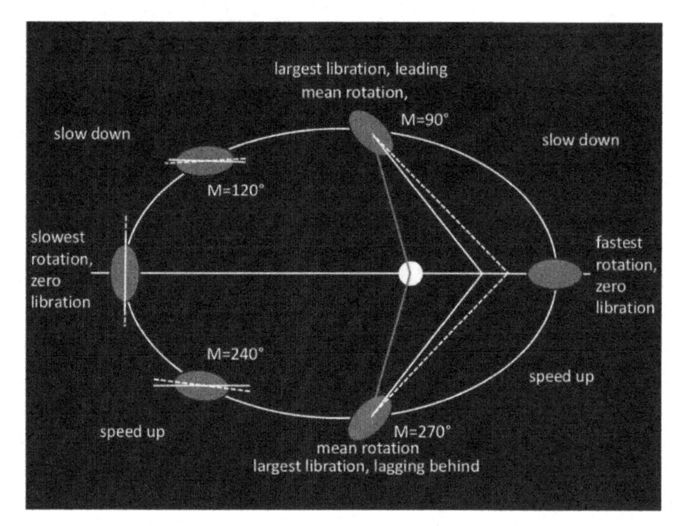

Figure 1. Schematic of the phase of the annual libration of Mercury. White continuous lines represent the approximate orientation of the long axis without libration, dashed lines the orientation including libration.

obtain the libration equation

$$\frac{d^2\gamma}{dM^2} + \frac{3}{4}\frac{B-A}{C}e\left(7 - \frac{123}{8}e^2\right)\sin 2\gamma = -\frac{3}{2}\frac{B-A}{C}\left(1 - 11e^2 + \frac{959}{48}e^4 + \cdots\right)\sin M \tag{2.4}$$

(Peale 2005). Here harmonic forcing terms of the diurnal frequencies and time-dependent torque components proportional to the small libration angle have been neglected. This equation has the form of a forced harmonic oscillator and the libration will oscillate with a period equal to the orbital period.

The phase of the libration is illustrated in Fig. 1. For mean anomalies between 0 and 180 degrees, the long axis of Mercury is ahead of the direction to the sun in the sense of the rotation of Mercury. Therefore, the gravitational torque exerted by the sun on Mercury will tend to decelerate the rotation. For mean anomalies larger than 180 degrees, the torque tends to accelerate the rotation. The rotation speed is thus largest at perihelion and smallest at aphelion. As for a pendulum passing through its equilibrium position at highest velocity, Mercury passes through the equilibrium of zero libration at the perihelion and its long axis there is in the direction to the sun. Although the rotation slows down for mean anomalies increasing from 0 to 90 degrees, the libration increases as the rotation is faster than average. At a mean anomaly of 90 degrees, Mercury reaches its largest libration leading with respect to a constant mean rotation and rotates at the mean rotation rate. Since Mercury's rotation continues to slow down when further advancing in its orbit, the libration decreases but remains positive until it becomes zero at aphelion. In the second half of its orbit, the long axis of Mercury lags behind the orientation for constant rotation until it again reaches zero libration at the perihelion.

The libration amplitude can immediately be obtained from libration Eq. (2.4), but we first discuss the case of a liquid core. When the core is in addition assumed to be spherically symmetric and not interacting with the mantle, the total torque of the sun is exerted on the solid silicate shell and only the shell with polar moment of inertia C^m

librates with amplitude g_M given by

$$g_M = \frac{3}{2}\frac{B-A}{C^m}\left(1 - 11e^2 + \frac{959}{48}e^4 + \cdots\right). \tag{2.5}$$

The libration is linearly dependent on the equatorial moment of inertia difference $B - A$, which is related to the torque, and is inversely proportional to the shell's polar moment of inertia, which represents the rotational inertia (resistance to rotational forcing). When the core is solid, both the silicate shell and core librate with an amplitude that is given by the same equation in which the polar moment of the shell must be replaced by the total polar moment of inertia of Mercury. This difference is key to the interpretation of the libration. For a solid core, the expected libration amplitude is about 190 m (the polar moment of inertia of Mercury is determined from the same radar observations which also yield an accurate estimate of the obliquity, Margot *et al.* 2012), less than half the observed value. This demonstrates that Mercury's core must be (at least partially) liquid. Since all other quantities on which the libration amplitude depends are sufficiently well known, libration is thus seen to provide an accurate estimate of the polar moment of inertia of the mantle. This, in turn, can be used to estimate the radius of the core.

Before describing results about the interior structure, we first need to assess the importance of up to now neglected effects on the libration. First, Mercury's core interacts with the mantle in different ways. However, Peale *et al.* (2002) and Rambaux *et al.* (2007) (see also Van Hoolst 2007) have shown that all direct interactions between the liquid core and the mantle, such as viscous coupling, electromagnetic coupling, gravitational coupling and pressure coupling related to topography of the core-mantle boundary, have a negligibly small effect on the libration amplitude that is well below the current measurement precision.

An inner core interacts directly with the mantle through gravitational coupling if the long axes of inner core and mantle are not aligned. The global magnetic field observed by Mariner 10 and MESSENGER, if caused by dynamo action, suggests that Mercury contains a growing solid inner core in a liquid outer core. The size of the inner core is currently not constrained but it cannot be excluded that it might possibly even be relatively large although its size is limited by the global radial contraction of about 7 km estimated from Mercury's tectonics (Byrne *et al.* 2014). For small inner cores, the effect of an inner core on Mercury's 88 days libration amplitude is below the observational precision. If Mercury's inner core is larger than at least 1000 km it could, however, be of the order of the current observational uncertainty and be larger than the future BepiColombo observational precision expected to be about 10 m or somewhat smaller (Pfyffer *et al.*, 2011) (see e.g. Fig. 2, Van Hoolst *et al.* 2012, Dumberry *et al.* 2013). Gravitational coupling between the inner core and the mantle slightly reduces the mantle libration by up to 20 m and therefore the mantle moment of inertia derived from the libration is also somewhat smaller than if derived from Eq. (2.5) (Van Hoolst *et al.* 2012).

The observed libration amplitude and obliquity, together with the recent determination of the gravity field of Mercury by MESSENGER provide strong constraints on the radial mass distribution in Mercury. Libration stands out because it provides an estimate of a subsystem of the planet (the silicate part) and not, as for the other data, of the whole planet. The geodesy data set strong limits on the size of the core in particular. According to Hauck *et al.* (2013) and Rivoldini & Van Hoolst (2013) the core radius is about 2000 km with an uncertainty of only about 2%.

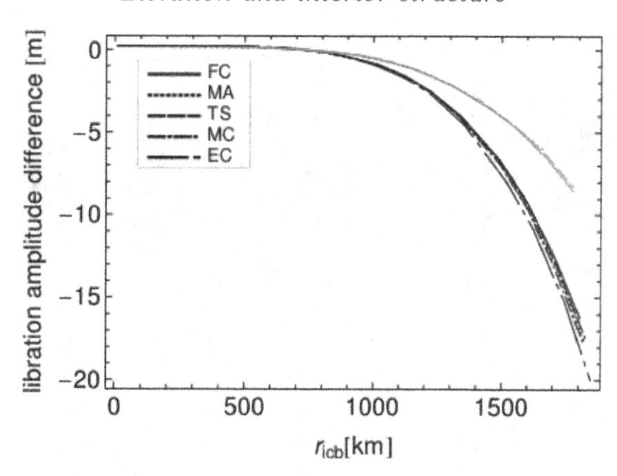

Figure 2. Difference in libration amplitude (in m) due to the inner core as a function of inner core radius for different models of the interior structure of Mercury (Van Hoolst *et al.* 2012)

3. Large icy satellites

Several lines of evidence strongly indicate that the largest icy satellites of the solar system (Ganymede, Titan, Callisto, and Europa) all have a subsurface ocean beneath their surface (Khurana *et al.*, 1998, Kivelson *et al.*, 2002, Béghin *et al.* 2010, Baland *et al.*, 2012, Iess *et al.* 2012). Essential properties of the subsurface oceans and overlying ice shell, including their size, are not well known. Further insight into the ocean and ice shell characteristics can be obtained from several methods, including those already used for the ocean detection: magnetic observations, observations of the satellite tides (e.g. Wahr *et al.*, 2006) and rotation (e.g. Baland *et al.*, 2011). In particular, the forced diurnal librations, which have not yet been unambiguously detected so far for large icy satellites, might provide useful constraints.

By using an angular momentum approach as for Mercury, the diurnal libration amplitude g_s of an entirely solid and rigid satellite is seen to depend on the moments of inertia of the satellite and on the orbital eccentricity:

$$g_s = 6\frac{B-A}{C}e. \tag{3.1}$$

Of the four largest icy satellites, Europa has the largest rigid libration amplitude of about 134 m. Titan's amplitude is about 50 m, and the libration amplitude of Callisto and Ganymede is about 12 m and 10 m, respectively (Van Hoolst *et al.* (2013), see also Comstock & Bills (2003)).

The phase of the libration is illustrated in Fig. (3) and can be explained as follows. The direction of the long axis is approximately into the direction of the empty focus (Murray and Dermott 2000). Therefore, the long axis of the satellite is lagging behind the direction to the planet in the sense of the rotation for mean anomalies between 0 and 180 degrees, and is ahead of that direction for mean anomalies larger than 180 degrees. This is opposite to the case of Mercury. The gravitational torque exerted by the planet on the satellite will tend to speed up the rotation in the first half of the orbit and to slow it down in the second half. The rotation speed is thus smallest at pericenter and largest at apocenter. The largest libration will be reached at a mean anomaly of 270 degrees when the rotation is decelerating and reaches the mean rotation. The libration is smallest and lagging the rotation at a mean anomaly of 90 degrees when the rotation is accelerating and reaches the mean rotation rate.

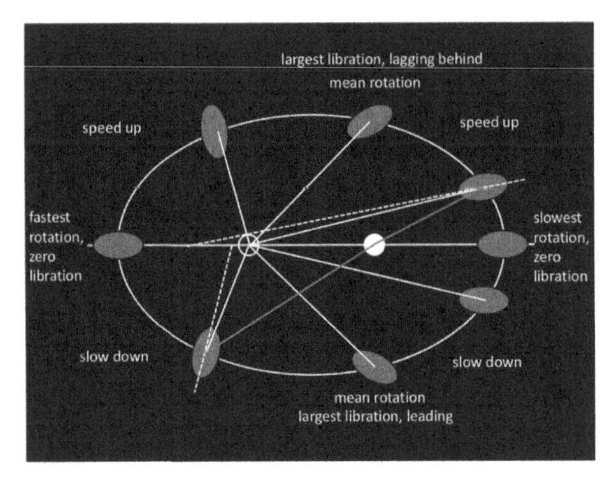

Figure 3. Schematic of the phase of the diurnal libration of synchronous satellites. White continuous lines represent the approximate orientation of the long axis without libration, dashed lines the orientation including libration.

Since the subsurface ocean is close to the surface, one might expect that the real libration amplitude would be much larger if the shell could be considered to be librating almost independently of the deeper interior (Van Hoolst *et al.* 2008). However, the thin ice shell responds to the gravitational forcing of the central planets not only by libration but also by deformation (tides). As pointed out by Goldreich and Mitchell (2010) this strongly limits the libration amplitude. This reduction can be understood by considering the limit case of a satellite with an ocean that reaches up to the surface. The gravitational and, to a lesser extent, the geometric shape of the satellites is strongly determined by the static tides and the centrifugal acceleration which result in synchronously rotating satellites with a triaxial ellipsoidal shape. On top of these static tides are dynamic tides, mainly with an orbital period. If the ocean extends to the surface (or better if the satellite is entirely liquid) the total tidal bulge (static + periodic) would be in the direction to the central planet at any moment. In that situation, the central planet cannot exert a gravitational torque on the satellite and there would be no libration.

The libration of an icy satellite with a liquid layer can be studied by considering the change in angular momentum of each solid and liquid layer. The angular momentum of a layer changes due to the gravitational torque of the planet on the static shape and on the periodic tidal bulges but also by interaction torques between the different layers. Viscous and electromagnetic torques can probably be neglected because the spin up times associated with these torques are expected to be many orders of magnitude longer than the libration period as in the case of Mercury. The torques to be considered then are the gravitational torques between the static and tidal bulges of the different layers and the pressures torques of the liquid layers on the surrounding solid layers induced by the changing gravitational field resulting from orientation changes and periodic tidal bulges of the different layers. Figures 4 and 5 illustrate the possible different orientation and periodic tidal bulges of the layers resulting in interlayer torques for a satellite (likely representative of Ganymede) with two liquid layers - a subsurface ocean and a liquid iron outer core - and three solid layers - the outer ice shell, a mantle, and a solid iron inner core.

Periodic zonal tides are a separate class of tides to be considered since they change the polar moment of inertia and therefore the rotational response of the layers to forcing. The polar moment of inertia is largest at pericenter and smallest at apocenter. It therefore decreases in the first half of the orbit, effectively contributing to the acceleration of the

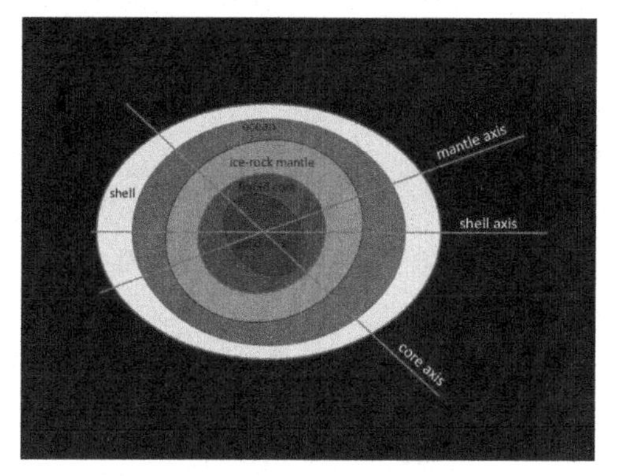

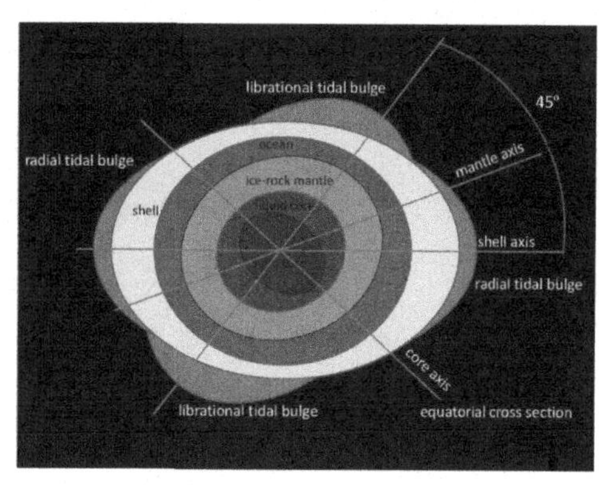

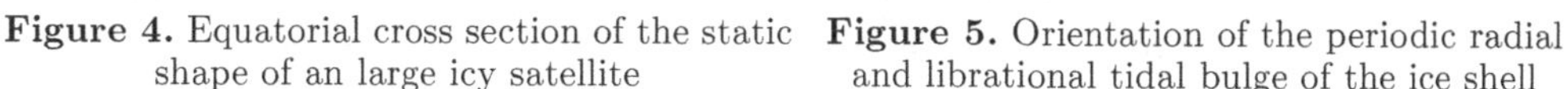

Figure 4. Equatorial cross section of the static shape of an large icy satellite

Figure 5. Orientation of the periodic radial and librational tidal bulge of the ice shell

rotation, and increases in the second half, contributing to the deceleration. As a result, the zonal tides increase the libration.

The libration amplitude of the ice shell above an ocean is inversely proportional to the shell moment of inertia because it represents the rotational inertia and proportional to a forcing term which consists of three contributions: the external gravitational and associated pressure forcing of the shell, the forcing of the shell due to coupling between the librational bulges and the static bulges of different layers, and the effect of the zonal tides (Van Hoolst *et al.* 2013). The last term is at most 10% of the external torque due to the planet, and the first two terms oppose each other. Interaction between the periodic and static bulges of different layers, in particular between the static bulge of the solid interior and the librational bulge of the ice shell (see Fig. 5), can periodically force the shell libration, in contrast to the torque between static tidal bulges, which cannot force libration and is zero when the libration angles are zero. This interlayer forcing of the libration opposes the forcing due to the central planet as can be seen from the geometry of the bulges, with the librational bulge approximately 45 degrees ahead of the static bulges (see Fig. 5).

The libration amplitude is rather insensitive to the thickness of the ice shell, as was suggested by Goldreich and Mitchell (2010), and is close to the value of the libration for a satellite without subsurface ocean. It mainly depends on the rigidity and the viscosity of the ice shell (see also Jara-Orué and Vermeersen 2014). Figs. 6 and 7 show the dependence on rigidity and viscosity for a model of the interior structure of Ganymede with a subsurface ocean at a depth of 50 km below the surface. If the ocean is assumed to be freezed out and the whole interior were to be rigid, the libration amplitude would be 8.86 m for the particular model considered. With an ocean and considering elastic behavior, the libration amplitude increases almost linearly with increasing rigidity of the ice shell by about 2m/GPa (Fig. 6). For viscosities of the ice shell below about 10^{15} Pa s (for an ice shell rigidity of 3.3×10^9 Pa), the libration amplitude decreases to about 1 m. The Maxwell timescale for a viscosity of 10^{15} Pa s and an ice shell rigidity of 3.3×10^9 Pa is close to the orbital period of Ganymede. Therefore, the behavior will be be essentially elastic for larger viscosities and tend to that of a viscous fluid for lower viscosities.

Acknowledgements

I am grateful to my collaborators on these topics: Rose-Marie Baland, Attilio Rivoldini, and Antony Trinh.

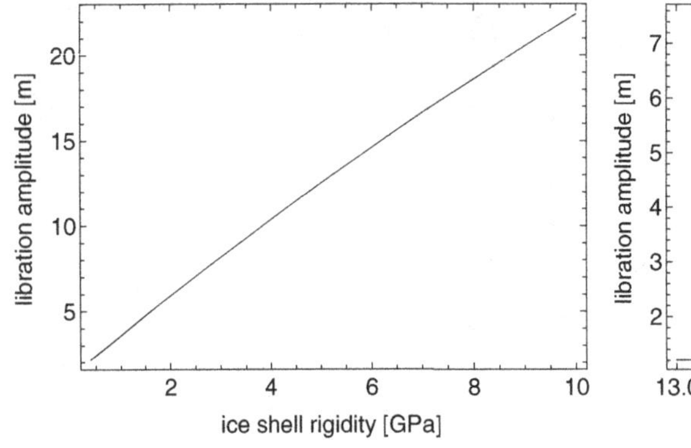

Figure 6. Libration amplitude of Ganymede as a function of the rigidity of the ice shell

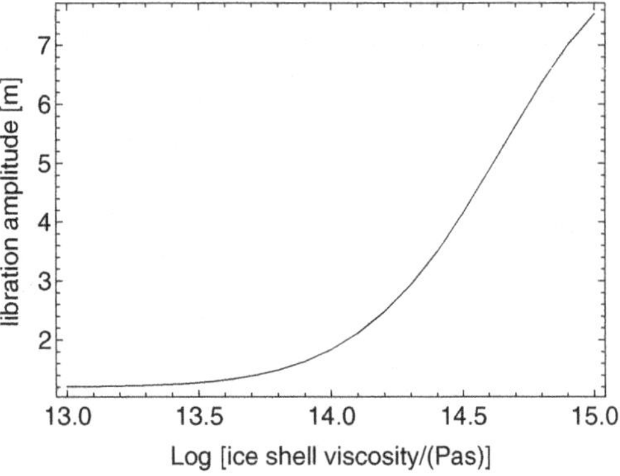

Figure 7. Libration amplitude of Ganymede as a function of the ice viscosity of the shell

References

Baland, R.-M., Van Hoolst, T., Yseboodt, M., & Karatekin, O., 2011, *A&A*, 530, A141

Béghin, C., Sotin, C., & Hamelin, M., 2010, *C. R. Geoscience*, 342, 425

Byrne, P. K., Klimczak, C., Celal Sengor, C. A., Solomon, S. C., Watters, T. R., & Hauck II, S. A., 2014, *Nature Geoscience*, 7, 301

Comstock, R. L. & Bills, B. G. 2003, *J. Geophys. Res.*, 108(E9), 5100

Dumberry, M., Rivoldini, A., Van Hoolst, T., & Yseboodt, M., 2010, *Icarus*, 225, 62

Goldreich, P. M. & Mitchell, J. L. 2010, *Icarus*, 209, 631

Hauck, S. A., Margot, J.-L., Solomon, S. C., Phillips, R. J., Johnson, C. L., Lemoine, F. G., Mazarico, E., McCoy, T. J., Padovan, S., Peale, S. J., Perry, M. E., Smith, D. E., & Zuber, M. T., 2013. *J. Geophys. Res.*, 118, 1

Iess, L., R. A. Jacobson, M. Ducci, D. J. Stevenson, J. I. Lunine, J. W. Armstrong, S. W. Asmar, P. Racioppa, N. J. Rappaport, P. Tortora, 2012, *Science*, 337, 457

Jara-Orué, H. & Vermeersen, B. L. A. 2014, *Icarus*, 229, 31

Khurana, K. K., Kivelson, M. G., Stevenson, D. J., Schubert, G., Russell, C. T., Walker, R. J., & Polanskey, C. 1998, *Nature*, 395, 777

Kivelson, M. G., Khurana, K. K., & Volwerk, M., 2002, *Icarus*, 157, 507

Margot, J.-L., Peale, S. J., Jurgens, R. F.,. Slade, M. A., & Holin, I. V., 2007, *Science* 316, 710

Margot, J.-L., Peale, S. J., Solomon, S. C., Hauck, S. A., Ghigo, F. D., Jurgens, R. F., Yseboodt, M., Giorgini, J. D., Padovan, S., & Campbell, D. B., 2012, J. Geophys. Res., 117, E00L09

Murray, C. D. & Dermott, S. F., 2000, Solar System Dynamics, Cambridge University Press

Peale, S. J., 1976, *Nature* 262, 765

Peale, S. J., 2005, *Icarus*, 178, 4

Peale, S. J., Phillips, R. J., Solomon, S. C., Smith, D. E., & Zuber, M. T., 2002, *Meteoritics and Planetary Science*, 37, 1269

Pettengill, G. H. & Dyce, R. B., 1965, *Nature*, 206, 1240

Pfyffer, G., Van Hoolst, T., & Dehant, V., 2011, *Planetary and Space Science*, 59, 848

Rambaux, N., Van Hoolst, T., Dehant, V., & Bois, E., 2007, *A&A*, 468, 711

Rivoldini, A., & Van Hoolst, T., 2013, *Earth and Planetary Science Letters*, 377-378, 62

Smith, D. E., & 16 co-authors, 2012, *Science*, 336, 214-217

Van Hoolst, T., 2007, *Treatise on Geophysics*, ed. G. Schubert, Vol.10: Planets and Moons (ed. T. Spohn), 123

Van Hoolst, T., Rambaux, N., Karatekin, Ö., Dehant, V., & Rivoldini, A. 2008, *Icarus*, 195, 386

Van Hoolst, T., Rivoldini, A., Baland, R.-M., & Yseboodt, M., 2012, *Earth and Planetary Science Letters*, 333-334, 83

Van Hoolst, T., Baland, R.-M., & Trinh, A. 2013, *Icarus*, 226, 299

Wahr, J. M., Zuber, M. T., Smith, D. E., & Lunine, J. I., 2006, *J. Geophys. Res.*, 111, E12005

Complex Planetary Systems
Proceedings IAU Symposium No. 310, 2014
Z. Knežević & A. Lemaitre, eds.

doi:10.1017/S1743921314007704

Spin-orbit resonances and rotation of coorbital bodies in quasi-circular orbits

Philippe Robutel[1], Alexandre, C. M. Correia[2,1] and Adrien Leleu[1]

[1]IMCCE, Observatoire de Paris, UPMC Univ. Paris 06, Univ. Lille 1, CNRS,
77 Av.Denfert-Rochereau, 75014 Paris, France
email: robutel@imcce.fr

[2]Departamento de Física, I3N, Universidade de Aveiro, Campus de Santiago, 3810-193 Aveiro, Portugal
email: correia@ua.pt

Abstract. The rotation of asymmetric bodies in eccentric Keplerian orbits can be chaotic when there is some overlap of spin-orbit resonances. Here we show that the rotation of two coorbital bodies (two planets orbiting a star or two satellites of a planet) can also be chaotic even for quasi-circular orbits around the central body. When dissipation is present, the rotation period of a body on a nearly circular orbit is believed to always end synchronous with the orbital period. Here we demonstrate that for coorbital bodies in quasi-circular orbits, stable non-synchronous rotation is possible for a wide range of mass ratios and body shapes. We further show that the rotation becomes chaotic when the natural rotational libration frequency, due to the axial asymmetry, is of the same order of magnitude as the orbital libration frequency.

Keywords. celestial mechanics, coorbitals, rotation, resonance, spin-orbit resonance

1. Rotation on a Keplerian orbit

Considering that the spin axis of a body is perpendicular to the orbital plane, the orientation of the long axis in an inertial reference frame is given by the angle θ satisfying the differential equation (e.g. Goldreich & Peale, 1966):

$$\ddot{\theta} + \frac{\sigma^2}{2}\left(\frac{a}{r}\right)^3 \sin 2(\theta - f) = 0, \text{ with } \sigma = n\sqrt{\frac{3(B-A)}{C}}, \tag{1.1}$$

where $A < B < C$ are the moments of inertia of the body, (r, f) the polar coordinates of the center of the body, a and n its instantaneous semi-major axis and mean motion. One of the simplest situations is the well-known case of a rotating body orbiting the primary in Keplerian motion. Then, the equation (1.1) can be expanded in Fourier series of the time, given the expression:

$$\ddot{\theta} + \frac{\sigma^2}{2}\sum_{k=-\infty}^{+\infty} X_k^{-3,2}(e)\sin(2\theta - knt) = 0, \tag{1.2}$$

where the Hansen coefficients $X_k^{-3,2}(e)$ are of order $e^{|k-2|}$ as the eccentricity e tends to zero. The equation (1.2) imposes that the main resonances, namely the spin-orbit resonances, are centered at $\dot{\theta} = kn/2$ and that their width, in terms of frequency, is of order $\sigma\sqrt{e}^{|k-2|}$. As a consequence, the three largest spin-orbit resonances are, at least for small to moderate eccentricity, the 1:1, 3:2 and 1:2. For small values of σ or e, the islands are isolated and the global dynamics is regular in the concerned region. If the width of the resonance islands increases, for example by increasing the axial asymmetry, and consequently σ, or the eccentricity of the Keplerian orbit, the resonances partially

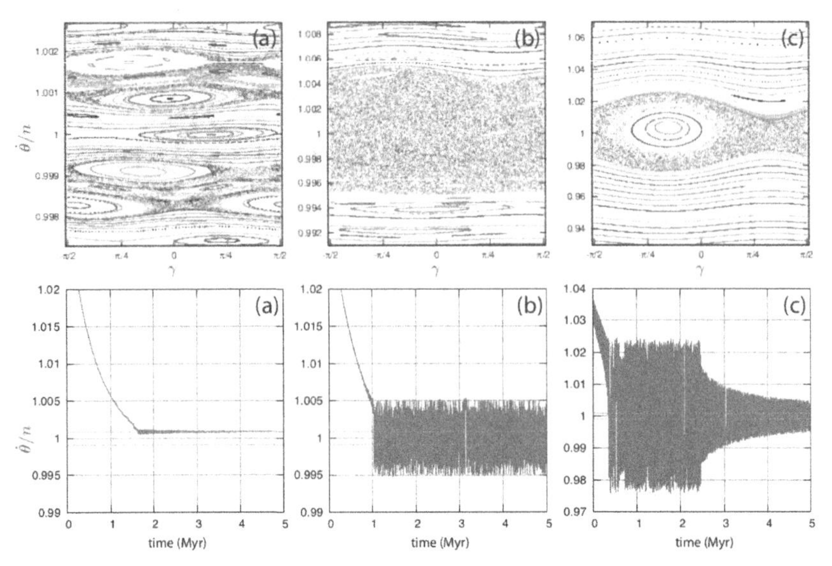

Figure 1. Top: Poincaré surface of section in the plane of coordinates $(\gamma, \dot{\theta}/n)$ for $\beta = 10^{-0.5}$ (a), $10^{0.2}$ (b) and $10^{1.2}$ (c). Bottom: examples of tidal evolution of the rotation rate versus time.

overlap to give rise to a huge chaotic sea. This kind of phenomenon was described by Wisdom *et al.* (1984) in the case of Hyperion. In contrast, if the eccentricity is decreased to zero the width of all spin-orbit resonances, except the 1:1 which corresponds to the synchronous rotation, shrinks to zero. In this case, the Hansen coefficients $X_k^{-3,2}(0)$ with $k \neq 2$ vanish, and the rotational differential equation becomes equivalent to the one of the simple pendulum. Adding tidal dissipations to the rotation model, the rotator can be captured in a spin-orbit resonance, where the probability of the capture depends on the width of each resonance and also on the tidal model (see Goldreich & Peale, 1966). Of course, for circular orbits, the only possible final state is the synchronous rotation.

If the circular orbital motion is now perturbed by an additional companion body, the possible rotational final states can be very different from the Keplerian case, as it was previously shown by Correia & Robutel (2013).

2. The coorbital case

Let us denote m_0 the mass of the central body, m_1 the mass of the rotating body and m_2 the mass of its perturbing coorbital companion. Adapting the theory developed by Érdi (1977) for the restricted 3-body problem to the case of three massive bodies, and neglecting the quantities of second order and more in the small quantity $\mu = (m_1 + m_2)/(m_0 + m_1 + m_2)$, the mean longitudes and the semi-major axes of the coorbitals can be approximated by the expressions (see Robutel *et al.*, 2011):

$$\begin{aligned} \lambda_1(t) &\approx nt + \delta\zeta(t) + \lambda^0, & a_1(t) &\approx \bar{a}\left(1 - \delta\tfrac{2}{3}\tfrac{\dot{\zeta}(t)}{n}\right) \\ \lambda_2(t) &\approx nt - (1-\delta)\zeta(t) + \lambda^0, & a_2(t) &\approx \bar{a}\left(1 + (1-\delta)\tfrac{2}{3}\tfrac{\dot{\zeta}(t)}{n}\right) \end{aligned} \tag{2.1}$$

where λ^0 is a constant, $\delta = m_2/(m_1 + m_2)$, the mean semi-major axis $\bar{a}$ is a constant equal to $(1-\delta)a_1 + \delta a_2$ at the fist order in μ, and the variable $\zeta = \lambda_1 - \lambda_2$ satisfies the second order differential equation:

$$\ddot{\zeta} = -3\mu n^2 \left[1 - (2 - 2\cos\zeta)^{-3/2}\right] \sin\zeta. \tag{2.2}$$

This differential equation, which is one of the most common representations of the coorbital motion (see Morais 1999 and references therein), possesses two stable equilibrium points at $(\zeta, \dot{\zeta})$ equal to $(\pi/3, 0)$ and $(-\pi/3, 0)$ and an unstable one at $(\pi, 0)$, corresponding respectively to the two Lagrangian equilateral configurations (L_4 and L_5) and to the Eulerian collinear configuration of the type L_3. The separatrices emanating from this last unstable point split the phase space in three different regions: two corresponding to the tadpole trajectories surrounding one of the two Lagrange's equilibria, and another one corresponding to the horseshoe orbits which surround the three above-mentioned fixed points. As the solutions of (2.2) are periodic functions of frequency ν (although this frequency depends on the considered trajectory, it is always of order $\sqrt{\mu}$ and consequently small with respect to n), we have:

$$\left(\frac{a_1}{r_1}\right)^3 \mathrm{e}^{-\mathrm{i}2\delta\zeta} = \sum_{k\in\mathbb{Z}} \rho_k \mathrm{e}^{\mathrm{i}(k\nu t+\phi_k)}, \tag{2.3}$$

where the amplitudes ρ_k and the phases ϕ_k depend on the chosen trajectory. As in the Keplerian case, the term $(a/r)^3 \sin 2(\theta - f)$ appearing in (1.1) can be approximated at the first order in μ by the expression:

$$a_1^3 r_1^{-3} \mathrm{Imag}\left[\mathrm{e}^{2i(\theta-f)}\right] = a_1^3 r_1^{-3} \mathrm{Imag}\left[\mathrm{e}^{2i(\theta-nt-\lambda^0)} \mathrm{e}^{-2i\delta\zeta}\right], \tag{2.4}$$

the rotational equation (1.1) can be expanded in Fourier's series in the form:

$$\ddot{\gamma} = -\frac{\sigma^2}{2} \sum_{k\in\mathbb{Z}} \rho_k \sin(2\gamma + k\nu t + \phi_k), \tag{2.5}$$

where $\gamma = \theta - nt - \lambda^0$ represents the orientation of the rotating body with respect to a reference frame rotating at the angular velocity n. It turns out that the main resonances can be found at $\dot{\theta} = n + k\nu/2$, with $k \in \mathbb{Z}$. The respective width of each associated resonant island is equal to $\sigma\sqrt{\rho_k}$. At one of the two Lagrange's equilibria ($\zeta = \pm\pi/3, \dot{\zeta} = 0$) the motion of the two coorbital bodies is circular, then $a_1 = r_1 = \bar{a}$ and the coefficients of the expansion (2.5) verify $\rho_0 = 1$ and $\rho_k = 0$ for $k \neq 0$. In this case, the equation (2.5) becomes equivalent to the pendulum's equation, and as for a circular Keplerian orbit, the only possible spin-orbit resonance is the synchronous one. For small orbital librations around L_4 (resp. L_5) the ρ_k with $k \neq 0$ become strictly positive and the sequence of these coefficients, which satisfies the relations $\rho_0 > \rho_{\pm 1} > \cdots > \rho_{\pm k} > \cdots$, is rapidly decreasing. This phenomenon gives rise to new resonant islands at both sides of the synchronous spin-orbit resonance. The widest islands are located at $\dot{\theta} = n + \nu/2$ and $\dot{\theta} = n - \nu/2$, which corresponds respectively to a super-synchronous and to a sub-synchronous rotation. The distance between these resonances being small, of the order of ν (that is, of order $\sqrt{\mu}$), an increase in the orbital libration amplitude causes an overlapping of these resonances which can lead to chaotic rotations. We show in Correia & Robutel (2013) that the global dynamics of the problem is mainly controlled by two parameters: the amplitude of the orbit around the fixed points and the quantity $\beta = \sigma/(\sqrt{\mu}n)$ which is proportional to the frequency ration σ/ν.

For large tadpole orbits and more specifically for horseshoe orbits, the situation becomes more complicated than the one described above. First of all, the sequence of the coefficients ρ_k does not decrease rapidly. It follows that the width of the islands centered at $\dot{\theta} = n \pm k\nu/2$ with moderate k are comparable, which greatly enriches the rotational dynamics. In addition, ρ_0 is not necessarily the dominant coefficient and the synchronous island is no longer the widest one, as it is shown in Fig. 1-a (top).

The top panel of Fig. 1 represents the Poincaré sections at the time $t = 2\pi/\nu$ (the coorbital period) of the differential equation (1.1) for three different values of β. In these three cases, the bodies whose masses verify the relation $m_1 = m_2 = 10^{-6}m_0$ are on a horseshoe orbit of amplitude $\text{Max}|\lambda_1 - \lambda_2| = 320°$. This figure is equivalent to those in Correia & Robutel (2013), although in previous work we only show the phase space for tadpole orbits. In Fig. 1-a, where $\beta = 10^{-0.5}$, the phase space is dominated by 5 islands approximatively centered at $\dot{\theta} = n, n \pm \nu/2, n \pm \nu$. As mentioned above, the synchronous island is not the widest one. The islands are isolated and consequently, the dynamics is globally regular. For a larger values of β (typically of order 1) the main islands begin to overlap, generating a broad chaotic zone around the synchronous rotation (Fig. 1-b). Finally, when $\beta \gg 1$ (Fig. 1-c), the proximity of the main resonant islands leads to their complete overlapping. The dynamical structure is comparable to that of a modulated pendulum (see Morbidelli 2002) where the single island, which corresponds to the synchronous rotation, is surrounded by a small chaotic layer.

3. Discussion

By studding the rotation of a body in a quasi-circular orbit perturbed by a coorbital, we highlight a new mechanism generating spin-orbit resonances, where the role played by the eccentricity in the case of Keplerian orbits is replaced by the amplitude of libration around the coorbital equilibria. For a zero amplitude (Lagrange's configurations), the only possible spin-orbit resonance is the synchronization. The increase of this amplitude results in new resonant islands, associated to super-synchronous and sub-synchronous rotation, whose width becomes maximal in the horseshoe region. If tidal dissipation is added to the equations (1.1), various final states can be reached. In the bottom panel of the figure 1 we show examples of the rotational final states for coorbital bodies using the viscous linear model (see Mignard, 1979), whose contribution to the equation of the rotation is given by: $\ddot{\gamma} = -Ka_1^6 r_1^{-6}(\dot{\gamma} - \delta\dot{\zeta})$, with $K = 250\,\text{yr}^{-1}$. In these three simulations, the rotation rate that begins slightly above the synchronization ($\dot{\theta}/n = 1.033$), decreases before reaching its final state in a few million years. In the case (a), the rotation ends in a super-synchronous rotation of rate $\dot{\theta} = n + \nu/2$. In (b), the rotation enters the large chaotic region generated by the marginal overlapping of the main spin-orbit islands, while in (c), the rotation reaches the synchronous island which is the only significant resonance for late values of β.

References

Correia, A. C. M. & Robutel, P. 2013, *AJ*, 779, 20
Érdi, B. 1977, *Celestial Mechanics*, 15, 367
Goldreich, P. & Peale, S. 1966, *ApJ*, 71, 425
Mignard, F. 1979, *Moon and Planets*, 20, 301
Morbidelli, A. 2002, *Modern Celestial Mechanics* (London: Taylor and Francis)
Morais, M. H. M. 1999, *A&A*, 350, 318
Robutel, P., Rambaux, N., & Castillo-Rogez, J. 2011, *Icarus*, 211, 758
Wisdom, J., Peale, S. J., & Mignard, F. 1984, *Icarus*, 58, 137

Complex Planetary Systems
Proceedings IAU Symposium No. 310, 2014
Z. Knežević & A. Lemaitre, eds.

doi:10.1017/S1743921314007716

A supplementary note on constructing the general Earth's rotation theory

Victor A. Brumberg and Tamara V. Ivanova

Institute of Applied Astronomy,
St.–Petersburg, 191187, Russia
email: vabrumberg@verizon.net, itv@ipa.nw.ru

Abstract. Representing a post-scriptum supplementary to a previous paper of the authors Brumberg & Ivanova (2011) this note aims to simplify the practical development of the Earth's rotation theory, in the framework of the general planetary theory, avoiding the non–physical secular terms and involving the separation of the fast and slow angular variables, both for planetary–lunar motion and Earth's rotation. In this combined treatment of motion and rotation, the fast angular terms are related to the mean orbital longitudes of the bodies, the diurnal and Euler rotations of the Earth. The slow angular terms are due to the motions of pericenters and nodes, as well as the precession of the Earth. The combined system of the equations of motion for the principal planets and the Moon and the equations of the Earth's rotation is reduced to the autonomous secular system with theoretically possible solution in a trigonometric form. In the above–mentioned paper, the Earth's rotation has been treated in Euler parameters. The trivial change of the Euler parameters to their small declinations from some nominal values may improve the practical efficiency of the normalization of the Earth's rotation equations. This technique may be applied to any three-axial rigid planet. The initial terms of the corresponding expansions are given in the Appendix.

Keywords. celestial mechanics, methods: analytical, Earth: rotation

---

This note represents a post-scriptum remark to our paper Brumberg & Ivanova (2011) (later denoted by BI) to improve the practical efficiency of the normalization of the Earth's rotation equations. The seven first–order differential equations of the Earth's rotation treated in this paper in the formula (2.45) typical for General Planetary Theory (GPT) are:

$$\dot{X} = i\mathcal{N}[PX + R(X,t)] \tag{1}$$

where X and R stand for 7–vectors of the variables and right–hand members, respectively, $X = (X_i) = (u,\, \bar{u},\, v,\, \bar{v},\, m',\, \bar{m}',\, m_3)$, and $R = (R_i)$, $\quad (i = 1, 2, \ldots, 7)$.
The first four components of X represent the Euler parameters related to the ordinary Euler angles φ, ψ, θ by means of the relations : $u = -\sin\frac{\theta}{2}\, e^{-i\frac{\psi+\varphi}{2}}$, $v = i\cos\frac{\theta}{2}\, e^{i\frac{\psi-\varphi}{2}}$ and interrelated by the identity:

$$u\bar{u} + v\bar{v} \equiv 1. \tag{2}$$

The last three components of X are small dimensionless quantities determining the components of the angular Earth's rotation velocity $\omega = (\omega_i)$, $(i = 1, 2, 3)$ by means of $\omega_1 = \Omega\, m_1$, $\omega_2 = \Omega\, m_2$, $\omega_3 = \Omega\,(1 + m_3)$, Ω being the mean Earth's rotation velocity. In consistency with our previous papers $\Omega = -2n, n = \text{const}$.

With I_1, I_2, I_3 denoting the Earth's principal inertia moments, $k_1 = \frac{I_3 - I_1}{2I_2}$ and $k_2 = \frac{I_3 - I_2}{2I_1}$, we write:

$$m_1 = 2\sqrt{k_2}\, m_1', \quad m_2 = 2\sqrt{k_1}\, m_2' \quad m = m_1 + im_2 \quad m' = m_1' + im_2'\,. \tag{3}$$

The right–hand members are:

$$
\begin{aligned}
R_1 &= m_3 u - m\bar{v} & R_2 &= -\bar{R}_1, \\
R_3 &= m_3 v + m\bar{u}, & R_4 &= -\bar{R}_3 \\
R_5 &= -4\sqrt{k_1 k_2}\, m_3 m' - \frac{1}{4n^2}\left(\frac{1}{\sqrt{k_1}}\frac{\mathcal{M}_2}{I_2} - \frac{i}{\sqrt{k_2}}\frac{\mathcal{M}_1}{I_1}\right) & R_6 &= -\bar{R}_5 \\
R_7 &= 2\sqrt{k_1 k_2}\,\frac{I_1 - I_2}{I_3}(m'^2 - \bar{m}'^2) + \frac{i}{2n^2}\frac{\mathcal{M}_3}{I_3},
\end{aligned}
$$

$\mathcal{M}_1$, $\mathcal{M}_2$, $\mathcal{M}_3$ being the components of the torque vector. Their expressions and expansion techniques are given in BI.

At last, $\mathcal{N}$ and P are 7×7 diagonal matrices with the structure $\mathcal{N} = \mathrm{diag}(n, n, n, n, n, n, n)$, $P = \mathrm{diag}(1, -1, 1, -1, \epsilon, -\epsilon, 0)$ with $\epsilon = -4\sqrt{k_1 k_2}$. The main frequencies of u, v and m' are n, n and $n\epsilon$, respectively (the last frequency corresponds to the Euler period of the Earth's rotation). The frequency for m_3 is zero.

In BI, the Earth's rotation system (1) was treated together with the system describing the planetary–lunar motion in the framework of GPT. Due to the identity (2) and the specific features of the right–hand members, there were no difficulties related to the peculiarities of P. However, from the practical point of view, it might be more adequate not to deal with variables u, v themselves but rather to use their small declinations from some nominal values. The simplest option seems to choose, as a nominal initial approximation, $\underset{0}{X}$ satisfying the equations: $\underset{0}{\dot{X}} = i\mathcal{N}P\underset{0}{X}$. The components of $\underset{0}{X}$ are:

$$
\underset{0}{u} = a\, e^{int}, \underset{0}{v} = b\, e^{int}, \underset{0}{m'} = c\, e^{i\epsilon nt}, \underset{0}{m_3} = 0, \tag{4}
$$

with constant a, b, c (the zero value for $\underset{0}{m_3}$ can be always provided by the choice of Ω).

In terms of the Euler angles, this nominal approximation corresponds to $\varphi = -2nt + \varphi_0$, $\psi = \psi_0$, $\theta = \theta_0$, φ_0, ψ_0, θ_0 being constants.

Moreover, similarly to (2), we obtain: $a\bar{a} + b\bar{b} \equiv 1$ with

$$
a = -\sin\frac{\theta_0}{2}\, e^{-i\dfrac{\psi_0 + \varphi_0}{2}}, b = i\cos\frac{\theta_0}{2}\, e^{i\dfrac{\psi_0 - \varphi_0}{2}}, c = -\frac{1}{4n}\left(\frac{(\omega_1)_0}{\sqrt{k_2}} + i\frac{(\omega_2)_0}{\sqrt{k_1}}\right),
$$

$(\omega_j)_0$ $(j = 1, 2)$ being the components of the angular Earth's rotation velocity for the initial moment of time.

Returning now to the original equations (1), let us define $X = \underset{0}{X} + \delta X$ reducing (1) to:

$$
\delta\dot{X} = i\mathcal{N}[P\delta X + R(\underset{0}{X} + \delta X, t)]. \tag{5}
$$

Equations (1) and (5) have the same form but, in contrast to X, the components δX are small quantities. The right–hand members in (1) are trigonometric functions of the mean longitudes of the planets and the Moon. In addition, the right–hand members in (5) have two trigonometric arguments nt and ϵnt related to the Earth's rotation. With numerical values for φ_0, ψ_0, θ_0 and c, one can easily get the right–hand member expansions generalizing those treated in BI. Afterwards, one may solve (5) in traditional manner, by using Picard–type iterations, in powers of δX or in more sophisticated manner, by combining (5) with the planetary–lunar GPT equations and finding the normalizing transformation to the combined secular system, as it was done in BI. It may be noted that the right–hand member R in (5) is holomorphic with respect to δX with the zero– and first–degree terms proportional to the small perturbing parameters. Such terms involve the corresponding contributions into the normalization transformation described by (4.8)–(4.12) of BI but do not interfere with the resulting secular system.

References

Brumberg, V. & Ivanova, T. 2011, *Cel. Mech. Dyn. Astron.* 109, 385

Appendix

In Appendix we give a set of formulas to compute the initial terms of the expansions for the Earth's rotation variables (or any other three–axial rigid planet). Using (3) for the variable $\underset{0}{m}$ and the nominal solution (4), we obtain:

$$\delta\dot{u} = in(\delta u - \underset{0}{m}\,\underset{0}{\bar{v}}) = in\delta u - in\bar{b}\,[(\sqrt{k_2}+\sqrt{k_1}\,)ce^{i(-1+\epsilon)nt} + (\sqrt{k_2}-\sqrt{k_1}\,)\bar{c}e^{i(-1-\epsilon)nt}],$$

$$\delta\dot{v} = in(\delta v + \underset{0}{m}\,\underset{0}{\bar{u}}) = in\delta v + in\bar{a}\,[(\sqrt{k_2}+\sqrt{k_1}\,)ce^{i(-1+\epsilon)nt} + (\sqrt{k_2}-\sqrt{k_1}\,)\bar{c}e^{i(-1-\epsilon)nt}],$$

$$\delta\dot{m}' = in\epsilon\,\delta m' - \frac{1}{4n}\left(\frac{1}{\sqrt{k_2}}\frac{\mathcal{M}_1}{I_1} + \frac{i}{\sqrt{k_1}}\frac{\mathcal{M}_2}{I_2}\right),$$

$$\delta\dot{m}_3 = -\frac{1}{2}in\epsilon\,\frac{I_1 - I_2}{I_3}\left(c^2 e^{2i\epsilon nt} - \bar{c}^2 e^{-2i\epsilon nt}\right) - \frac{1}{2n}\frac{\mathcal{M}_3}{I_3}$$

resulting to

$$\delta u = \bar{b}\Big[c\frac{\sqrt{k_2}+\sqrt{k_1}}{2-\epsilon}e^{i(-1+\epsilon)nt} + \bar{c}\frac{\sqrt{k_2}-\sqrt{k_1}}{2+\epsilon}e^{i(-1-\epsilon)nt}\Big],$$

$$\delta v = -\bar{a}\Big[c\frac{\sqrt{k_2}+\sqrt{k_1}}{2-\epsilon}e^{i(-1+\epsilon)nt} + \bar{c}\frac{\sqrt{k_2}-\sqrt{k_1}}{2+\epsilon}e^{i(-1-\epsilon)nt}\Big],$$

$$\delta m' = -\frac{1}{4n}\left[\int\left(\frac{1}{\sqrt{k_2}}\frac{\mathcal{M}_1}{I_1} + \frac{i}{\sqrt{k_1}}\frac{\mathcal{M}_2}{I_2}\right)e^{-i\epsilon nt}dt\right]e^{i\epsilon nt},$$

$$\delta m_3 = -\frac{1}{4}\frac{I_1 - I_2}{I_3}\left(c^2 e^{2i\epsilon nt} + \bar{c}^2 e^{-2i\epsilon nt}\right) - \frac{1}{2n}\int\frac{\mathcal{M}_3}{I_3}\,dt.$$

In addition, we demonstrate the initial terms of $\delta m'$ and δm_3 expansions provided that the Sun and the Moon move in planar circular orbits.

$$\begin{aligned}
\delta m' = {} & (1+\epsilon_1)(1-2a\bar{a})\left[ab\frac{KC_1+LC_2}{2-\epsilon}e^{2int} + \bar{a}\bar{b}\frac{KC_2+LC_1}{2+\epsilon}e^{-2int}\right] \\
& + \sum_{j=3,9}(\delta_{j3}+\epsilon_1\delta_{j9}) \times \Bigg[a\bar{b}^3\frac{KC_2+LC_1}{2+2n'_j+\epsilon}e^{-2i(nt+\lambda_j)} - \bar{a}^3 b\frac{KC_2+LC_1}{2-2n'_j+\epsilon}e^{-2i(nt-\lambda_j)} \\
& - a^3\bar{b}\frac{KC_1+LC_2}{2-2n'_j-\epsilon}e^{2i(nt-\lambda_j)} + \bar{a}b^3\frac{KC_1+LC_2}{2+2n'_j-\epsilon}e^{2i(nt+\lambda_j)}\Bigg] \\
& + \frac{1}{2}\sigma C_1\Bigg\{a\bar{b}^3\,L\left[\frac{e^{-i(2nt+\lambda_3+\lambda_9)}}{2+n'_3+n'_9+\epsilon} + 5\frac{e^{-i(2nt+3\lambda_3-\lambda_9)}}{2+3n'_3-n'_9+\epsilon}\right] \\
& + \bar{a}b^3\,K\left[\frac{e^{i(2nt+\lambda_3+\lambda_9)}}{2+n'_3+n'_9-\epsilon} + 5\frac{e^{i(2nt+3\lambda_3-\lambda_9)}}{2+3n'_3-n'_9-\epsilon}\right] \\
& - \bar{a}^3 b\,L\left[\frac{e^{-i(2nt-\lambda_3-\lambda_9)}}{2-n'_3-n'_9+\epsilon} + 5\frac{e^{-i(2nt-3\lambda_3+\lambda_9)}}{2-3n'_3+n'_9+\epsilon}\right]
\end{aligned}$$

$$-a^3\bar{b}\,K\left[\frac{e^{i(2nt-\lambda_3-\lambda_9)}}{2-n_3'-n_9'-\epsilon}+5\frac{e^{i(2nt-3\lambda_3+\lambda_9)}}{2-3n_3'+n_9'-\epsilon}\right]$$
$$+3(1-2a\bar{a})\bar{a}\bar{b}\,L\left[\frac{e^{-i(2nt+\lambda_3-\lambda_9)}}{2+n_3'-n_9'+\epsilon}+\frac{e^{-i(2nt-\lambda_3+\lambda_9)}}{2-n_3'+n_9'+\epsilon}\right]$$
$$+3(1-2a\bar{a})ab\,K\left[\frac{e^{i(2nt+\lambda_3-\lambda_9)}}{2+n_3'-n_9'-\epsilon}+\frac{e^{i(2nt-\lambda_3+\lambda_9)}}{2-n_3'+n_9'-\epsilon}\right]\Bigg\}, \tag{6}$$

$$\delta m_3 = B+\overline{B}, \tag{7}$$

$$B = L\Bigg\{-\frac{1}{2}(1+\epsilon_1)a^2b^2e^{4int}+\sum_{j=3,9}\frac{1}{2}(\delta_{j3}+\epsilon_1\delta_{j9})\left[a^4\frac{e^{2i(2nt-\lambda_j)}}{2-n_j'}+b^4\frac{e^{2i(2nt+\lambda_j)}}{2+n_j'}\right]$$
$$+\frac{1}{2}\sigma\Bigg\{a^4\left[\frac{e^{i(4nt-\lambda_3-\lambda_9)}}{4-n_3'-n_9'}+5\frac{e^{i(4nt-3\lambda_3+\lambda_9)}}{4-3n_3'+n_9'}\right]+b^4\left[\frac{e^{i(4nt+\lambda_3+\lambda_9)}}{4+n_3'+n_9'}+5\frac{e^{i(4nt+3\lambda_3-\lambda_9)}}{4+3n_3'-n_9'}\right]$$
$$-6a^2b^2\left[\frac{e^{i(4nt+\lambda_3-\lambda_9)}}{4+n_3'-n_9'}+\frac{e^{i(4nt-\lambda_3+\lambda_9)}}{4-n_3'+n_9'}\right]\Bigg\}\Bigg\}-\frac{1}{4}\frac{I_1-I_2}{I_3}c^2e^{2i\epsilon nt}.$$

In these expressions,

$$C_1 = \frac{I_3\left(I_1\sqrt{k_2}+I_2\sqrt{k_1}\right)}{2I_1I_2\sqrt{k_1k_2}}, \qquad C_2 = \frac{I_3\left(I_2\sqrt{k_1}-I_1\sqrt{k_2}\right)}{2I_1I_2\sqrt{k_1k_2}},$$
$$K = -\frac{3}{4}\left(1-\frac{I_1+I_2}{2I_3}\right)\frac{GM_0}{A_3^3n^2}, \qquad L = -\frac{3}{8}\frac{I_1-I_2}{I_3}\frac{GM_0}{A_3^3n^2},$$
$$\varepsilon_1 = \tfrac{M_9}{M_0}\left(\tfrac{A_3}{A_9}\right)^3, \qquad \sigma = \tfrac{M_9}{M_3}\tfrac{A_9}{A_3}.$$

$n_j' = \dfrac{n_j}{n}$ $(j = 3, 9)$, M_0 is the mass of the Sun, M_j, A_j, λ_j denote masses, semi–major axes and mean longitudes of the Earth–Moon barycentre (for $j = 3$) and the Moon (for $j = 9$), respectively, δ_{ij} being the Kronecker symbol.

The resulting expansions (6) and (7) are obtained in the literal form with respect to two small parameters K and L. Needless to say, they have the trigonometric form with respect to time.

Complex Planetary Systems
Proceedings IAU Symposium No. 310, 2014
Z. Knežević & A. Lemaitre, eds.

doi:10.1017/S1743921314007728

New clues on the interior of Titan from its rotation state

Benoît Noyelles[1] **and Francis Nimmo**[2]

[1]NAmur Center for CompleX SYStems (naXys), University of Namur, Rempart de la Vierge 8, B-5000 Namur, Belgium
email: benoit.noyelles@gmail.com

[2]Department of Earth and Planetary Sciences, University of California at Santa Cruz, 1156 High Street, Santa Cruz, California 95064, USA
email: fnimmo@es.ucsc.edu

Abstract. The Saturnian satellite Titan is one of the main targets of the *Cassini-Huygens* mission, which revealed in particular Titan's shape, gravity field, and rotation state. The shape and gravity field suggest that Titan is not in hydrostatic equilibrium, that it has a global subsurface ocean, and that its ice shell is both rigid (at tidal periods) and of variable thickness. The rotational state of Titan consists of an expected synchronous rotation rate and an unexpectedly high obliquity (0.3°) explained by Baland *et al.* (2011) to be a resonant behavior. We here combine a realistic model of the ice shell and interior and a 6-degrees of freedom rotational model, in which the librations, obliquity and polar motion of the rigid core and of the shell are modelled, to constrain the structure of Titan from the observations. We consider the gravitational pull of Saturn on the 2 rigid layers, the gravitational coupling between them, and the pressure coupling at the liquid-solid interfaces.

We confirm the influence of the resonance found by Baland *et al.* that affects between 10 and 13% of the possible Titans. It is due to the 29.5-year periodic annual forcing. The resonant Titans can be obtained in situations in which a mass anomaly at the shell-ocean boundary (bottom loading) is from 80 to 92% compensated. This suggests a 250 to 280 km thick ocean below a 130 to 140 km thick shell, and is consistent with the degree-3 analysis of Hemingway *et al.* (2013).

Keywords. celestial mechanics, planets and satellites: individual: Titan

---

The *Cassini-Huygens* space mission has significantly improved our knowledge of Titan. In particular, there are more and more clues suggesting the presence of a global internal ocean. This contribution aims at modeling the rotation of Titan in considering a rigid shell surrounding a global ocean and a rigid core. We here extend the studies by Baland *et al.* (2011) and Baland *et al.* (2014) to a system having 6 dynamical degrees of freedom, i.e. three each (longitude, obliquity and polar motion) for the shell and the inner core.

1. Our knowledge of Titan

Several studies since Lunine & Stevenson (1987) have suggested that Titan's large size results in a global ocean beneath its icy shell. This ocean is consistent with the high tidal Love number measured by *Cassini* (Iess *et al.* (2012)).

Hints on the interior can be obtained from the gravity field of Titan that depends on the mass distribution. Unfortunately, two methods of solving for the gravity field do not exactly converge (Tab.1). While the SOL2 solution, using all available data on Titan, could be consistent with a hydrostatic body (J_2/C_{22} close to 10/3), the SOL1 solution, based only on 4 flybys of *Cassini*, is inconsistent with hydrostatic equilibrium. Anyway,

Table 1. The two solutions for the gravity field of Titan, from Iess *et al.* (2010).

	SOL1	SOL2
J_2	$(3.1808 \pm 0.0404) \times 10^{-5}$	$(3.3462 \pm 0.0632) \times 10^{-5}$
C_{22}	$(9.983 \pm 0.039) \times 10^{-6}$	$(1.0022 \pm 0.0071) \times 10^{-5}$
J_2/C_{22}	3.186 ± 0.042	3.339 ± 0.067

neither of these two solutions is consistent with a hydrostatic Titan when compared to the shape (Zebker *et al.* (2009)). This discrepancy led to the suggestion of mass anomalies in Titan, resulting from either lateral variations of the shell thickness (Nimmo & Bills (2010)) or of the shell density (Choukroun & Sotin (2012)). The recent degree-3 analysis of the topography by Hemingway *et al.* (2013) supports lateral variations.

Observations of the rotation of Titan suggest that it rotates synchronously (Meriggiola & Iess (2012)), as usually expected for the main natural satellites. However, an unexpectedly high obliquity of 18.6 ± 3 arcmin at the mean date March 11th, 2007 has been measured (Stiles *et al.* (2008), Meriggiola & Iess (2012)), that cannot be explained with a rigid Titan (Bills & Nimmo (2008)). Baland *et al.* (2011) have shown that the presence of a global ocean could explain this obliquity.

2. Our model

We consider Titan as a triaxial body composed of 3 layers: a rigid inner core, a global ocean, and a shell that is completely rigid at tidal periods but deformable at much longer periods, as appropriate for a viscoelastic material like ice. We then build realistic model-Titans. For that, we start from a hydrostatic Titan in which we add a mass anomaly, either at the surface (top loading), or at the ocean-crust boundary (bottom loading). Because the duration of loading is likely tens of Myr at least, we then consider a partial isostatic compensation of this mass anomaly. We assume Airy isostasy, in which lateral variations of the shell thickness occur. Only model Titans that are consistent with the measured shape and gravity fields are kept. Since two solutions exist for the gravity field, we have two sets of model Titans. The resulting model Titans depend on 12 parameters, i.e. a density and 3 radii defining the triaxial outer boundary of each of the three layers. We can reproduce the observations only with bottom loading, and require between 80 and 85% isostatic compensation for the gravity field SOL1, and between 85 and 92% for the gravity field SOL2.

We then simulate the rotation of each of these model Titans at the dynamical equilibrium, i.e. synchronous rotation of the two rigid layers and equilibrium obliquity corresponding to the Cassini State 1 for rigid bodies. As Baland *et al.* (2011) did, we consider the gravitational torque of Saturn on Titan, the gravitational coupling between the two rigid layers, and the pressure coupling at the fluid-solid interfaces. We extended their equations to 6 degrees of freedom, simulating simultaneously, for the two rigid layers, the longitudinal motion, the obliquity, and the polar motion, as Szeto & Xu (1997) did but without considering the pressure coupling at the two rigid-fluid boundaries. We then extracted the rotational outputs, and in this study we particularly focus on the obliquity. A weakness in our physical model is that we assume that the shell is completely rigid at tidal periods. The elasticity can be critical for the longitudinal librations (Goldreich & Mitchell (2010), Van Hoolst *et al.* (2013), Richard *et al.* (2014)); the extent to which it matters for obliquity is not yet clear.

We model the orbit of Titan around Saturn with TASS1.6 ephemerides (Vienne & Duriez (1995)), that have the advantage of being available over 9000 years. This is re-

Table 2. Quasiperiodic decomposition of the orbital quantity $I \exp \imath\Omega$ from TASS1.6 (Vienne & Duriez (1995)), I being the inclination of Titan and Ω the longitude of its ascending node. The variations of the shell obliquity resonate with the annual forcing (Term 5).

N	Amplitude (arcmin)	Period (yr)	Cause
1	38.52	∞	Sun
2	19.18	-703.51	Saturn's J_2
3	0.90	-3263.07	Iapetus
4	0.77	14.73	Semi-annual
5	0.13	-29.46	Annual

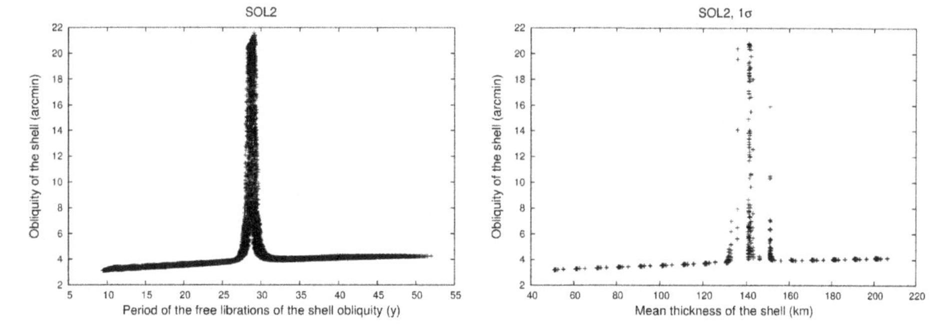

Figure 1. Mean obliquity of Titan vs. the period of the free oscillations (left) and the shell thickness (right). This illustrates how the resonance constrains the interior.

quired to accurately study the obliquity, which results in a slow motion.

3. A resonant behavior

Fig.1 shows that two regimes can be expected for the shell obliquity. Away from the peak, we see a small obliquity, close to the one to be expected for a rigid body. However, we see near the peak much larger obliquities. As Baland *et al.* (2011) already noticed, these larger obliquities are in fact due to a resonance between the annual forcing (see Tab.2) and the free oscillations of the obliquity of the shell, their period being denoted T_4. Baland *et al.* (2011) also suggest a possible resonance with the Term 2 of Tab.2 with a wider range of parameters than we have. They also find it in Baland *et al.* (2014) with a non-hydrostatic Titan and consider it more probable than a resonance with the annual forcing.

For model Titans consistent with the gravity solution SOL1, the high obliquity suggests a shell thickness of ≈ 130 km, and ≈ 140 km for SOL2. 10 to 13% of our model Titans fall into this resonance, making it an acceptable explanation.

To deduce the period of the free oscillations T_4 consistent with the obliquity measurement, we have to keep in mind that the measured obliquity is an instantaneous value, different from the mean obliquity (see Fig. 2). Actually two periods are possible, we have either $T_4 = 29.3 \pm 0.03$ years, or $T_4 = 29.572^{+0.019}_{-0.015}$ years.

4. A time-varying obliquity

A numerical simulation of the time evolution of this obliquity predicts a variation of 7 arcmin over the duration of the *Cassini* mission. The slope should be either positive or

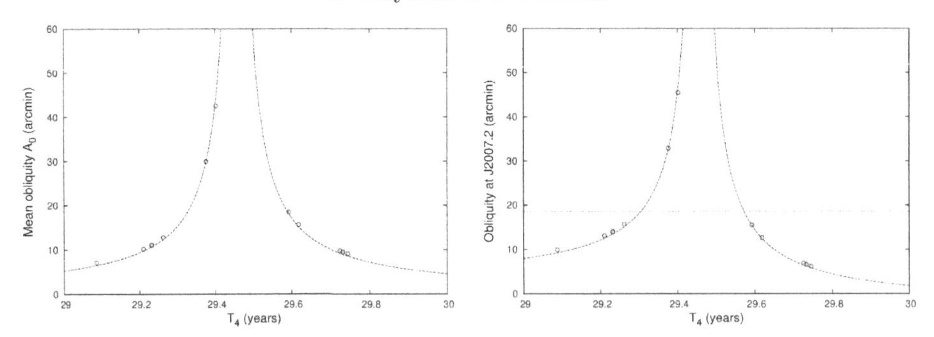

Figure 2. Mean (left) and instantaneous (right) obliquity of Titan. The asymmmetry present for the instantaneous obliquity but absent for the mean one illustrates the difference between the two quantities. The measurement corresponds to the instantaneous obliquity.

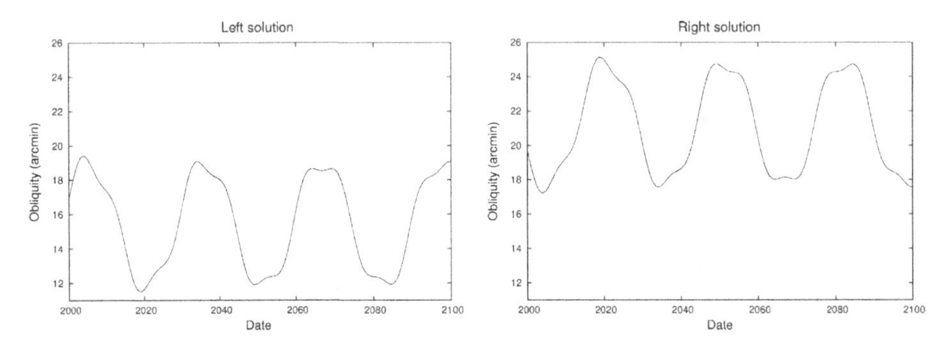

Figure 3. Possible time variations of the obliquity of Titan, with $T_4 = 29.3$ years (left) and $T_4 = 29.572$ years (right).

negative, depending whether T_4 is smaller or bigger than 29.46 years. So, observing this slope would give an additional constraint on the state of the interior.

References

Baland, R.-M., Van Hoolst, T., Yseboodt, M., & Karatekin, Ö. 2011, *A&A*, 530, A141
Baland, R.-M., Tobie, G., Lefèvre A., & Van Hoolst, T., 2014, *Icarus*, 237, 29
Bills, B. G. & Nimmo, F. 2008, *Icarus*, 196, 293
Choukroun, M. & Sotin, C. 2012, *Geophysical Research Letters*, 39, L04201
Goldreich, P. M. & Mitchell, J. L. 2010, *Icarus*, 209, 631
Hemingway, D., Nimmo, F., Zebker, H., & Iess, L. 2013, *Nature*, 500, 550
Iess, L., Rappaport, N. J., & Jacobson, R. A. *et al.* 2010, *Science*, 327, 1367
Iess, L., Jacobson, R. A., & Ducci, M. *et al.* 2012, *Science*, 337, 457
Lunine, J. I. & Stevenson, D. J. 1987, *Icarus*, 70, 61
Meriggiola, R. & Iess, L. 2012, *European Planetary Science Congress*, id. EPSC2012-593
Nimmo, F. & Bills, B. G. 2010, *Icarus*, 208, 896
Noyelles, B., & Nimmo, F., submitted
Richard, A., Rambaux, N., & Charnay, B. 2014, *Planetary and Space Science*, 93-94, 22
Stiles, B. W., Kirk, R. L., & Lorenz, R. D. *et al.* 2008, *AJ*, 135, 1669
Szeto, A. M. K. & Xu, S. 1997, *Journal of Geophysical Research*, 102, 27651
Van Hoolst, T., Baland, R.-M., & Trinh, A. 2013, *Icarus*, 226, 299
Vienne, A. & Duriez, L. 1995, *A&A*, 297, 588
Zebker, H. A., Stiles, B., & Hensley, S. *et al.* 2009, *Science*, 324, 921

Complex Planetary Systems
Proceedings IAU Symposium No. 310, 2014
Z. Knežević & A. Lemaitre, eds.

doi:10.1017/S174392131400773X

Complements to the longitudinal librations of an elastic 3-layer Titan on a non-Keplerian orbit

Andy Richard[1,2] **and Nicolas Rambaux**[1,2]

[1]IMCCE, Observatoire de Paris, UMR 8028, CNRS, UPMC
77 avenue Denfert-Rochereau, 75014 Paris, France
email: arichard@imcce.fr

[2]Université Pierre et Marie Curie, UPMC - Paris 06,
4 Place Jussieu, 75005 Paris
email: Nicolas.Rambaux@imcce.fr

Abstract. Titan longitudinal librations are dependent on the satellite internal structure and the elastic behavior of the surface. The elastic deformation of the surface is related to the perturbing potential through the Love theory. In a previous paper, we described the deformation as a response to the tidal potential exerted by Saturn at orbital frequency. Here we improve the tidal deformation reponse by including the effect of the libration angle and the orbital perturbations. We then provide the libration amplitudes associated with the rotational model of a tidally deformed three-layer Titan evolving on a non-Keplerian orbit.

Keywords. planets and satellites: individual (Titan), librations, rotation, elasticity, perturbations

1. Introduction

Goldreich and Mitchell (2010) suggested that the librational motion of icy satellites should be strongly reduced when the icy shell behaves elastically. In the case of an elastic icy shell, the surface should deform instead of rotate under the influence of the gravitational potential of the planet. In the case of Titan, the reduction of the longitudinal libration has been assessed by Van Hoolst *et al.* (2013) and Jara-Orué and Vermeersen (2014) at the orbital frequency and Richard *et al.* (2014) for elastic three-layered Titan evolving on a non-Keplerian orbit. Richard *et al.* (2014) computed the deformations of the surfaces by using the Love theory and the perturbing tidal potential detailed by Giampieri (2004). We analyzed the Horizons ephemerides (Giorgini *et al.* 1996) of Titan to obtain the frequencies of the perturbed orbit present in the gravitational torque exerted by Saturn. By taking into account the elasticity, two new libration frequencies were identified. Here we complete the results obtained in the paper of Richard *et al.* (2014) by including the libration angle and the orbital perturbations in the deformation process and evaluating the libration amplitude resulting at different orbital forcing frequencies in the case of a non-Keplerian orbit.

2. Rotational model

The rotational dynamics of Titan composed of an icy shell, an internal ocean and a solid inner core can be given by the Euler equations written for each layer in the Tisserand reference frame of the satellite by taking into account the external and internal gravitational torques and the ocean pressure torque (see *e.g.* Richard *et al.* 2014).

Under the tidal potential of Saturn, the surfaces of the layers composing Titan are deformed. Thus, the mass distribution depends on the time and the inertia tensor can be decomposed into a static and a dynamic part such as $\mathcal{I} = \mathcal{I}_0 + \Delta\mathcal{I}(t)$. The dynamic part of the inertia tensor can be obtained by using the Love theory describing the deformation under the influence of a perturbing potential.

The perturbing tidal potential W for a synchronous satellite is expressed by Giampieri (2004) when the libration angle is neglected. As a complement to the paper of Richard *et al.* (2014), the libration angle and the orbital perturbations are included in the potential which is then written in spherical coordinates (r_o, θ, ϕ) with r_o the mean radius of the layer considered, θ and ϕ the colatitude and longitude respectively

$$W(r_o,\theta,\phi) = -\frac{GM}{a^3}r_o^2\left(\frac{a}{r}\right)^3\left[-\frac{1}{2}P_2^0(\cos\theta) + \frac{1}{4}P_2^2(\cos\theta)\cos 2\phi\cos 2(\nu - L - \gamma) + \frac{1}{4}P_2^2(\cos\theta)\sin 2\phi\sin 2(\nu - L - \gamma)\right], \quad (2.1)$$

where G is the gravitational constant, M is the mass of the planet, a is the semi-major axis of Titan orbit, P_2^j are the associated Legendre polynomials of degree 2, ν is the true longitude, L is the mean longitude and γ is the libration angle.

The gravitational torque exerted by Saturn on Titan surface depends on the position of the planet expressed in the reference frame of the satellite. We use the rotation matrices R_1 and R_3 to define cartesian coordinates (x, y, z) of the planet expressed in Titan frame such as $(x, y, z)^T = R_3(\varphi)R_1(\Theta)R_3(\Psi)R_3(-\Omega)R_1(-i)R_3(-\omega)(r\cos f_p, r\sin f_p, 0)^T$ where f_p is the true anomaly of the planet, ω, i and Ω are the classical orbital elements, and φ, Θ and Ψ are the Euler angles. By using the Saturn equatorial plane as the inertial reference frame, Θ and i can be neglected.

The longitude of the planet in the reference frame of the satellite is given by $\phi_p = \Omega + \omega + f_p - \Psi - \varphi$. We use the definition of Eckhardt (1965) $\Psi + \varphi = \pi + L_s + \gamma$, where L_s is the mean longitude of the satellite defined as $L = Nt + L_0$ with N being the mean mean motion (Vienne and Duriez 1995) and L_0 its phase. Thus, the libration angle γ includes the periodic variations of the orbital motion. By using the true longitude of the satellite defined as $\nu = \Omega + \omega + f = \pi + \nu_p$, ν_p being the true longitude of the planet expressed in the satellite reference frame, the set of planet coordinates can be written as functions of $\nu - L$ and γ.

By including the libration angle in the deformations, the external and internal gravitational torques exerted on the icy shell provided by the expressions (27) and (29) of Richard *et al.* (2014) can be generalized as

$$\Gamma_{ext} = \frac{1}{2}\left(\frac{a}{r}\right)^3\left(K_s - \frac{K_s^p}{3} + K_{int}^{s/p} - \frac{K_{int}^{p/p}}{8}\right)\sin 2(\nu - L - \gamma_s), \quad (2.2)$$

$$\Gamma_{int} = \left(K_{int} - \frac{K_{int}^{s/p}}{2} - \frac{K_{int}^{p/s}}{12} + \frac{K_{int}^{p/p}}{16}\right)\sin 2(\gamma_i - \gamma_s) - \left(\frac{a}{r}\right)^3\left(\frac{K_{int}^{p/s}}{12} - \frac{K_{int}^{p/p}}{16}\right)\sin 2(\nu - L - \gamma_i), \quad (2.3)$$

with γ_s and γ_i the libration angles of the icy shell and the inner core, K_s and K_s^p the amplitudes of the torques exerted on the static and dynamic figures of the shell respectively, K_{int} and $K_{int}^{s/p}$ the amplitudes of the internal gravitational torques exerted by the static figure of the shell on the static and dynamic figure of the inner core respectively, $K_{int}^{p/s}$ and $K_{int}^{p/p}$ the amplitudes of the internal torques exerted by the dynamic figure of the shell on the static and dynamic figure of the inner core respectively. All these amplitudes

Table 1. Frequency analysis of $\nu - L$ for Titan obtained with the JPL Horizons ephemerides taken over 400 years. The identification is made with the TASS ephemerides. The magnitudes are truncated at 30". The initial date is J2000.

i	**Frequency (rad/days)**	**Period (days)**	**Magnitude (")**	**Phase (degrees)**	**Identification**
1	0.394018	15.9464	11899.3237	163.3693	$L_6 - \phi_6$
2	0.788036	7.9732	212.5868	-32.7941	$2L_6 - 2\phi_6$
3	0.394081	15.9439	56.6941	-68.1211	$L_6 - \Phi_6$
4	0.001169	5376.6331	43.7313	-66.0428	$2L_9$
5	0.000584	10750.3648	37.5508	138.4821	L_9
6	0.392897	15.9919	31.5673	10.8789	$L_6 - \Phi_6 - 2L_9$

are defined in Richard *et al.* (2014). The presence of the libration angle in the torques will induce variations of the proper frequencies, especially due to the elastic term K_s^p which is of the same order of magnitude as the rigid torque amplitude K_s.

3. Orbital motion

The Saturn gravitational torque exerted on the multi-layered satellite depends on the orbital motion of the satellite expressed through $\nu - L$ in (2.2). Due to interaction with the Sun, the planets and other satellites, the orbital motion of Titan is perturbed and $\nu - L$ can be approximated by a Fourier series such as (Vienne and Duriez 1995) $\nu - L = \sum_{j=0}^{\infty} H_j \sin(\omega_j t + \alpha_j)$, where ω_j are the frequencies of the orbital motion, H_j the magnitudes and α_j the phases.

In Table 1, the frequency analysis of $\nu - L$ obtained with the Horizons ephemerides (Giorgini *et al.* 1996) provided over 400 years is detailed. We used the notation of Vienne and Duriez (1995) where ϕ is close to the longitude of the periapsis, Φ is close to the longitude of the node and subscripts 6 and 9 refer to Titan and the Sun respectively. In the paper of Richard *et al.* (2014), the short period of the ephemerides compared to the evolution rate of the longitude of the node and the periapsis argument ($\sim$ 703 years) lead us to an incorrect identification of a frequency that we attributed to a perturbation of Iapetus. The comparison with frequency analysis of the TASS ephemerides of Vienne and Duriez (1995) taken over 30000 years allowed to identify a better candidate.

Indeed, the determination of the third term of the frequency analysis (Table 1) is problematic due to the proximity of its frequency with the first term. With Horizons, the frequency is determined at 0.394081 rad/days and has a difference of ± 0.000014 rad/days with the frequency determined with TASS. This difference is due to the short time scale of 400 years of ephemerides which is insufficient to precisely determine the third term frequency. However, TASS permitted to identify the argument of this term as $L_6 - \Phi_6$.

4. Librational motion

The librational motion in longitude of Titan is decomposed at the different orbital forcing frequencies. The variations of the inertia tensor due to the perturbed orbital motion and the librations are included in the libration equations for the ice shell and the solid inner core. The solutions are then obtained by decomposing the system over each perturbing frequency.

The amplitudes of the system are obtained analytically and listed in Table 2. Here we used the interior model CA10 described in Richard *et al.* (2014) and derived from Castillo-Rogez and Lunine (2010). The presence of terms coming from the torques exerted

Table 2. Analytical libration amplitudes of the icy shell (expressed as equator deviation in meters) for the CA10 interior model of Titan at different forcing frequencies. The last column describes the difference of amplitudes between the librational models.

Freq. (rad/days)	**Period** (days)	**CA10** (Richard *et al.* (2014))	**CA10** (this paper)	**Amp. diff.** (%)
0.394018	15.9464	-72.751 m	-73.142 m	0.54
0.001169	5376.63	552.265 m	574.051 m	3.94
0.000584	10750.4	470.171 m	474.673 m	0.96
0.000063	99,027.4	51.963 m	30.523 m	41.26
0.001121	5606.25	29.185 m	17.792 m	39.04

on the dynamic figure of the different layers modifies the proper frequencies ω_1 and ω_2 of the system obtained in the rigid case by about 52% for ω_1 and 2% for ω_2 compared to the values of Richard *et al.* (2014). These variations of the proper frequencies are of the same order as those obtained in the model of libration at orbital frequency of Van Hoolst *et al.* (2013).

The libration amplitudes at orbital frequency are strongly reduced when compared to the rigid case where the amplitudes are higher by about a factor 5. The librational angle included in the deformation process modifies the amplitudes at the orbital frequency by less than 1%. However, by including the librations and orbital perturbations in the deformations, the responses at low frequencies are increased, as for the dominant saturnian annual term increased by about 15 meters, or about 4%. These variations can be attributed to the modification of the proper frequencies due to the tidal perturbing potential and the deformation. Finally, the two frequencies of 0.000063 and 0.001121 rad/days described by Richard *et al.* (2014) and resulting from the orbital forcing exerted on the dynamic figure of the satellite are less dependent on the internal ocean. Their amplitudes for the oceanic models are reduced by about 40% when the libration angle is included in the deformations and are now of the same order as for the solid model. The variations between models are below one meter.

The signature of the internal structure exists in the high frequency forced libration but its amplitude is strongly reduced when compared to the rigid case. The detection of such a signature is thus challenging.

Acknowledgments

The authors wish to thank Alain Vienne for the fruitful discussions on the frequency analysis of the TASS ephemerides. The authors are also thankful to the IAU and the CS of Paris Observatory for their financial support.

References

Castillo-Rogez, J. & Lunine, J. I. 2010, *Geophys. Res. Lett.*, 37, 20205
Eckhardt, D. H. 1965, *AJ*, 70, 466
Giampieri, G. 2004, *Icarus*, 167, 228
Giorgini, J. D., *et al.* 1996, *AAS*, 28, 1158
Goldreich, P. M. & Mitchell, J. L. 2010, *Icarus*, 209, 631
Jara-Orué, H. M. & Vermeersen, B. L. A. 2014, *Icarus*, 229, 31
Richard, A., Rambaux, N., & Charnay, B. 2014, *Planet. Space Sci.* 93, 22
Van Hoolst, T., Baland, R.-M., & Trinh, A. 2013, *Icarus*, 226, 299
Vienne, A. & Duriez, L. 1995, *A&A*, 297, 588

Complex Planetary Systems
Proceedings IAU Symposium No. 310, 2014
Z. Knežević & A. Lemaitre, eds.

doi:10.1017/S1743921314007741

The long-period forced librations of Titan

Marie Yseboodt and Tim Van Hoolst
Royal Observatory of Belgium
email: m.yseboodt@oma.be

Abstract. A moon in synchronous rotation has longitudinal librations because of its non-spherical mass distribution and its elliptical orbit around the planet. We study the librations of Titan with periods of 14.7y and 29.5y and include deformation effects and the existence of a subsurface ocean. We take into account the fact that the orbit is not Keplerian and has other periodicities than the main period of orbital motion around Saturn due to perturbations by the Sun, other planets and moons. An orbital theory is used to compute the orbital perturbations due to these other bodies.

We numerically evaluate the amplitude of the long-period librations for many interior structure models of Titan constrained by the mass, radius and gravity field. Measurements of the librations may give constraints on the interior structure of the icy satellites.

Keywords. planets and satellites: individual (Titan), dynamics, rotation, planetary interiors

1. Introduction

The orbital motion of the icy moon Titan is not Keplerian, therefore Titan experiences forced librations with different frequencies. Here we study the longitudinal librations at the semi-annual period of Saturn around the Sun (14.7y) and its annual period (29.5y). These two periods are long with respect to the diurnal period (16 days). We also take into account the atmospheric torque at the Saturn semi-annual period.

We include deformation effects and the existence of a subsurface ocean. We investigate how the internal structure changes these forced librations.

2. Interior models and tides

We calculate the long-period libration for a lot of different density profiles for Titan with a liquid ocean and 3 solid layers: a solid core (c) composed of a mixture of ice and rocks, a high-pressure solid ice mantle (m), a liquid ocean (o) and an outer solid icy shell (s). We refer to the interior layer (i) as the layer composed of the solid core and the solid mantle. The layers are assumed to have an homogeneous density.

The interior models are constrained by the mass, the radius, and the moment of inertia (MOI $= 0.3431 \pm 0.0004$, Iess (2012)) of Titan.

We use the Clairaut equation for a synchronous satellite deformed by rotation and static tides in order to compute the flattenings and the principal moments of inertia A_l,

Layer	Thickness (km)/Radius (km)	Density (kg m^{-3})
Ice shell (s)	$h_s = 5$–200	800–1200
Ocean (o)	$r_o = 2375$–2570	1000–1400
Ice mantle (m)	$r_m = 2000$–2550	1200–1400
Ice/Rock core (c)	$r_c = 1600$–2200	2430–3419

Tab. 1: The numerical range for the densities and the minimum and maximal size of the 4 layers of Titan (Baland *et al.* 2011).

B_l and C_l ($A_l \leqslant B_l \leqslant C_l$) for each layer l. The polar flattening is defined as $\alpha = \frac{(a+b)/2-c}{(a+b)/2}$ where a, b and c are the radii of the principal axes of the ellipsoid ($a > b > c$) and the equatorial flattening is defined as $\beta = (a-b)/a$.

We include the elastic effects due to tides according to Van Hoolst *et al.* (2013): (1) there is a modification of the gravitational torque exerted by Saturn because of the periodic tidal bulges, (2) additional torques exist between the periodic tidal bulges and the static shape of another solid layer, and (3) there is a zonal tide.

3. Forcings

3.1. *Orbital perturbations*

In order to evaluate the orbital perturbations, we use the SAT360 orbital satellite ephemeris of Titan from R. A. Jacobson covering 400y around J2000. The orbital elements of Titan are given in an inertial frame based on the equatorial plane of Saturn at J2000. The dependence of the equation of motion to the orbital elements will be through the mean longitude $\lambda = M + \omega + \Omega$ where M, ω and Ω are the traditional orbital elements. We use the TRIP software (Gastineau & Laskar 2011) for a decomposition into a quasiperiodic series:

$$\lambda(t) = \sum \lambda_A \cos(\omega_f\, t + \phi_\lambda) \tag{3.1}$$

where λ_A is the amplitude of the orbital perturbation and ω_f its frequency.

The main perturbing frequencies here are due to the annual and semi-annual motion of Saturn around the Sun.

By using another orbital theory valid on a longer interval, the orbit of Titan can be seen to have additional long-period variations, with are much longer periods than the diurnal period (about 16d). The TASS 1.7 orbital model for Titan (Vienne & Duriez 1995) decomposes the mean longitude over 36 frequencies. The 5 largest perturbations have periods of 3263.1y, 703.5y, 14.7y, 29.5y and 914.8y.

We neglect the obliquity of Titan (0.3°, Stiles *et al.* (2008, 2010)) and the inclination of the orbit of Titan on the equatorial plane of Saturn (0.3-0.4°). Therefore only the z-component of the quantities has to be considered.

We neglect the direct torques of the bodies other than Saturn. Indirect effects of the orbital perturbations have a much larger influence on the rotational motion than these direct torques.

3.2. *Atmospheric torque*

There is an exchange of Atmospheric Angular Momentum (AAM) between the surface of Titan and the dense atmosphere (by friction of winds), and therefore a coupling between these two layers. The dynamic is driven by the insolation, therefore the motion is mostly at the Saturnian semi-annual frequency. The expression for the atmospheric torque is $\Gamma_A \cos(\omega_f\, t + \phi_\lambda + \Delta\phi)$. The amplitude of the torque at the semi-annual frequency computed by Tokano & Neubauer (2005) from a Global Circulation Model (GCM) is $\Gamma_A = 1.6\, 10^{17}$ kg m^2/s^2. The phase $\Delta\phi$ at the Saturnian semi-annual frequency is evaluated to be about 1.9 rad at J2000 (computed from Van Hoolst *et al.* (2009)). Richard *et al.* (2014) uses a much smaller atmospheric torque amplitude.

4. Equations of motion

We express the equations of motion of the shell and of the interior in terms of the small libration angle γ. The rotation angle ϕ is the angle between the X-axis of the inertial frame to the long axis of Titan. The difference between these two angles is equal to the mean longitude $\lambda = \phi - \gamma$.

We write the equation of conservation of angular momentum for each layer l by taking into account all the torques acting on the layers, both the external gravitational and internal torques; including the pressure torques. The two equations of rotation for the shell (s) and for the interior (i) are:

$$C_s\ddot{\gamma}_s + C_s\ddot{\lambda} + K_1\,\gamma_s + K_2\,\gamma_i = \Gamma_A \tag{4.1}$$

$$C_i\ddot{\gamma}_i + C_i\ddot{\lambda} + K_4\,\gamma_s + K_5\,\gamma_i = 0 \tag{4.2}$$

We neglect the short-periodic variations related to the orbital period of Titan since we are interested in the long-period variations. We have used the small angle approximation and we neglected dissipation inside the satellite.

The strengths of the couplings K_1, K_2, K_4 and K_5 depend on the coupling between the layers, both the static and periodic parts (Van Hoolst *et al.* 2013).

5. Long-period libration results

5.1. *Saturn annual libration*

At the Saturn annual libration (29.5y), the orbital perturbation on the mean longitude is $\lambda_{An} = 37$ as, which corresponds to 466 m on the surface of Titan.

The frequency is far away from the period of the 2 eigenmodes which vary between 58d and 1.2y and between 2.5y and 5y, depending on the interior structure of Titan. Therefore, the amplitude of libration in longitude of the icy shell (γ_s angle) is very small (< 10m).

For this frequency, the rotation angle of the shell (ϕ_s) has an amplitude between 469 and 475m, depending on the interior model.

5.2. *Saturn semi-annual libration*

At the Saturn semi-annual libration (14.7y), the orbital perturbation on the mean longitude is $\lambda_{SAn} = 43$ as, which corresponds to 548 m on the surface. If there is no atmospheric torque, the corresponding libration amplitude (γ_s) of the icy shell is very small (18-48m, depending on the interior model, see Fig. 1).

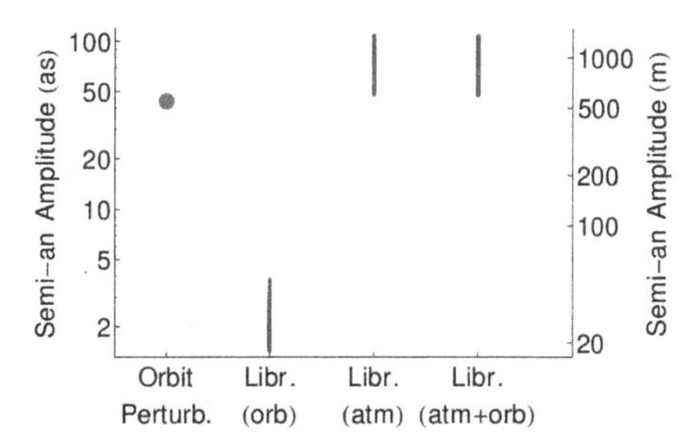

Figure 1. The amplitude of the orbital perturbation (the red point) at the Saturn semi-annual period of Saturn ($T = 14.7$y). The blue points show the libration amplitudes of the ice shell (γ_s) for different interior models by taking into account the orbital perturbation only, the atmospheric torque only, and the orbital perturbation and the atmospheric torque together. The scale is logarithmic.

This is much smaller than the semi-annual libration due to to the atmospheric torque, which can have amplitudes between 600 m and 1350 m as shown by Van Hoolst *et al.* (2013). Therefore, adding the orbital perturbation if the atmospheric torque is considered makes a very small difference (amplitude between 591 m and 1335 m).

For a 10 times smaller atmospheric torque (as for example in Richard *et al.* (2014)), the amplitude γ_s is about 10 times smaller (= 69–159m)

The amplitude of the shell rotation angle (ϕ_s angle) is between 837 and 1477 m.

If we do not take into account the tides (rigid planet), the amplitude γ_s is 3 times smaller (201–245m).

The libration of the ice shell may be observed in the future, but not the libration of the interior.

5.3. *The libration at the very long period*

Using the amplitudes given in the TASS orbital theory for the very long periods (3263.1y, 703.5y and 914.8y), we compute the libration amplitude for the γ_s angle. Since these periods are much longer than the periods of two eigenmodes (less than 5y), their libration amplitude is smaller than 0.1m and may be neglected.

6. Conclusion

Since the Saturn annual and semi-annual frequencies are far away from the two eigenfrequencies, there are no large amplification for the libration. The Saturn annual (29y) libration will be too small (<10m) to be detected in a near future. The Saturn semi-annual (14.7y) libration may have a large amplitude (up to 1.5km), depending on the interior model and the value of the atmospheric torque.

Acknowledgments

This work was financially supported by the Belgian PRODEX program managed by the European Space Agency in collaboration with the Belgian Federal Science Policy Office.

References

Baland, R.-M., Van Hoolst, T., Yseboodt, M. & Karatekin, Ö. 2011. *A&A*, 530, A141.

Gastineau, M. & Laskar, J. 2011. *ACM Commun. Comput. Algebra*, 44, 194.

Iess, L., Jacobson, R. A., Ducci, M., Stevenson, D. J., Lunine, J. I., Armstrong, J. W., Asmar, S. W., Racioppa, P., Rappaport, N. J., & Tortora, P. 2012. *Science*, 337, 457.

Richard, A., Rambaux, N., & Charnay, B. 2014. *Planetary and Space Science*, 93, 22.

Stiles, B. W., Kirk, R. L., Lorenz, R. D., Hensley, S., Lee, E., Ostro, S. J., Allison, M. D., Callahan, P. S., Gim, Y., Iess, L., & 5 coauthors 2008. *AJ*, 135, 1669.

Stiles, B. W., Kirk, R. L., Lorenz, R. D., Hensley, S., Lee, E., Ostro, S. J., Allison, M. D., Callahan, P. S., Gim, Y., Iess, L., & 5 coauthors 2010. *AJ*, 139, 311.

Tokano, T. & Neubauer, F. M. 2005. *Geophysical Research Letters*, 32, 24203.

Van Hoolst, T., Baland, R.-M., & Trinh, A. 2013. *Icarus*, 226, 299.

Van Hoolst, T., Rambaux, N., Karatekin, Ö., & Baland, R.-M. 2009. *Icarus*, 200, 256.

Vienne, A. & Duriez, L. 1995. *A&A*, 297, 588.

Complex Planetary Systems
Proceedings IAU Symposium No. 310, 2014
Z. Knežević & A. Lemaitre, eds.

doi:10.1017/S1743921314007753

Scaling laws to understand tidal dissipation in fluid planetary layers and stars

Pierre Auclair-Desrotour[1,2], Stéphane Mathis[2,3] and Christophe Le Poncin-Lafitte[4]

[1]IMCCE, Observatoire de Paris, CNRS UMR 8028,
77 Avenue Denfert-Rochereau, 75014 Paris, France
email: `pauclair-desrotour@imcce.fr`

[2]Laboratoire AIM Paris-Saclay, CEA/DSM - CNRS - Université Paris Diderot,
IRFU/SAp Centre de Saclay, F-91191 Gif-sur-Yvette, France
email: `stephane.mathis@cea.fr`

[3]LESIA, Observatoire de Paris, CNRS UMR 8109, UPMC, Univ. Paris-Diderot,
5 place Jules Janssen, 92195 Meudon, France

[4]SYRTE, Observatoire de Paris, CNRS UMR 8630, UPMC,
61 Avenue de l'Observatoire, 75014 Paris, France
email: `christophe.leponcin@obspm.fr`

Abstract. Tidal dissipation is known as one of the main drivers of the secular evolution of planetary systems. It directly results from dissipative mechanisms that occur in planets and stars' interiors and strongly depends on the structure and dynamics of the bodies. This work focuses on the mechanism of viscous friction in stars and planetary layers. A local model is used to study tidal dissipation. It provides general scaling laws that give a qualitative overview of the different possible behaviors of fluid tidal waves. Furthermore, it highlights the sensitivity of dissipation to the tidal frequency and the roles played by the internal parameters of the fluid such as rotation, stratification, viscosity and thermal diffusivity that will impact the spins/orbital architecture in planetary systems.

Keywords. hydrodynamics, waves, turbulence, planet-star interactions

1. Introduction

The bodies that compose a planetary system often present one or several fluid regions. For example, stars are completely fluid, gaseous giant planets are mainly constituted of large fluid envelopes, and some rocky planets and icy satellites such as the Earth, Europa or Enceladus have internal and external fluid layers. Submitted a tidal perturbation, these fluid regions move and deform periodically like solid ones. These tidal motions are affected by different kinds of dissipation mechanisms: the viscous friction, thermal diffusion and Ohmic dissipation in the presence of a magnetic field. Theoretical studies of these processes have emerged after the first study of a tidally deformed body performed by Lord Kelvin (Kelvin 1863) and the developments made in 1911 by Love, who introduced the so-called Love-numbers (Love 1911). In the 1960's, Goldreich established the link that bounds the long-term evolution of planetary systems to the internal dissipation (see Goldreich & Soter 1966). He introduced the commonly used Q tidal quality factor that intervenes in the dynamical equations. Then, several works, such as Efroimsky & Lainey (2007) for rocky bodies, illustrate the impact of dissipation on the evolution of spin and orbital parameters.

Over the past decades, studies have been carried out on the tidal dissipation in fluid bodies, these studies being mainly about stars and the envelopes of gaseous giant planets

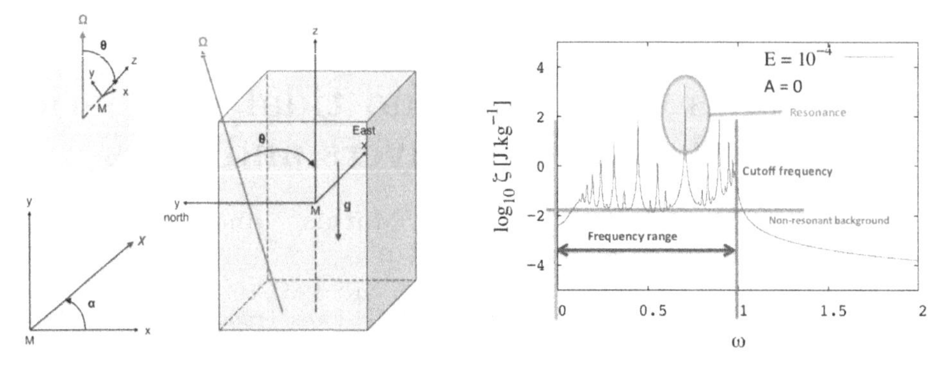

Figure 1. Left: The studied model: an inclined rotating Cartesian fluid box where the viscosity, thermal diffusion, and stratification of the fluid are taken into account. **Right:** A typical dissipation frequency spectrum computed from the analytical model. ζ is the mass energy dissipated by viscous friction per rotation period in the box ; $\omega = \chi/2\Omega$ is the tidal frequency of the perturbation normalized by the inertial frequency 2Ω. The Ekman number (E) and the stratification parameter (A) are defined in the text hereafter.

(Ogilvie & Lin 2004, 2007; Remus *et al.* 2012). Given the great diversity of existing tidal motions and dissipation mechanisms, the nature of the response varies, taking very different possible complex forms. For example, the internal fluid layers of a planet and the interiors of stars dissipate energy through the mechanism of viscous friction, but their physical parameters vary over several orders of magnitude. This is the reason why a local approach, described further (and fully developed in Auclair-Desrotour, Mathis & Le Poncin Lafitte, in preparation), is privileged here: it allows to explore the physics using simplified tractable models (see also Ogilvie 2005; Jouve & Ogilvie 2014) and comes as a complement to global models by providing qualitative results.

In a fluid, a perturbation can generate a large spectrum of waves. The acoustic waves are the first kind but their characteristic frequency is too high to allow a significative response to a low-frequency tidal perturbation. This later will rather excite inertial waves and gravity waves, the combination of the two kinds being gravito-inertial waves. Inertial waves have the Coriolis force as restoring force. Their typical frequency is the inertial frequency, 2Ω, where Ω is the spin frequency of the body. Gravity waves, which have the buoyancy as restoring force, propagate in stably stratified fluid layers. They are characterized by the Brunt-Väisälä frequency, often denoted N in the literature, which results from radial variations of the specific entropy.

These physical parameters, Ω and N, like the viscosity ν and the thermal diffusivity κ of the fluid have a real impact on the tidal kinetic energy dissipated by viscous friction. The goal of this study is to understand this dissipation. Thus, through the simplified model of a local fluid box, scaling laws are obtained, characterizing the dissipation spectra, to give simplified models that can be used as a first step in celestial mechanics studies. In this way, it is demonstrated that dissipation strongly depends on the forcing frequency and internal parameters in the case of fluids, and so do the spin and orbital evolution of corresponding planetary system (Auclair-Desrotour *et al.* 2014).

2. Model and results: waves in a box

Consider a local section of a fluid region in any star or planet. It can be represented as a rotating Cartesian box of side length L inclined with respect to the spin axis of the body $\mathbf{\Omega}$ with a colatitude θ (Fig. 1). This model generalizes the first local model

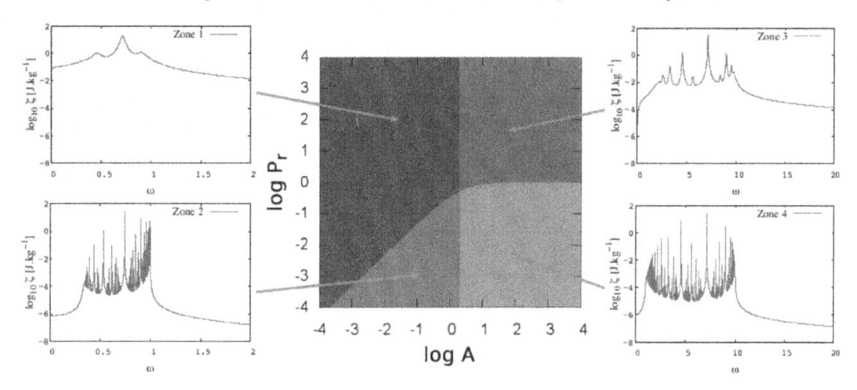

Figure 2. Asymptotical behaviors of the tidal waves. Zones colored in blue and purple correspond to inertial waves, the two other to gravity waves ; zones colored in blue and red correspond to the case where viscous diffusion predominates over thermal diffusion, the two zones below corresponding to the opposite case.

presented in Ogilvie & Lin (2004). The local coordinates (x, y, z) are defined so that z corresponds to the radial direction, and x and y to the West-East and South-North directions respectively. The fluid is supposed to be locally homogeneous of background density ρ, kinetic viscosity ν, and thermal diffusivity κ. This implies three control parameters, denoted A, E, and K. The behavior of the waves is given by $A = (N/2\Omega)^2$: inertial waves correspond to $A \ll 1$, and pure gravity waves to $A \gg 1$. Then, the so-called Ekman number $E = 2\pi^2\nu/\left(\Omega L^2\right)$, and $K = 2\pi^2\kappa/\left(\Omega L^2\right)$ weight the effects of viscous and thermal diffusions with respect to the Coriolis effect. Their ratio forms the Prandlt number $Pr = E/K$. The fluid is submitted to a tidal perturbative force of frequency χ.

The energy dissipated by viscous friction in the box is computed analytically. Expressed as a function of $\omega = \chi/2\Omega$, the normalized frequency of the perturbation, it yields the dissipation spectrum of the system (Figs. 1, 2 and 3). Note that this latter takes the form of a batch of resonances located between $\chi_1 \approx 2\Omega\cos\theta$ and $\chi_2 \approx N$. Its properties, such as the positions ω_{mn}, widths l_{mn} and heights H_{mn} of the peaks, m and n being respectively the horizontal and vertical wave numbers, the number of resonances $N_{\rm kc}$, the height of the non-resonant background $H_{\rm bg}$ (corresponding to the equilibrium tide) and the sharpness ratio $\Xi = H_{11}/H_{\rm bg}$, depend on the regime of the waves highlighted by the model (Fig. 2). For each regimes, these properties are described by scaling laws obtained analytically as functions of the control parameters and synthesized in Table 1. An illustrative example is given in Fig. 3 for l_{11}.

3. Conclusions

The analytical local model of the fluid box appears as a useful and efficient way to explore a large domain of physical parameters. It allows to identify and to describe the asymptotical behaviors of the fluid qualitatively and can be used in parallel with more complex global models to understand the physics of dissipation that must be used in celestial mechanics' studies. In this study, it has been restricted to viscous friction but it is possible to take the other dissipations, enumerated in introduction, into account. One could also think about applying it in a near future to Alfvén waves, and not only to gravito-inertial waves, by adding a magnetic field to the model, stars and planets being magnetized bodies.

Table 1. Scaling laws of the properties of the energy viscously dissipated for the different regimes (we define $A_{11} \equiv 2\cos^2\theta$ and $Pr_{11} \equiv A/\left(A + A_{11}\right)$). **Top left:** Inertial waves dominated by viscosity. **Top right:** Gravity waves dominated by viscosity. **Bottom left:** Inertial waves dominated by heat diffusion. **Bottom right:** Gravity waves dominated by heat diffusion. F is the amplitude of the forcing.

Domain	$A \ll A_{11}$		$A \gg A_{11}$	
$Pr \gg Pr_{11}$	$\frac{\chi_{mn}}{2\Omega} \propto \frac{n}{\sqrt{m^2+n^2}}\cos\theta$	$N_{kc} \propto E^{-1/2}$	$\frac{\chi_{mn}}{2\Omega} \propto \frac{m}{\sqrt{m^2+n^2}}\sqrt{A}$	$N_{kc} \propto A^{1/4}E^{-1/2}$
	$l_{mn} \propto E$	$H_{mn} \propto F^2E^{-1}$	$l_{mn} \propto E$	$H_{mn} \propto F^2E^{-1}$
	$H_{bg} \propto F^2E$	$\Xi \propto E^{-2}$	$H_{bg} \propto F^2EA^{-1}$	$\Xi \propto AE^{-2}$
$Pr \ll Pr_{11}$	$\frac{\chi_{mn}}{2\Omega} \propto \frac{n}{\sqrt{m^2+n^2}}\cos\theta$	$N_{kc} \propto A^{-1/2}K^{-1/2}$	$\frac{\chi_{mn}}{2\Omega} \propto \frac{m}{\sqrt{m^2+n^2}}\sqrt{A}$	$N_{kc} \propto A^{1/4}K^{-1/2}$
	$l_{mn} \propto AK$	$H_{mn} \propto F^2A^{-2}EK^{-2}$	$l_{mn} \propto K$	$H_{mn} \propto F^2EK^{-2}$
	$H_{bg} \propto F^2E$	$\Xi \propto A^{-2}K^{-2}$	$H_{bg} \propto F^2EA^{-1}$	$\Xi \propto AK^{-2}$

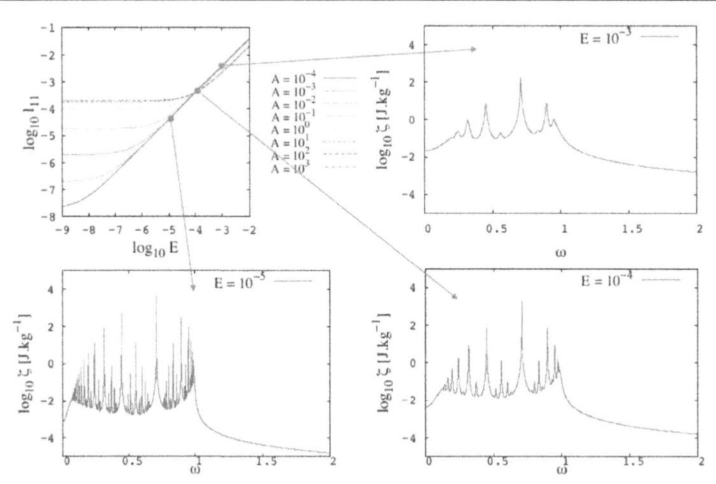

Figure 3. Width at mid-height of the main resonance as a function of E for different values of A and the corresponding dissipation spectra.

References

Auclair-Desrotour, P., Le Poncin-Lafitte, C., & Mathis, S. 2014, *A&A*, 561, L7

Efroimsky, M. & Lainey, V. 2007, *Journal of Geophysical Research (Planets)*,112, 12003

Goldreich, P. & Soter, S. 1966, *Icarus*, 5, 375

Jouve, L. & Ogilvie, G. I. 2014, *Journal of Fluid Mechanics*, 745, 223

Lord Kelvin 1863, *Phil. Trans. Roy. Soc. London, Treatise on Natural Philosophy*, 2, 837

Love, A. E. H. 1911, *Some Problems of Geodynamics*, Publisher: Cambridge University Press

Ogilvie, G. I. 2005, *Journal of Fluid Mechanics*, 543, 19

Ogilvie, G. I. & Lin, D. N. C. 2004, *ApJ*, 610, 477

Ogilvie, G. I. & Lin, D. N. C. 2007, *ApJ*, 661, 1180

Remus, F., Mathis, S., & Zahn, J.-P 2012, *A&A*, 544, A132

Complex Planetary Systems
Proceedings IAU Symposium No. 310, 2014
Z. Knežević & A. Lemaitre, eds.

doi:10.1017/S1743921314007765

Revisiting the capture of Mercury into its 3:2 spin-orbit resonance

Benoît Noyelles[1], Julien Frouard[2], Valeri V. Makarov[3] and Michael Efroimsky[3]

[1]naXys, University of Namur, Belgium,
email: benoit.noyelles@gmail.com

[2]CIERA, Northwestern University, Evanston IL 60208 USA,
email: frouard@imcce.fr

[3]United States Naval Observatory, Washington DC 20392 USA
email: {valeri.makarov,michael.efroimsky}@usno.navy.mil

Abstract. We simulate the despinning of Mercury, with or without a fluid core, and with a frequency-dependent tidal model employed. The tidal model incorporates the viscoelastic (Maxwell) rebound at low frequencies and a predominantly inelastic (Andrade) creep at higher frequencies. It is combined with a statistically relevant set of histories of Mercury's eccentricity. The tidal model has a dramatic influence on the behaviour of spin histories near spin-orbit resonances. The probabilities of capture into high-order resonances are greatly enhanced. Exploring several scenarios, we conclude that the present 3:2 spin state was achieved by entrapment of an initially prograde cold Mercury when its age was less than 20 Myr, i.e., well before differentiation.

Keywords. celestial mechanics, planets and satellites: individual (Mercury)

1. Previous studies

In the literature hitherto, three scenarios of Mercury's entrapment have been discussed.

(*a*) *A prograde rigid Mercury.* The probability of capture into the 3:2 spin-orbit state is $\approx 7\%$, with a constant eccentricity ≈ 0.206, see Goldreich & Peale (1966). The probability increases to $\approx 55\%$ due to multiple crossings induced by secular variations of the eccentricity, see Correia & Laskar (2004).

(*b*) *A prograde Mercury with a liquid core.* Within this scenario, Mercury was more likely to be trapped into the 2:1 resonance than into the 3:2 one, as demonstrated by Peale & Boss (1977), Correia & Laskar (2009).

(*c*) *A Mercury once synchronised.* Wieczorek *et al.* (2012) argued that the allegedly asymmetric distribution of impact craters was the signature of a past synchronous rotation destabilised later by an impact.

All these scenarios rely on the CTL (*Constant Time Lag*) tidal model, which cannot be applied to terrestrial planets of considerable viscosities. Among the mathematical consequences of that model is a *stable state of pseudosynchronous rotation* on which the previous studies are based. We revisit these scenarios, using a physics-based tidal model.

2. A more realistic tidal model

At low obliquities, the polar tidal torque reads as (Noyelles *et al.* (2014))

$$\mathcal{T}_{tide} \approx \frac{3}{2}\frac{GM_{\odot}^2}{a}\left(\frac{R}{a}\right)^5 \sum_{j,q=-\infty}^{\infty} G_{20q}(e)G_{20j}(e)k_2 \sin\epsilon_2 \cos\left[(q-j)\mathcal{M}\right]. \qquad (2.1)$$

$\mathcal{M}$ and $n \equiv \dot{\mathcal{M}}$ are the mean anomaly and mean motion; θ and $\dot{\theta}$ are the rotational angle and spin rate of the planet; $k_2 \sin \epsilon_2$ is a function the Fourier mode $\omega_{2m0q} \approx (2+q)n - m\dot{\theta}$. Its shape is determined by the planet's self-gravitation and rheology. Viscoelastic (Maxwell) at low frequencies, the rheology comprises both viscoelastic and inelastic reaction (Andrade creep) at higher frequencies (Efroimsky(2012)).

Kink-shaped, the quality function $k_2(\omega_{2m0q}) \sin \epsilon_2(\omega_{2m0q})$ goes continuously through zero in the resonance $\omega_{2m0q} = 0$. With all terms expressed as functions of $\dot{\theta}$, the series (2.1) is a superposition of kinks. Employment of this torque radically changes the entrapment probabilities and excludes pseudosynchronism, see Makarov & Efroimsky(2013).

3. Scenario 1: A prograde rigid Mercury

As soon as our tidal model is used, Mercury almost always gets trapped into the 3:2 resonance *on the first crossing*. Moreover, the absence of a stable pseudosynchronous rotation makes several crossings impossible. So, if Mercury is not trapped into a high-order resonance, it falls into the synchronous one. A hot Mercury (with a short Maxwell time τ_M) is more likely to fall into the 2:1 resonance than into the current 3:2.

4. Scenario 2: A prograde Mercury with a core

We also considered a differentiated Mercury with core-mantle friction, following Goldreich & Peale (1967). When our tidal model is used, the 2:1 resonance is certain for the current eccentricity (0.206). Only a past low eccentricity or a collision disrupting the 2:1 resonance (Correia & Laskar (2012)) could have made the current configuration possible.

5. Scenario 3: A once synchronous Mercury

The distribution of craters, according to the MESSENGER data (Fassett *et al.* (2012)), suggests an East-West asymmetry consistent with a past synchronous rotation. However, the absence of pseudosynchronous stable rotation requires the impact to be energetic enough to make Mercury reach the 3:2 resonance. This would leave a crater larger than 600 km, while the use of the CTL model would require only a crater of 300 km. For a detailed critical analysis of this scenario, see Noyelles *et al.* (2014).

6. Conclusion

Within the Scenario 1 of an initially prograde cold Mercury, the 3:2 resonance is the likeliest end state. The capture takes place in less than 20 Myr, well before differentiation.

References

Correia, A. C. M. & Laskar, J. 2004, *Nature*, 429, 848
Correia, A. C. M. & Laskar, J. 2009, *Icarus*, 201, 1
Correia, A. C. M. & Laskar, J. 2012, *ApJL*, 751:L43
Efroimsky, M. 2012. *ApJ*, **746**:150 ERRATA: *ApJ*, **763**:150 (2013)
Fassett, C. I., Head, J. W., & Baker, D. M. H., *et al.* 2012, *JGR*, 117, E00L08
Goldreich, P. & Peale, S. J. 1966, *AJ*, 71, 425
Goldreich, P. & Peale, S. J. 1967, *AJ*, 72, 662
Makarov, V. V., & Efroimsky, M. 2013. *ApJ*, Vol. **764**, id. 27
Noyelles, B., Frouard, J., Makarov, V. V., & Efroimsky, M. 2014, *Icarus*, 241, 26
Peale, S. J. & Boss, A. 1977, *JGR*, 82, 743
Wieczorek, M. A., Correia, A. C. M.., & Le Feuvre, M., *et al.* 2012, *Nature Geoscience*, 5, 18

Complex Planetary Systems
Proceedings IAU Symposium No. 310, 2014
Z. Knežević & A. Lemaitre, eds.

doi:10.1017/S1743921314007777

Variational Chaos Indicators: Application to the Restricted Three-Body Problem

Alexey M. Koksin and Vladimir A. Shefer

Tomsk State University, 634050 Tomsk, Russia
email: shefer@niipmm.tsu.ru

Abstract. A comparison of several known dynamical indicators of chaos based on the numerical integration of differential variational equations is performed. The comparison is implemented on the examples of studying dynamics in the planar circular restricted three-body problem.

Keywords. celestial mechanics, methods: n-body simulations; methods: numerical; instabilities

1. Introduction

The classical method for distinguishing between regular and chaotic motions in dynamical systems is the calculation of the maximal Lyapunov Characteristic Number (mLCN). For a continuous dynamical system with n degrees of freedom the mLCN is defined as

$$\mathrm{mLCN} \equiv \sigma = \lim_{t\to\infty} (\ln \delta / t)$$

using the solution of the system of motion equations and variational equations:

$$\dot{\mathbf{x}} = \mathbf{f}(\mathbf{x}(t)), \qquad \dot{\boldsymbol{\delta}} = \frac{\partial \mathbf{f}}{\partial \mathbf{x}}(\mathbf{x}(t))\boldsymbol{\delta}; \qquad \mathbf{x}, \boldsymbol{\delta} \in R^{2n}, \qquad t \in R. \tag{1.1}$$

Here, t is the independent variable (time), the vector $\mathbf{x}(t)$ sets a trajectory in the phase space of the dynamical system; $\mathbf{x}_0 \equiv \mathbf{x}(t_0)$, $t_0 = 0$ are the initial conditions for the equations of motion, $\boldsymbol{\delta}(t)$ is the tangent vector determining the evolution of initial unit deviation vector relative to the trajectory $\mathbf{x}(t)$; $\delta \equiv \|\boldsymbol{\delta}(t)\|$.

Usually the convergence of $\ln \delta / t$ to σ is slow, therefore the reliable estimation of the mLCN requires a long calculation time. It stimulated the emergence of so-called *fast Lyapunov indicators* allowing to study the phase space on relatively short time intervals.

The aim of this research was to compare some often-used fast chaos indicators (hereafter CIs) on the examples of their working in the case of the dynamical system specified by the equations of the planar circular restricted three-body problem (Szebehely 1967).

2. Numerical Experiments and Results

We selected the following chaos indicators: the Fast Lyapunov Indicator (FLI, Froeschlé *et al.* 1997), the Orthogonal Fast Lyapunov Indicator (OFLI, Fouchard *et al.* 2002), the Mean Exponential Growth factor of Nearby Orbits (MEGNO, Cincotta & Simo 2000), the Smaller Alignment Index (SALI, Skokos 2001), and the Average Power Law Exponent (APLE, Lukes-Gerakopoulos *et al.* 2008). Specific applications of some listed CIs to the circular restricted three-body problem were discussed, for example, in (Fouchard *et al.* 2002; Frouard *et al.* 2008; Morais & Giuppone 2012).

We used also the Orthogonal Mean Exponential Growth factor of Nearby Orbits (OMEGNO) introduced by us (Shefer & Koksin 2013). It is determined by

$$\mathrm{OMEGNO}(t) = 2(\theta - \zeta)/t, \tag{2.1}$$

where θ and ζ are calculated by integrating the differential equations

$$\dot{\theta} = \ln \delta_\perp, \qquad \dot{\zeta} = \theta/t \tag{2.2}$$

simultaneously with the equations (1.1). We use $\delta_\perp = \|\boldsymbol{\delta}_\perp(t)\|$, where $\boldsymbol{\delta}_\perp(t)$ is the vector component of $\boldsymbol{\delta}(t)$ orthogonal to the direction of the phase flow. The initial conditions for (2.2) are as follows: $\theta(0) = 0$, $\zeta(0) = 0$. At $t = 0$ the right-hand side of the second equation of system (2.2) is taken equal to zero. OMEGNO(t) $\to 0$ with $t \to \infty$ for any stable periodic orbit. Thus, this MEGNO's modification gives the indicator, which makes it possible to separate periodicity in the regular component of the phase space.

The problem considered in this work describes the motion of a massless body P perturbed by two massive bodies P_1 and P_2 (called primaries) with respective masses $1-\mu$ and μ ($\mu \leqslant 0.5$) moving around their barycenter O. The length unit is such that the distance between the primaries is unity, and the time unit τ is such that the orbital period of the primaries equals $2\pi\tau$.

In the rotating frame xOy, in which P_1 and P_2 are located at points with coordinates $(-\mu, 0)$ and $(1-\mu, 0)$, respectively, the equations of motion of P are written in the form $\ddot{x} = 2\dot{y} + x - (1-\mu)(x+\mu)/r_1^3 - \mu(x-1+\mu)/r_2^3$, $\quad \ddot{y} = -2\dot{x} + y - (1-\mu)y/r_1^3 - \mu y/r_2^3$, where $r_1^2 = (x+\mu)^2 + y^2$ and $r_2^2 = (x-1+\mu)^2 + y^2$.

We accepted that the primaries have equal masses ($\mu = 0.5$), and the Jacobi integral is equal to 4.

For the initial vector $\mathbf{x}_0 \equiv (x_0, \dot{x}_0, y_0, \dot{y}_0)$ we took the set of 961 values, where x_0-coordinate is uniformly distributed in the interval $[-0.49, -0.01]$ with the step of size 0.0005, $\dot{x}_0 = 0$, $y_0 = 0$, and $\dot{y}_0 > 0$ is determined from the Jacobi integral.

As a method of numerical integration of differential equations the Dormand–Prince algorithm of the 8(7)th order (Hairer *et al.* 1987) was chosen.

With above-mentioned initial conditions we have calculated the values of the CIs. The final values on the time $t = 10000\tau$ are represented graphically (Fig. 1).

In all the figures obtained the chaotic motion appears separated from the regular resonant one as a whole. But the MEGNO and the OMEGNO separate these motions most clearly because these indicators have the universal threshold value ($\cong 2$). The FLI, the OFLI, and the SALI have not reference values clearly discriminating the regular orbits from the weak chaotic ones.

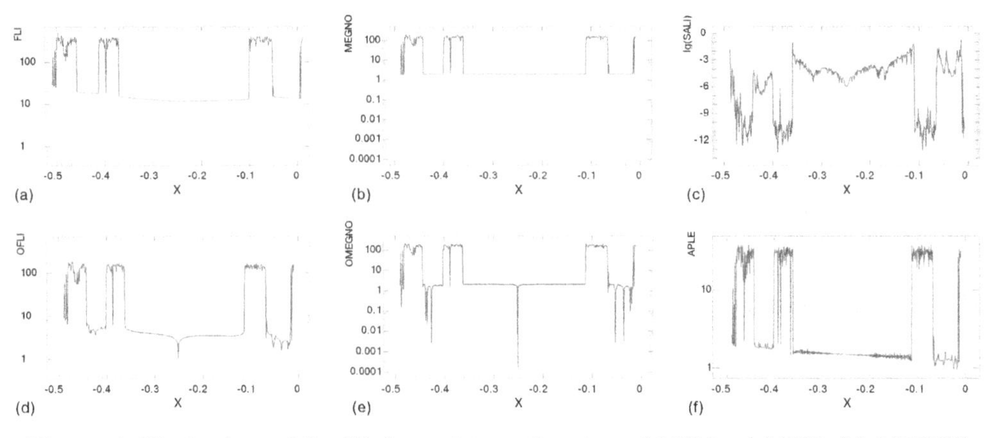

Figure 1. Final values of the CIs for an integration time of 10000τ: (a) FLI, (b) MEGNO, (c) SALI, (d) OFLI, (e) OMEGNO, (f) APLE.

There is a difference between the values of the FLI, the OFLI, the SALI, and the APLE found for x_0 in $[-0.44, -0.40]$ and $[-0.065, -0.015]$. As both intervals correspond to the same set of orbits, this difference can be explained only by the relative positions of the initial vectors $\mathbf{x}_0$ and $\boldsymbol{\delta}_0$ (analogous conclusions were made by Fouchard *et al.* (2002)).

A comparison shows that on the curves of the OFLI and the OMEGNO are several minima, which are not presented on the curves of the FLI and the MEGNO. These minima refer to the presence of periodic orbits in the neighbourhood of the corresponding initial conditions. The final values of the SALI and the APLE demonstrate minima in the regular motion areas too, but some of them are not related with the periodic component of the motion. By the final values of the FLI and the MEGNO it is not possible to identify a fine structure of the regular component of the motion.

For effective study of large sets of initial conditions with the help of the CIs' final values it is important to know the reliability on the CIs' thresholds that make a confident distinction between chaotic and regular motions.

We took the same initial conditions as in the previous experiment and identified the chaotic orbits by the Poincaré surface of section (there are found 263 chaotic orbits, i.e. 27,4%). The indicators we chose have the theoretical estimations of their thresholds (time-dependent or time-independent). We splitted the full time of integration (10000τ) on 20 sub-intervals per 500τ for each. We have computed the number of chaotic orbits in each sub-interval with each indicator on the basis of it threshold. We estimated the reliability on the corresponding threshold by comparing this number with the true count of chaotic orbits.

The thresholds associated with the FLI and the APLE define all 961 orbits as chaotic ones for the complete time interval. The thresholds for the MEGNO and the SALI give the number of chaotic orbits increased approximately 2.5 times. So they are inefficient too.

The thresholds for the orthogonal indicators are very close to the values allowing to achieve a stable fraction of the chaotic component, close to the true fraction.

Consequently, the thresholds' estimations for the FLI, the APLE, the MEGNO, and the SALI are not reliable and they need an empirical adjustment. The thresholds for the OFLI and the OMEGNO practically do not require any corrections.

To trace the evolution of the CIs' values with time for different types of motion, we considered four representative orbits with the following initial conditions for x_0: -0.25065550 (O_1), -0.35 (O_2), -0.28 (O_3), -0.08 (O_4). The values of $\dot{x_0}$, y_0, and $\dot{y_0}$ are the same as in the cited above set of initial conditions. The orbit O_1 is a periodic orbit, O_2 and O_3 are resonance orbits from the O_1's neighbourhood, O_4 belongs to chaotic orbits. The initial conditions for O_1, O_3, and O_4 were taken from (Fouchard *et al.* 2002). The variations of the CIs' values for four selected orbits in the time interval 10000τ are showed graphically in a logarithmic time scale (Fig. 2).

In this numerical experiment all the CIs clearly separate the regular resonance motion from the strong chaotic one.

In regard to separation of different levels of the regular motion, here the results of the CIs' working can be described as following.

The FLI, the MEGNO, and the APLE for the resonance libration orbits detect their stable quasi-periodic nature very quickly and show a power of their proximity to O_1. But in this case we can not say uniquely that O_1 is a periodic orbit. The rest of the CIs not only separate the quasi-periodic orbits between themselves, but they detect the periodic orbit. As this takes place, the closer the quasi-periodic orbit to the periodic one the later their essential distinction is detected.

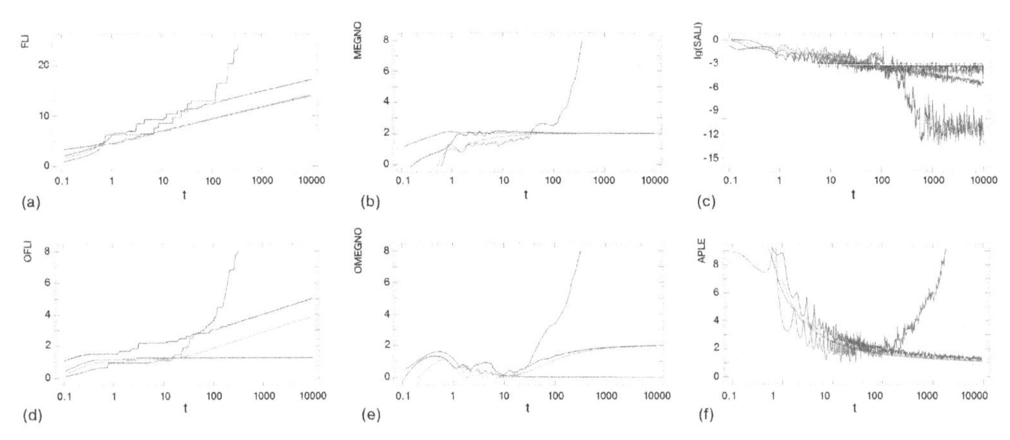

Figure 2. Evolution of the CIs with time for the four representative orbits O_1 (the red line), O_2 (the light-blue line), O_3 (the dark-blue line), and O_4 (the purple line): (a) FLI, (b) MEGNO, (c) SALI, (d) OFLI, (e) OMEGNO, (f) APLE.

For each quasi-periodic orbit the OFLI temporarily takes a constant value (the *plateau* in figure) but later it grows linearly with the same slope. In this case the length of the *plateau* increases with approaching the quasi-periodic orbit to the periodic one. Hence at the end of the selected integration interval the OFLI's value for the orbit more close to the periodic one is less compared with the OFLI's value for the more distant orbit.

The OMEGNO's values for the quasi-periodic orbits in very short time confidently go away from zero approaching the value 2 from below. The farther the quasi-periodic orbit is from the periodic one the earlier the OMEGNO's values fall into the nearest neighbourhood of 2.

The APLE's values for O_1, O_2, and O_3 approache unity from above with time. The closer the quasi-periodic orbit to the periodic one the closer the APLE's values to unity.

The function lg(SALI) for the quasi-periodic orbits reaches different constant values. Here, the closer is the quasi-periodic orbit to the periodic one the lower is the *plateau* and the later is reached this *plateau*.

Thus, the OFLI, the OMEGNO, and the SALI allow to detect and locate periodic orbits. But the OMEGNO is only CI from these having the universal reference value ($\cong 0$) for periodic motion. The OFLI and the SALI give only relative results, which require calibration with a known reference orbit.

The work was supported by the Ministry of Education and Science of the Russian Federation, project no. 2014/223(1766).

References

Cincotta, P. M., Simo, C. 2000, *Astron. Astrophys. Suppl. Ser.*, 147, 205

Fouchard, M., Lega, E., Froeschlé, Ch., & Froeschlé, C. 2002, *Celest. Mech. Dyn. Astr.*, 83, 205

Froeschlé, C., Lega, E., & Gonczi, R. 1997, *Celest. Mech. Dyn. Astr.*, 67, 41

Frouard, J., Fouchard, M., & Vienne, A. 2008, *SF2A-2008*, 121

Hairer, E., Norsett, S. P., & Wanner, G. 1987, *Solving Ordinary Differential Equations. Nonstiff Problems* (Berlin, Heidelberg: Springer–Verlag)

Lukes-Gerakopoulos, G., Voglis, N., & Efthymiopoulos, C. 2008, *Physica A*, 387, 1907

Morais, M. H. M. & Giuppone, C. A. 2012, *MNRAS*, 424, 52

Shefer, V. A. & Koksin, A. M. 2013, *Izv. Vusov. Fizika*, 56, No.6/3, 256

Skokos, Ch. 2001, *J. Phys. A*, 34, 10029

Szebehely, V. 1967, *Theory of Orbits* (New York: Academic Press)

Complex Planetary Systems
Proceedings IAU Symposium No. 310, 2014
Z. Knežević & A. Lemaitre, eds.

doi:10.1017/S1743921314007789

The parametric instability of resonance motion in restricted three body problem

Alexey E. Rosaev[1]

[1]NPC Nedra, Yaroslavl, Russia,
150000, Svobody str., 8/38
email: hegem@mail.ru

Abstract. A method of analyses of dynamical system is applied to the planetary restricted three-body problem (RTBP). It is well known, that equations of motion of restricted 3-body problem in rotating rectangular frame may be reduced to the second order differential equation with periodic coefficients (Hills equation). Here Hills equation in cylindrical coordinate frame is derived. It gives the possibility to estimate width and position of the unstable zones. The dependence of the position of unstable zones on orbital eccentricity of the test particle is derived. Some followings of this simple linear model are noted.

Keywords. celestial mechanics, methods: analytical, restricted three body problem, resonance, stability

---

1. Introduction

The three-body problem is a continuous source of study, since the discovery of its non-integrability due to Poincare (1892). However, some problems are still unresolved (Celletti *et al.*(2002)). In this paper, we consider the planar circular restricted three-body problem, when mass m revolves around M (M is much more than m) in a circular orbit and the third body is considered with negligible mass m_0. We assume that all bodies move on the same plane. The equations of planar restricted Hill's problem in rectangular frame may be reduced to a Hill equation for normal distance from variation orbit x see Szebehely(1967). In our view, cylindrical system has some advantages over other ones, because the variation of one of the coordinates - central distance R - is always restricted and may be considered as a small parameter at the problem.

The main equations for the planar circular 3-body problem in cylindrical frame are:

$$\frac{d^2}{dt^2}\mathrm{R} - \mathrm{R}\left(\frac{d}{dt}\lambda\right)^2 = \frac{\partial}{\partial R}U \tag{1}$$

$$\frac{d}{dt}\left(\mathrm{R}^2\frac{d}{dt}\lambda\right) = \frac{\partial}{\partial\lambda}U \tag{2}$$

where the perturbation function U corresponding to the heliocentric frame of reference is :

$$U = \frac{Gm}{D} + \frac{GM}{R} - \frac{Gm\,R\cos(\delta\,\lambda)}{r^2} = \mathrm{U}(R,\,\lambda,\,t) \tag{3}$$

where M - mass of the primary; m - mass of the perturbing body; R, r - distance from the mass center test particle and perturbing body accordingly(the vector r between the bodies of masses M and m being expressed as a function of t); G - constant of gravity, $\delta\lambda$ - angle between perturbed and perturbing bodys(differential longitude of perturbed body),ω, ω_s - mean motions perturbed and perturbing body accordingly, and:

$$D = \sqrt{R^2 + r^2 - 2\,R\,r\cos(\,\delta\,\lambda\,)} \tag{4}$$

There are two small parameters in problem: x - radial shift from intermediate (in variations) orbit and r/R - the ratio of mean distance of perturbing and perturbed body. Let R(t)=R_0+x(t), where R_0 is constant. In this paper, simple way to linearization of planar three body problem is considered. For study long-time evolution of orbits we consider averaged equations. In many averaged systems the second equation above is reduced to momentum L conservation. In this case we can rewrite first equation as:

$$\frac{d^2}{dt^2}\,R - \frac{L^2}{R^3} = -\frac{GM}{R^2} + \frac{\partial}{\partial R}\,U \tag{5}$$

2. Simple linearization method

Due to we have two small parameters, we can subsequently expand last expression by power x and R/r (or R/r and x) and restrict themselves by first order terms.

The principle part of expansion perturbation function may be written with using Legendres polynomials Stiefel&Scheifele (1971). In case of the outer perturbing body it is possible to expand perturbation function by power R/r

$$U = \frac{GM}{R} + \sum_{p=2}^{\infty} \frac{Gm}{r}\left(\frac{R}{r}\right)^p P_p(\,\cos(\,\delta\,\lambda\,)\,) \tag{6}$$

After that, it is possible to transform this expansion into Fourier series. We obtain Hill's equation:

$$\frac{d^2}{dt^2}\,x + \omega^2\,x = \mathrm{f}(\,t\,) \tag{7}$$

where f(t) and ω depends on time.

Now we can easy take into account the eccentricity of perturbed particle orbit and show, that such generalization leads to very interesting results. At the elliptic osculating orbit, after the entering osculating orbital elements a - semimajor axis, e - eccentricity, for angular momentum L approximately:

$$L = R^2\,\frac{d}{dt}\,\lambda = \sqrt{GM\,a\,(\,1 - e^2\,)} \tag{8}$$

R_0 depend on time. The phase-averaged value of R_0 approximately:

$$R_0 = \frac{1}{2}\,\pi \int_0^{2\,\pi} \frac{a\,(\,1 - e^2\,)}{1 + e\cos(\,\phi\,)}\,d\phi = a\,\sqrt{1 - e^2} \tag{9}$$

After angular momentum substitution and linearization:

$$3\,\frac{L_0{}^2}{R_0{}^4} - 2\,\frac{GM}{R_0{}^3} = \frac{GM}{a^3}\left(1 - \frac{3}{8}\,e^4\right) = \frac{GM}{a^3}\,(1-\alpha) \tag{10}$$

Finally, for the probe particle elliptic motion case ω depends on eccentricity:

$$\omega^2 = \frac{GM}{a^3}\left(1 - \frac{3}{8}\,e^4\right) + \sum_{p=2}^{\infty} \frac{Gm}{r^3}\, p\,(p-1)\left(\frac{R}{r}\right)^{(p-2)} P_p(\cos(\delta\,\lambda)) \tag{11}$$

Then we can group terms with equal p $\delta\lambda$ by using Laplace's coefficients. The main frequency may be expressed:

$$\omega^2 = \frac{GM}{a^3}\left(1 - \frac{3}{8}\,e^4\right) + \left(\sum_{p=2}^{\infty} \frac{Gm\, b_p \cos(\delta\,\lambda)}{r^3}\right) \tag{12}$$

where b_p is easy to calculate numerically.

For investigation of stability, x may be negligible small. So, we can restrict only 1-st order in expansion by power x. The Mathieu equation is a limit case of Hill's equation and it is more simply for studying. On the other side, because both equations (Hill and Mathieu) are linear, the main area of instability of Mathieu equation must be present in Hill's equation solution. Consequently, and in this case exist the areas of instability, complying with condition, see Landau&Liphshitz (1973):

$$\frac{\omega}{\omega_s} = \frac{n}{n-2+\alpha} \tag{13}$$

Thereby received that orbits near resonances (2n+1)/(2n−1) are unstable parametric. This conclusion completely coincides with results of the studies of the declared problem in the paper Hadjidemetriou(1982) by methods of matrix algebra. Directly, from view of linear equations in elliptic case, we can explain one interesting feature - centres of

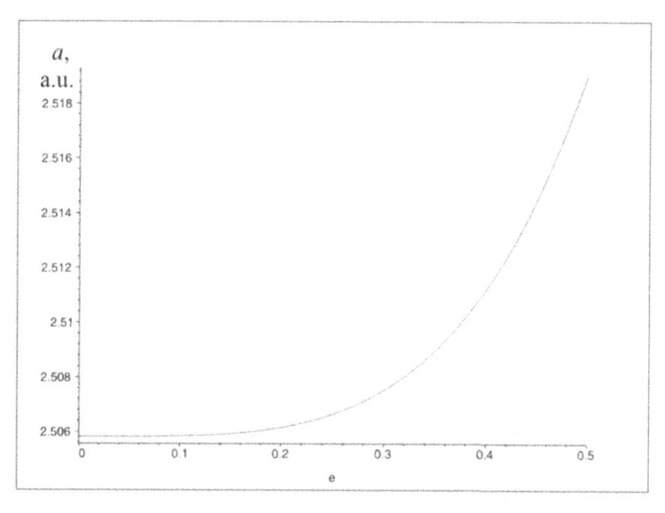

Figure 1. Position of exact resonance in dependence on the eccentricity for 3:1 commensurability with Jupiter.

resonant zones are shifted out relative exact commensurability (resonance). In simple case one perturbing body, centres of instable resonant zones moved away from exact commensurabilities toward a source of perturbation (Fig. 1). So, exact commensurability may be out of the relatively according instable zone!

There are few another interesting applications of the presented simple linear average model. We note two of them. First, magnitude of eccentricity variations at resonance related with variations of semimajor axis. When eccentricity increase, semimajor axis decrease. As it is known (see Morbidelli (2002)), such behaviour is very common in resonance celestial mechanical systems. This relation can be explained in simple averaged linear model, but details are require special careful consideration and can be a subject of separate paper. Second, it is possible overlapping of unstable zones at high order of resonance.

3. Conclusions

It is known, that some different restricted three body problem modifications can be described by linear equation with time-dependent coefficients. Here we give the additional simplest method to obtain linear equation of motion. It is very suitable equation in variation to study perturbed motion at resonance. Just in a very simple model we can explain some very important characteristics of resonant motion. The main results of this work may be formulated in such form:

1. There are not exist stable orbits with e=0 and i=0 in neighbourhood of mean motion resonances (2n-1)/(2n+1).

2. Positions and width of unstable zones depends on eccentricity. The results can be applied to a number of problems of Solar system dynamics. A number of minor planets on high elliptic resonance orbits show the regular behaviour of their orbits. We suppose that stability of these orbits can be provided due to large eccentricity.

In addition, some followings of this simple linear model are noted.

References

Celletti, A., Chessa, A., Hadjidemetriou, J., & Valsecchi, G. B. 2002 *Cel. Mech. Dyn. Astr.* 83,239-255.

Morbidelli, A. 2002 *Modern celestial mechanics*. Taylor and Francis, 356

Hadjidemetriou, J. D.1982 *Celestial Mechanics*, 27, 305-322.

Landau, L. D. & Liphshitz, E. M. 1973 *Mechanics*. Moscow, Nauka, 104-105.

Poincare, H. 1892 *Les Méthodes Nouvelles de la Mécanique Celeste*, Gauthier Villars, Paris.

Stiefel, E. L., & Scheifele, G. 1971 *Linear and Regular Celestial Mechanics*. Springer-Verlag Berlin Heidelberg New York, 305.

Szebehely, V. 1967 *Theory of Orbits*. Acad.press New York and London.

Complex Planetary Systems
Proceedings IAU Symposium No. 310, 2014
Z. Knežević & A. Lemaitre, eds.

doi:10.1017/S1743921314007790

Three-body problem, the measure of oscillating types. A short review

Christian Marchal

French National Office for Aerospace Studies and Researches (ONERA)
email:Christian.Marchal@onera.fr

Abstract. The theoretical three body problem, with three given non infinitesimal point masses, has two types of oscillating motions. In the first type at least two mutual distances are unbounded, but their inferior limit is bounded: there are an infinite number of larger and larger ejections, but without escape. In the second type, it is the velocities that are unbounded: there are an infinite number of nearer and nearer quasi-collisions, without exact collisions.

The first type has only a theoretical interest: its measure in phase space is zero. But the second type has a positive measure in phase space and a physical interest: it governs most of the collisions of stars.

Keywords. galaxies: evolution, three body problem, collisions

1. The mechanism of indirect collisions

Among the different types of three body motions there exists the exchange type: a single star meets a binary and their mutual gravitational interaction leads to the hyperbolic expulsion of one star of the binary and to the formation of a new binary with the two remaining stars. That exchange type is particularly frequent if, in the reference axis of the center of mass, the energy integral of the three body system of interest is negative.

Consider now our galactic system: a majority of stars are binary or multiple stars. If then a weak binary meets a strong binary the four body resulting motion is very close to a three body motion (the motion of the two stars of the weak binary and the center of mass of the strong binary) and motions of exchange type are certainly frequent. These exchange motions lead to the birth of new triple systems: the strong binary and one star of the weak one, these new triple systems that can sometimes be of the second oscillating type.

Of course motions of the second oscillating type lead real bodies to collision and thus the probability of collision through that indirect way is not negligible.

2. The probability of collisions

If we take account the interstellar distances, the average size of stars and their average velocities, the frequency of direct collisions of stars is extremely small: perhaps a few per million years in our galactic system. On the contrary, the formation (and the destruction) of new triple systems through the meeting of two binary or multiple systems is a frequent phenomenon, and thus we are interested into the probability of formation of triple systems having an oscillating motion of the second type.

The four presented references deal with that question that will be presented in the following way. We will use the usual Jacobi decomposition of the three-body problem into two two-body systems: the two nearest bodies are the inner system and the motion of the third mass with respect to the center of mass of the inner system is the outer system.

Let us call **A** the total angular momentum of the three-body system of interest, it is the vectorial sum of the angular momentums $\mathbf{A}_{in}$ and $\mathbf{A}_{out}$ of the inner and the outer systems. We will also call **A**, $\mathbf{A}_{in}$ and $\mathbf{A}_{out}$ the corresponding moduli. The total angular momentum **A** is a well known integral of motion and the perturbations of the inner and outer systems over each other have generally little effect on the modulus $\mathbf{A}_{out}$ of the outer angular momentum (but the direction of that momentum can have large slow variations). Since the quasi-collisions of the two bodies of the inner system occur when its own angular momentum is very small, a usual condition for an oscillating motion of the second type is the almost equality of the moduli **A** and $\mathbf{A}_{out}$. This condition requires generally a very large mutual inclination of the inner and outer systems. For three given masses, it possible to have a rough estimate of the probability P of an oscillating motion of the second type. Consider the inner and outer systems under the following conditions: the semi-major axes and the eccentricities are given (and then also the moduli $\mathbf{A}_{in}$ and $\mathbf{A}_{out}$) and the other parameters are free (with an isotropic repartition of the orientations). In these conditions there are two main cases:

I) If $\mathbf{A}_{in} > 2\ \mathbf{A}_{out}$, rare case requiring a small outer mass, in that case no orientation can give $\mathbf{A} \simeq \mathbf{A}_{out}$ and the probability P is zero or very small.

II) If, as usual, $\mathbf{A}_{in} < 2\ \mathbf{A}_{out}$, the probability P is generally a little larger than the ratio ($\mathrm{m_{out}}\ \mathrm{T_{in}}$ / M $\mathrm{T_{out}}$), where $\mathrm{m_{out}}$ is the outer mass, M the total mass, $\mathrm{T_{in}}$ the period of the inner system and $\mathrm{T_{out}}$ that of the outer system.

Of course that ratio ($\mathrm{m_{out}}\ \mathrm{T_{in}}$ / M $\mathrm{T_{out}}$) is generally small, but it is far larger than those corresponding to direct collisions and the number of indirect collisions is perhaps one hundred, or even one thousand times larger than that of direct collisions. Notice that the energy of the collision of two average stars is about 10^{42} joules, which is much less than that of the supernova phenomenon, but this corresponds roughly to an ordinary nova phenomenon. It is then possible than the indirect collisions are responsible for a large proportion of the novae.

3. Conclusion

The importance, in phase space, of the oscillating motions of the second type may give them a major role in the evolution of the galactic system, even if the present theoretical study neglects many phenomenons as for instance the tidal effects that are certainly essential for bodies undergoing quasi-collisions.

References

Chenciner, A. & Libre, J., 1988, *Ergodic theorem and Dynamical systems*, 8, 63–72
Fejoz, J., 2001 *Journal of Differential Equations*, 175, 175–187
Marchal, C., 1978 *Acta Astronautica*, 5, 745–764
Zhao, L., 2013, *Thèse de Doctorat en Mathématiques* Université Paris VII Diderot

Complex Planetary Systems
Proceedings IAU Symposium No. 310, 2014
Z. Knežević & A. Lemaitre, eds.

doi:10.1017/S1743921314007807

The special case of the three body problem, when gravitational potential is given as the Kislik potential.

A. Shuvalova[1] **and T. Salnikova**[2]

[1]Lomonosov Moscow State University, Moscow, 119991, Russia
email: a.shuvalova@yahoo.com

[2]Lomonosov Moscow State University, Moscow, 119991, Russia
email: tatiana.salnikova@gmail.com

Abstract. In this paper we consider the special case of the planar circular restricted three-body problem by the example of the problem of the Earth, the Moon and a point mass, where the gravitational potentials of the Earth and the Moon are given as the Kislik potential. The Kislik potential takes into account the flattening of a celestial body on the poles. We find the relative equilibria solutions for a point mass and analyze their stability. We describe the difference between the obtained points and the classical solution of the three-body problem.

Keywords. celestial mechanics, methods: analytical, three-body problem, libration point, flattening, gravitational potential.

---

1. Introduction

In classical celestial mechanics for most cases it is quite enough to research the problem when the actual attracting masses are considered as material points. This situation can be easily explained by the fact that the size of attracting bodies compared with the distances between them is usually small enough to neglect. Completely different situation is observed in celestial ballistics. Sometimes the spacecraft is located near the body, and the distance between them is comparable with the size of the body. In this case it is necessary to consider the shape of the body. Therefore the ways of representing the gravitational potential of the planet are of interest from a practical point of view. A variety of main problems of celestial ballistics has been described in the middle of the 20th century. Those problems are connected with the motion of a spacecraft in the non-central gravitational field. The differential equations of motion are the following:

$$\ddot{x}-2\omega\dot{y}-\omega^2x=U'_x,\quad \ddot{y}+2\omega\dot{x}-\omega^2x=U'_y,\quad \ddot{z}=U'_z,\quad U=\frac{\gamma m}{r}+\mu R(x,y,z,t,\mu),\tag{1.1}$$

where γ is the gravitational constant, r is the position vector of a moving point, ω is the angular velocity of the rotating coordinate system, μR is the perturbation function, μ is a small parameter. Potential of the planet U can be written as

$$U=\frac{\gamma m}{r}+\sum_{k=2}^{\infty}J_k(R/r)^kP_k(z/r),\tag{1.2}$$

where P_k is the Legendre polynomial, R is the average equatorial radius of the planet, J_k is the constant. J_2 characterizes the flattening of the planet, J_3 characterizes the asymmetry of the planet relative to the equatorial plane ($J_2\sim-10^{-3}$, $J_3\sim10^{-5}$).

It is known that L. Euler has reduced to the quadratures the problem of the motion of the point mass in the gravitational field of two fixed points. This idea gets a new important application for the theory of the artificial satellites of the Earth. M. D. Kislik was the first who proposed to use the problem of two fixed masses to describe the gravitational potential. In 1961 V. G. Dyomin has showed that the potential of the two fixed points can represent the gravitational field of flattened planet with high accuracy (Dyomin (1968)). The potential can be determined by the following formula

$$U = \frac{\gamma m}{2}\Big(\frac{1+i\delta}{r_1} + \frac{1-i\delta}{r_2}\Big), \quad r_{1,2} = \sqrt{x^2+y^2+(z-c(\delta \pm i))^2}. \tag{1.3}$$

Constants c and δ characterize the shape of the planet. Thus, if a celestial body is compressed along its axis of dynamic symmetry, we can assume that the gravitational potential is approximated by the potential of two points with complex conjugate masses located at the imaginary distance, the gravitational potential is still a real value. If we know the constants J_2 and J_3 from (1.2), then c and δ can be expressed as

$$c = \frac{\sqrt{-J_3^2 - 4J_2^3}}{2J_2}R, \quad \delta = \frac{J_3}{\sqrt{-J_3^2 - 4J_2^3}}. \tag{1.4}$$

When $\delta = 0$, the formula (1.3) does not take into account the asymmetry of the planet relative to the equatorial plane. In this case the form of gravitational potential corresponds to the Kislik potential.

2. Overview

In this paper we consider the special case of the planar circular restricted three-body problem by the example of the problem of the Earth (E), the Moon (M) and a point mass (P), where the gravitational potentials of E and M are given as the Kislik potential.

Let us consider the planar circular restricted three-body problem. We neglect the mass of the P. E and M have masses m and $m-\mu$, respectively, and perform circular motion around their barycenter O. The rotating coordinate system is as usual: O is the origin of coordinates, x-axis goes through E and M. E, M and P move in the plane $z=0$. $\vec{\omega} = \omega\vec{e_z}$ is the angular velocity of the moving coordinate system. If we take the Kislik potential in the Earth-Moon problem, ω satisfies the following equation:

$$\omega^2 = \frac{\gamma(m-\mu)}{((x_E+x_M)^2-c^2)^{3/2}} + \frac{\gamma\mu}{((x_E+x_M)^2-\lambda^2c^2)^{3/2}}.$$

To simplify the analytical calculus let us take the classical solution of the two-body problem:

$$\omega^2 = \gamma m/(x_E+x_M)^3.$$

Then the gravitational potentials of E and M take the form

$$U_E = \frac{\gamma(m-\mu)}{\sqrt{(x+x_E)^2+y^2-c^2}}, \quad U_M = \frac{\gamma\mu}{\sqrt{(x-x_M)^2+y^2-\lambda^2c^2}}, \tag{2.1}$$

where λ is the constant coefficient. Motion of P is determined by the system of equations (1.1). Constant c for E is equal to 209.9 km. Ellipticity of M is 3 times less than ellipticity of E, the radius of M is equal to 0.273 radius of E. So in the numerical calculations we assume that $\lambda = 0$. We define the effective potential energy as

$$V_\omega = -\frac{1}{2}\omega^2(x^2+y^2) - U_E - U_M. \tag{2.2}$$

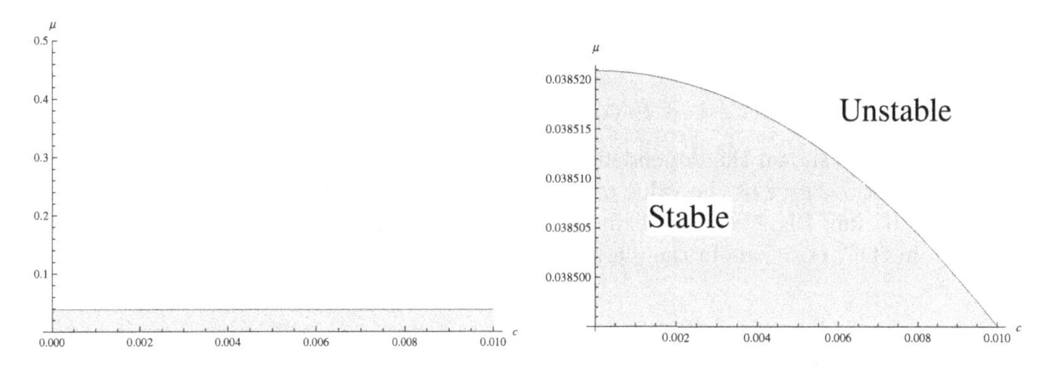

Figure 1. On the figure there is shown dependence of μ on c, if $\lambda = 0$. If $c = 0$, the condition for the stabilization L_4 and L_5 is the same as the condition for the stabilization in the classical three-body problem, i.e. $\mu < 0,0385209$.

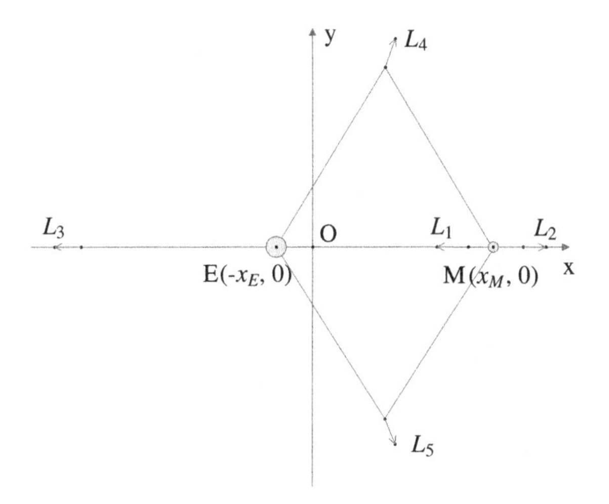

Figure 2.

Libration point. Relative equilibrium equations are

$$\omega^2 x = \frac{\gamma(x + x_E)(m - \mu)}{((x + x_E)^2 + y^2 - c^2)^{3/2}} + \frac{\gamma(x - x_M)\mu}{((x - x_M)^2 + y^2 - \lambda^2 c^2)^{3/2}},$$

$$\omega^2 y = \frac{\gamma y(m - \mu)}{((x + x_E)^2 + y^2 - c^2)^{3/2}} + \frac{\gamma y \mu}{((x - x_M)^2 + y^2 - \lambda^2 c^2)^{3/2}}.$$

If $y = 0$, we have 3 unstable libration points $L_{1,2,3}$ as in the usual three-body problem. L_1, L_2 and L_3 can be obtained from the equation:

$$\omega^2 x = \frac{\gamma(x + x_E)(m - \mu)}{((x + x_E)^2 - c^2)^{3/2}} + \frac{\gamma(x - x_M)\mu}{((x - x_M)^2 - \lambda^2 c^2)^{3/2}}. \tag{2.3}$$

The second member function has 4 vertical asymptotes ($x = -x_E \pm c$, $x = x_M \pm \lambda c$) and decreases monotonically.

If $y \neq 0$, then the coordinates of P can be found analytically, the coordinates of points L_4 and L_5 satisfy the equations: $(x + x_E)^2 + y^2 = (x_E + x_M)^2 + c^2$, $(x - x_M)^2 + y^2 = (x_E + x_M)^2 + \lambda^2 c^2$. In the classical solution of the three-body problem EML_4 and EML_5 satisfy the equations: $(x + x_E)^2 + y^2 = (x_E + x_M)^2$, $(x - x_M)^2 + y^2 = (x_E + x_M)^2$.

If we assume that $x_E + x_M = 1$, $m = 1$, $\mu = \widetilde{\mu}$, the condition for the stabilization of L_4 and L_5 in the first approximation takes the following form:

$$1 - 27\widetilde{\mu} + 27\widetilde{\mu}^2 + 9c^4(1 + (-1 + \lambda^4)\widetilde{\mu}) + 6c^2(-1 - 2(1 + 2\lambda^2)\widetilde{\mu} + 3(1 + \lambda^2)\widetilde{\mu}^2) > 0.$$

On Fig.1 there is shown the dependence of $\widetilde{\mu}$ on c, if $\lambda = 0$.

Our calculations give us the value of the shift of each libration point. It approximately equals to 0.05 km. Fig.2 schematically shows the directions of the shifts of the libration points from their positions in the classical problem.

3. Conclusion

Thus, we can conclude that in the problem considered above, the shape of the bodies does not provide substantial changes in the solutions. But the Earth is not the most flattened planet of the Solar system.

The value of c was calculated for several planets:

	c, km
Earth	209.9
Mars	150.013
Jupiter	8461.57
Saturn	7547.368

The polar flattening of Jupiter is equal to 0.065 (0.0033 for the Earth), the flattening of Saturn is equal to 0.1 . These values are greater, so in the problem, where we use a gravitational potential according to our algorithm for these planets, we can expect more visible effect.

References

Demin, V. G., 1968, *The motion of an artificial satellite in the noncentral gravitational field*, Nauka, Moscow.

Markeev, A. P., 1978, *The libration points in celestial mechanics and space dynamics*, Nauka, Moscow.

Worthington, J., 2012, *A Study of the Planar Circular Restricted Three Body Problem and the Vanishing Twist*, Thesis, Univ. Sydney.

Complex Planetary Systems
Proceedings IAU Symposium No. 310, 2014
Z. Knežević & A. Lemaitre, eds.

doi:10.1017/S1743921314007819

On the orbital structure of the HD 82943 multi-planet system

Roman V. Baluev[1,2] and Cristian Beaugé[3]

[1]Central Astronomical Observatory at Pulkovo of Russian Academy of Sciences, Pulkovskoje shosse 65, St Petersburg 196140, Russia

[2]Sobolev Astronomical Institute, St Petersburg State University, Universitetskij prospekt 28, Petrodvorets, St Petersburg 198504, Russia
email: r.baluev@spbu.ru

[3]Instituto de Astronomía Teórica y Experimental, Observatorio Astronómico, Universidad Nacional de Córdoba,
Laprida 854, (X5000BGR) Córdoba, Argentina

Abstract. HD 82943 hosts a mysterious multi-planet system in the 2:1 mean-motion resonance that puzzles astronomers for more than a decade. We describe our new analysis of all radial velocity data currently available for this star, including both the most recent Keck data and the older but more numerous CORALIE measurements.

Here we pay a major attention to the task of optimal scheduling of the future observation of this system. Applying several optimality criteria, we demonstrate that in the forthcoming observational season of HD 82943 (the winter 2014/2015) rather promising time ranges can be found. Observations of the near future may give rather remarkable improvement of the orbital fit, but only if we choose their time carefully.

Keywords. stars: planetary systems - stars: individual: HD 82943 - techniques: radial velocity - methods: data analysis - methods: statistical

---

1. Introduction

This paper can be treated as an addition to our recent work (Baluev & Beaugé 2014) devoted to a reanalysis of the radial velocity (RV) data for a unique multi-planet host star HD 82943. Here we only briefly summarize the most important of our previous results (Sect. 2), and present new ones, related to seeking the optimal observation dates for this star (Sect. 3).

2. Main results of the RV data analysis

In our work we used the entire set of the RV data currently available for HD 82943 in the public literature. These include the old CORALIE (Mayor *et al.* 2004) and the recent Keck (Tan *et al.* 2013) data. The Keck data were separated in two independent subsets that were acquired before and after a hardware upgrade. The primary results concerning our re-analysis of these data are described in (Baluev & Beaugé 2014). Thus we do not duplicate this discussion here, except for a brief summary of the conclusions:

(*a*) The Keck and CORALIE data are not in a good agreement with each other: fitting the entire data set plainly leads to a severely unstable orbital configuration of the two major planets *b* and *c*.

(*b*) One of the reasons for this mutual inconsistency is the likely presence of an additional systematic variation in the CORALIE (but not Keck) data with a period close

Table 1. Prescribed observations scheduling goals for HD 82943.

scheduling goal	critical parameters to refine	their number
goal 1	parameters of all three planets	14
goal 2	common orbital inclination	1
goal 3	location relatively to the two-planet ACR (see text)	4
goal 4	parameters of the third planet	3

to a year. Likely, this variation appeared due to some imperfections of the spectrum processing algorithm used for CORALIE.

(*c*) After removal of the CORALIE annual variation, the RV data still contain a significant periodicity with a period of $\sim$ 1000 d. We interpret this periodicity as a hint of the third planet that was previously suspected by Goździewski & Konacki (2006) and Beaugé *et al.* (2008).

(*d*) An RV fit implying a stable planetary configuration can be obtained only by including both the third planet and the CORALIE annual term in the RV model. Without these terms, the nominal (best fitting) solution appears unstable due to an antialigned initial apsidal state of the two major planets, and forcing this configuration to be stable would infer an unsuitably large shift of the fit from its nominal position.

(*e*) The planets in the best fitting configuration lie near the three-planet resonance with the periods ratio $P_c : P_b : P_d \approx 1 : 2 : 5$. The motion of the first two planets is truly resonant (a libration) in the vicinity of an aligned Apsidal Corotation Resonance (ACR), while the orbit of the third planet is rather uncertain and its orbital evolution can represent a libration (i.e. a true resonance) as well as a circulation (not a genuine resonance). The structure of the dynamical space in the vicinity of the third planet's orbit is pretty complicated and involves resonant domains intervening with the non-resonant ones.

3. Optimal planning of the future RV observations

Clearly, the orbital and dynamical structure of the HD 82943 system is still rather uncertain. In view of this, it may be useful to apply some optimal planning routines, in order to predict the time segments in future in which the new RV observations would improve or knowledge about the system, as well as to identify the time ranges where the new observations would be almost useless.

We solve this task by means of the optimal planning approaches described in (Baluev 2008a). In this method we should select the entire set of the fitted parameters, or any their subset, or even a set of some other quantities expressed by smooth functions of the original parameters. Our goal is to find an optimal time for a new observation in the future, in order to achieve a maximum reduction of the uncertainties in the targeted quantities. Here we adopt the so-called D-optimality criterion, in which the "reduction" of a multi-dimensional uncertainty is treated in terms of the volumes ratio for the relevant uncertainty ellipsoids (or determinants of the relevant covariance matrices).

In this work we consider the three-planet fit with the eccentricity of the third planet always fixed at zero. Otherwise this eccentricity is ill determined and generates dramatic non-linearity effects, which are not desirable. Four sets of target quantities to refine were considered in this work, defining four scheduling "goals". These goals are described in Table 1. The goal 3 from this table is defined in a rather complicated manner. Its purpose is to refine our knowledge about the position of the dynamical system relatively to the ACR of the two major planets. This information is important for the long-term dynamics and the stability of the system (Beaugé *et al.* 2003). In this case the set of target quantities includes 4 partial derivatives of the averaged Hamiltonian of the two major planets. When all these derivatives vanish, we deal with an exact ACR, so by reducing their uncertainties we can improve our knowledge about whether the system is close to or far from the ACR. The derivatives are computed using the method from (Baluev 2008b).

The results of the computation are shown in Fig. 1, where we plot the graphs of a function that indicates how much the uncertainty of the target quantities would reduce, provided that we make a single observation at the time given in the abscissa. This relative reduction is normalized so that it corresponds to a single degree of freedom in the set of target quantities. The vertical fringes label the yearly seasons when the star can (darker bands) or cannot (white bands) be actually observed.

We can see series peaks in these plots, marking the position of optimal times. In Table 2 we show more details concerning these optimal times for the nearest three observation seasons.

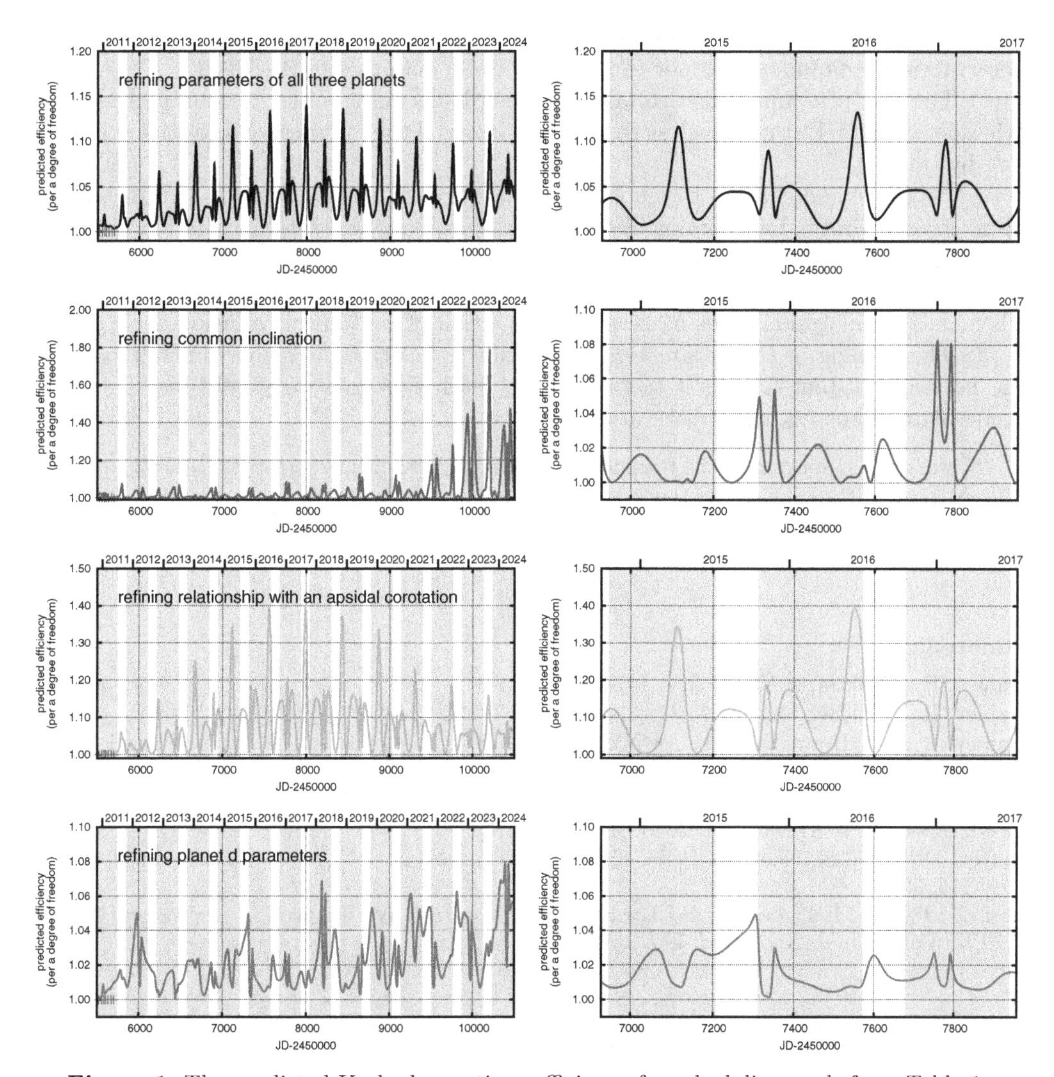

Figure 1. The predicted Keck observations efficiency for scheduling goals from Table 1.

Table 2. Optimal observation dates for HD 82943.

scheduling goal 1		scheduling goal 2		scheduling goal 3		scheduling goal 4	
JD-2450000	max. eff.	JD-2450000	max. eff.	JD-2450000	max. eff.	JD-2450000	max. eff.
observational season of 2014/2015							
begin – 6988	1.038	7024 ± 27	1.016	begin – 6985	1.123	7055 ± 32	1.029
7113 ± 15	1.116	7182 ± 17	1.018	7113 ± 17	1.343	7140 – end	1.029
7193 – end	1.035	-	-	7201 – end	1.095	-	-
observational season of 2015/2016							
7334 ± 10	1.090	7313 ± 8	1.049	7334 ± 10	1.187	begin – 7315	1.049
7392 ± 33	1.051	7351 ± 5	1.054	7393 ± 32	1.175	7357 ± 8	1.030
7553 ± 16	1.133	7454 ± 26	1.022	7551 ± 21	1.394	-	-
observational season of 2016/2017							
begin – 7743	1.047	7755 ± 7	1.082	begin – 7739	1.145	7748 ± 11	1.027
7773 ± 10	1.102	7790 ± 5	1.080	7773 ± 10	1.200	7793 ± 6	1.026
7831 ± 33	1.057	7891 ± 28	1.032	7831 ± 31	1.174	-	-

4. Conclusions

A few interesting matters can be noticed in Fig. 1 and Table 2:

(*a*) The peaks of the efficiency functions are rather narrow, meaning that allocating observation time randomly is not the best course of actions for HD 82943.

(*b*) The task of refining the orbital inclination looks antagonistic to refining the most other parameters. But nonetheless the relevant optimal time ranges tend to stick together side-by-side.

(*c*) We have good chances to refine the accuracy of the usual planetary parameters by up to $30-40\%$ in the forthcoming observing seasons. But the orbital inclination, which is only constrained thanks to the gravitational planet-planet perturbations, is an exception. It looks unrealistic to drastically improve the accuracy of this inclination before 2020s, when the orbital apsidal lines make a larger fraction of a secular revolution.

(*d*) The refining of the parameters of the third planet seems rather difficult both in the near and distant future. It seems that to reach this goal we should just patiently accumulate more and more observations.

The work was supported by the President of Russia grant for young scientists (MK-733.2014.2), by the Russian Foundation for Basic Research (project 14-02-92615 KO_a), and by the programme of the Presidium of Russian Academy of Sciences "Non-stationary phenomena in the objects of the Universe".

References

Baluev, R. V. 2008a, *MNRAS*, 389, 1375

Baluev, R. V. 2008b, *Cel. Mech. Dyn. Astron.*, 102, 297

Baluev, R. V. & Beaugé, C. 2014, *MNRAS*, 439, 673

Beaugé, C., Ferraz-Mello, S., & Michtchenko, T. A. 2003, *ApJ*, 593, 1124

Beaugé, C., Giuppone, C., Ferraz-Mello, S., & Michtchenko, T. A. 2008, *MNRAS*, 385, 2151

Goździewski, K. & Konacki, M. 2006, *ApJ*, 647, 573

Mayor, M., Udry, S., Naef, D., Pepe, F., Queloz, D., Santos, N. C., & Burnet, M. 2004, *A&A*, 415, 391

Tan, X., Payne, M. J., Lee, M. H., Ford, E. B., Howard, A. W., Johnson, J. A., Marcy, G. W., & Wright, J. T. 2013, *ApJ*, 777, 101

Complex Planetary Systems
Proceedings IAU Symposium No. 310, 2014
Z. Knežević & A. Lemaitre, eds.

doi:10.1017/S1743921314007820

Dynamics and Habitability in Binary Star Systems

Siegfried Eggl[1], Nikolaos Georgakarakos[2] and Elke Pilat-Lohinger[3]

[1]IMCCE, Observatoire de Paris, 77 Avenue Denfert-Rochereau, F-75014, Paris, France
email: siegfried.eggl@imcce.fr

[2] Higher Technological Educational Institute of Central Macedonia, Terma Magnesias, Serres 62124, Greece
email: georgakarakos@hotmail.com

[3] University of Graz, Institute of Physics, IGAM, Universitätsplatz 5, 8010 Graz, Austria
email: elke.pilat-lohinger@univie.ac.at

Abstract. Determining planetary habitability is a complex matter, as the interplay between a planet's physical and atmospheric properties with stellar insolation has to be studied in a self consistent manner. Standardized atmospheric models for Earth-like planets exist and are commonly accepted as a reference for estimates of Habitable Zones. In order to define Habitable Zone boundaries, circular orbital configurations around main sequence stars are generally assumed. In gravitationally interacting multibody systems, such as double stars, however, planetary orbits are forcibly becoming non circular with time. Especially in binary star systems even relatively small changes in a planet's orbit can have a large impact on habitability. Hence, we argue that a minimum model for calculating Habitable Zones in binary star systems has to include dynamical interactions.

Keywords. stars: binaries, stars: planetary systems, habitability, celestial mechanics

1. Introduction

One of the most intriguing aspects of exoplanet science is the prospect of finding worlds around other stars that might be capable of hosting life as we know it. This translates into a search for terrestrial planets in the so-called Habitable Zones (HZs), i.e. circumstellar regions where an Earth-analog is capable of sustaining liquid water on its surface. Finding planets in such zones does not automatically mean that they are habitable, though. The complex interplay between insolation, atmosphere, lithosphere and hydrosphere necessitates a determination of most planetary, stellar and orbital properties in order to judge potential habitability. Hence, each newly discovered candidate has to be assessed on an individual basis. This, of course, raises questions on the purpose and usefulness of the HZ concept. Yet, we believe that providing planet hunters with guidelines on where to look for potentially habitable worlds remains important, as long as observational resources are limited. This implies that predicted HZ borders have to be translatable into detectability domains for the respective exoplanet detection methods. HZ estimates should, therefore, contain reasonably detailed assumptions on the physical processes underlying planetary habitability. We argue, for instance, that dynamically induced changes in a planet's orbit can substantially change habitability conditions. If quantifiable, they should not be neglected in HZ calculations.

2. Insolation and Orbital Dynamics

So far, considerable efforts have been spent on investigating climatic stability of Earth-like planets leading to largely accepted estimates on the amount and spectral distribution of insolation that can render a world uninhabitable (Kasting *et al.* 1993; Selsis *et al.* 2007; Kopparapu *et al.* 2013). Recent so-called "effective insolation" limits can be found, for instance, in Kopparapu *et al.* (2014). Naturally, globally averaged climate models such as used by the previously named authors cannot account for all effects (e.g. Wang *et al.* 2014; Leconte *et al.* 2013), but they can be considered a reasonable first approximation for Earth-like planets that are far from the tidal-lock or tidal-heating regime. Alas, HZ borders can be calculated by solving the simultaneous equations:

$$S/I \leqslant 1 \text{ and } S/O \geqslant 1, \quad \text{leading to} \quad r_{\{I,O\}} = \left(\frac{L_\star}{4\pi S_c\{I,O\}}\right)^{1/2}, \tag{2.1}$$

where S is the insolation at the top of the atmosphere, $L_\star$ is the luminosity of the star, $I = S_{eff}^{inner}$ and $O = S_{eff}^{outer}$ denote the effective insolation limits for the inner and outer edge of the HZ, and $S_c = 1367\,[W/m^2]$ refers to the Solar constant. For the Sun and an Earth-like planet Kopparapu *et al.* (2014) predict $I \sim 1.11$ and $O \sim 0.36$, resulting in the following HZ limits expressed as distances r from the host star: $r_I \sim 0.95$ au and $r_O \sim 1.67$ au. Equation (2.1) describes a spherical shell around the host star where planets can populate orbits that respect the insolation limits necessary for habitability. Yet, what if the planet leaves this shell from time to time? This may happen, for instance, when the planet's orbit is elliptic. Then, the top of the atmosphere insolation can vary considerably with time, see Figure 1! Williams & Pollard (2002) could show that planets with Earth-like oceans can compensate climatic extremes due to excursion outside the classical HZs (henceforth CHZ, not to be confused with the Continuous Habitable Zone), as long as the average insolation stays within habitable limits. Later studies indicate, however, that the role of planetary eccentricity and its influence on insolation cannot easily be discarded (e.g. Spiegel *et al.* 2010; Dressing *et al.* 2010). We, thus, propose a way to include available information on the variability of planetary insolation directly into habitability considerations.

3. A Minimum Dynamical Model for HZs

Our aim is to acquire self consistent HZ estimates with a minimum of complexity that still retain the relevant physics. Hence, we use the effective insolation limits of a globally averaged atmosphere model, where variations in the planet's rotation state can be ignored, as long as its atmospheric properties do not change radically. Alterations in the activity of the star could be included in theory, but they are difficult to model and shall be neglected for the moment. In such a scenario, changes that are related to the planet's orbit can be considered the dominating cause for insolation variability. In order to include such information into our HZ model, we follow Eggl *et al.* (2012) and introduce the concepts of the Permanently (PHZ), Extended (EHZ) and the Averaged Habitable Zones (AHZ), see Figure 1. The PHZ is the region where a planet *always* stays within the effective insolation limits (I, O), i.e. $S/I \leqslant 1$ and $S/O \geqslant 1$. This is the "classical" definition of a HZ. The EHZ allows parts of the planetary orbit to lie beyond the classical HZ: $\langle S/I\rangle_t + \sigma_I \leqslant 1$ and $\langle S/O\rangle_t - \sigma_O \geqslant 1$, where $\langle S\rangle_t$ denotes the time-averaged effective insolation and σ^2 the effective insolation variance. High planetary eccentricity may not be prohibitive for habitability since the atmosphere can act as a buffer. Consequently, the AHZ encompasses all configurations that support the planet's time-averaged

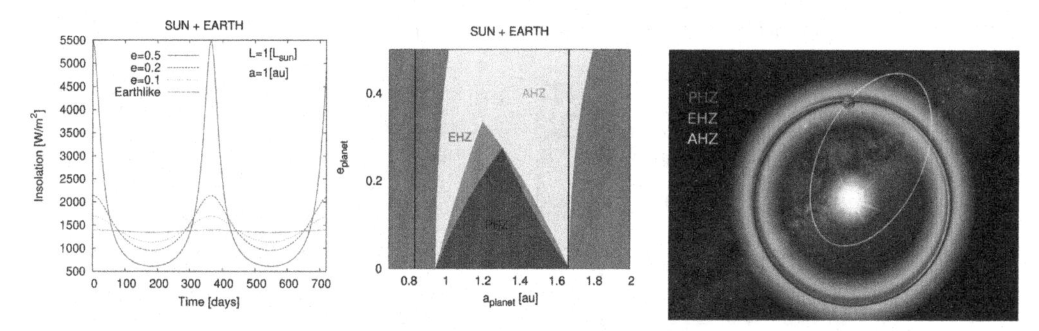

Figure 1. *left:* Insolation variability in the two body problem Sun + Earth for mutual orbits with various eccentricities. *mid:* A habitability map (HM) showing different kinds of Habitable Zones for an Earth-twin orbiting a Sun-like star on constant orbits with various semimajor axes and eccentricities. The different shades denote: PHZ, EHZ, AHZ and non-habitable regions. The black vertical lines correspond to the standard CHZ borders after Kasting *et al.* (1993) while the shaded HZs have been calculated using the values by Kopparapu *et al.* (2014). One can see that the PHZ and EHZ decrease rapidly with mounting planetary eccentricities while the AHZ remains practically constant up to high e_{planet} values. The difference between Kasting's and Kopparapu's HZ borders ($e_{planet} = 0$) also shows the importance of a classification scheme that can adapt to alterations in the climate model. *right:* A concept of the proposed HZ scheme that includes information on the variability of planetary insolation.

effective insolation to be within classical limits, e.g. also orbits with very high eccentricities: $\langle S/I \rangle_t \leqslant 1$ and $\langle S/O \rangle_t \geqslant 1$. Figure 1 shows a realization of this classification scheme for an Earth-like planet orbiting a Sun-like star.

4. Habitable Zones in Binary Star Systems

The possible existence of terrestrial planets in and around binary star systems (Hatzes 2013; Dumusque *et al.* 2012) has rekindled the scientific community's interest in determining Habitable Zones (HZ) in and around binary star systems (Forgan 2012, 2014; Kane & Hinkel 2013; Haghighipour & Kaltenegger 2013; Kaltenegger & Haghighipour 2013; Eggl *et al.* 2012, 2013b; Cuntz 2014; Jaime *et al.* 2014). Given the strong gravitational perturbations planets experience in such environments, significant variations in planetary orbits can be expected. Even if planets formed on circular orbits, for instance, the interaction with the binary would force the planet's orbital eccentricity become non-zero over time. As was shown in Eggl *et al.* (2012) the 'momentary' HZ for binary star systems are solutions of the two simultaneous planetary insolation equations normalized with respect to effective insolation limits. In compact notation that is

$$\bar{S}_I \leqslant 1 \text{ and } \bar{S}_O \geqslant 1, \quad \text{where} \quad \bar{S}(t)_{\{I,O\}} = \sum_{j=1}^{N} \frac{L_j}{4\pi S_c\{I_j, O_j\}} r_j^{-2}(t). \tag{4.1}$$

Here I_j and O_j denote the respective effective insolation limits that serve as spectral weighting factors and L_j are the stellar luminosities; $\bar{S}_{\{I,O\}}$ is the normalized total insolation at the inner ($\bar{S}_I$) and outer ($\bar{S}_O$) border of the combined HZ, and $r_j(t)$ represent the time dependent planet-to-star distances. For double star systems we have $N = 2$. Müller & Haghighipour (2014) derived a similar expression. Note, however, that two different sets of $\{a, b, c, d\}$ parameters, one for the inner and one for the outer HZ border are required to simultaneously fulfill the inequalities in their equation (5). Momentary HZs depend on the current relative positions of the stars and the planet. They tend to fluctuate on orbital and secular timescales. Deriving observational constraints from time

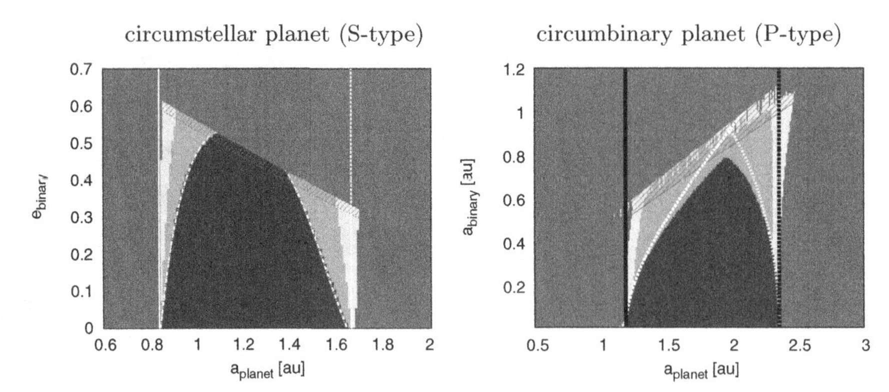

Figure 2. PHZ, EHZ, AHZ, non-habitable and dynamically unstable regions (upper diagonal areas) are shown for a G2V-G2V S-type system with a binary semi major axis of $a_b = 10$ au (*left*), and a set of G2V-G2V P-type systems on initially circular orbits with different semimajor axes (*right*). The coloring is the same as in Figure 1. The borders of the expected CHZ (to get classical estimates for P-type systems the radiation of both stars was combined in the binary's barycenter) are denoted by vertical lines. The white curves indicate analytical estimates for the extent of the PHZ. Effective insolation values by Kasting *et al.* (1993) were used. Evidently, the injected planetary eccentricities grow with the binary's eccentricity leading to a decrease in the width of the PHZs. Even for planets around binaries on circular orbits the harsh restrictions in the PHZ become evident with growing semimajor axis of the binary star.

dependent HZs can, thus, be difficult (e.g. Eggl *et al.* 2013a). To include information on the variability of plantary insolation, we apply the HZ scheme defined in section 2 and predict planetary insolation and HZ limits analytically using recent results from perturbation theory (Georgakarakos 2003, 2005, 2009). This allows us to model the Earth-analog's orbit and insolation evolution as a function of the system's initial parameters and time (Eggl *et al.* 2012). As can be seen in Figure 2, PHZ and EHZ and AHZ behave quite differently for various orbital configurations of a terrestrial planet in a binary star system. In all studied binary systems a clear decrease in the extent of the PHZ and EHZ can be observed for growing eccentricities of the binary's orbit. Of course, not only the binary's eccentricity, also its semimajor axis greatly influences the extent of permanently habitable regions. The right panel in Figure 2 shows that the PHZ can be considerably smaller than the CHZ for circumbinary planets (P-type), even if both the binary's and the planet's orbit are initially circular! The AHZ are mostly identical with CHZs. Only in regions close to the system's stability limits deviations in S-type as well as P-type systems occur. Our HZ estimates have, furthermore, been applied to well characterized S-type binary stars close to the Solar System (Eggl *et al.* 2013b). It was found that most systems do not only allow for habitability in an 'average sense', they even retain zones of permanent habitability.

5. Conclusions

The complex interplay between dynamical and radiative influences can turn the determination of Habitable Zones (HZ) in gravitationally interacting systems into a challenging affair. Flexibility and adaptability should, thus, be considered important properties of any HZ calculation scheme. We argue that a minimum dynamical model should be included in HZ calculations to ensure realistic estimates. The proposed classification scheme (PHZ, EHZ and AHZ) retains information on the variability of planetary insolation while providing HZ boundaries that remain valid up to stellar evolution timescales. It is flexible with respect to changes in atmospheric models as well as the choice of the underlying dynamical model. An alternative to time consuming numerical insolation calculations, the analytical estimates presented in Eggl *et al.* (2012) can help to shed light onto the diverse kinds of habitability occurring for terrestrial planets in gravitationally active systems. Their application to nearby S-type binary stars has shown that most of these systems allow for HZs in spite of their strong gravitational interactions and variable insolation conditions.

Acknowledgments

The authors would like to acknowledge the support of the Austrian FWF projects S11608-N16 (subproject of the NFN S116) and P22603-N16, the European Union Seventh Framework Program (FP7/2007-2013) under grant agreement no. 282703, as well as the IAU Symposium no. 310 travel grant.

References

Cuntz M., 2014, *ApJ*, 780, 14
Dressing C. D., Spiegel D. S., Scharf C. A., Menou K., & Raymond S. N., 2010,*ApJ* 721, 1295
Dumusque X. et al., 2012, *Nature*, 491, 207
Eggl S., Haghighipour N., & Pilat-Lohinger E., 2013a, *ApJ*, 764, 130
Eggl S., Pilat-Lohinger E., Funk B., Georgakarakos N., & Haghighipour N., 2013b, *MNRAS*, 428, 3104
Eggl S., Pilat-Lohinger E., Georgakarakos N., Gyergyovits M., & Funk B., 2012, *ApJ*, 752, 74
Forgan D., 2012, *MNRAS*, 422, 1241
Forgan D., 2014, *MNRAS* 437, 1352
Georgakarakos N., 2003, *MNRAS* 345, 340
Georgakarakos N., 2005, *MNRAS* 362, 748
Georgakarakos N., 2009, *MNRAS*, 392, 1253
Haghighipour N. & Kaltenegger L., 2013, *ApJ*, 777, 166
Hatzes A. P., 2013, *ApJ*, 770, 133
Jaime L. G., Aguilar L., & Pichardo B., 2014, *MNRAS*, 443, 260
Kaltenegger L., Haghighipour N., 2013, *ApJ*, 777, 165
Kane S. R. & Hinkel N. R., 2013, *ApJ*, 762, 7
Kasting J. F., Whitmire D. P., & Reynolds R. T., 1993, *Icarus*, 101, 108
Kopparapu R. K. et al., 2013, *ApJ*, 770, 82
Kopparapu R. K., Ramirez R. M., SchottelKotte J., Kasting J. F., Domagal-Goldman S., & Eymet V., 2014, *ApJ*, 787, L29
Leconte J., Forget F., Charnay B., Wordsworth R., Selsis F., Millour E., & Spiga A., 2013, *A&A*, 554, A69
Müller T. W. A. & Haghighipour N., 2014, *ApJ*, 782, 26
Selsis F., Kasting J. F., Levrard B., Paillet J., Ribas I., & Delfosse X., 2007, *ApJ*, 476, 1373
Spiegel D. S., Raymond S., Dressing C. D., Scharf C. A., Mitchell J. L., & Menou K., 2010, in *Astronomical Society of the Pacific Conference Series, 430, Pathways Towards Habitable Planets,* Coudé Du Foresto V., Gelino D. M., Ribas I., eds., 109
Wang, Y., Tian, F., & Hu, Y., 2014, *The Astrophysical Journal Letters*, 791, L12
Williams D. M., & Pollard D., 2002, *International Journal of Astrobiology*, 1, 61

Complex Planetary Systems
Proceedings IAU Symposium No. 310, 2014
Z. Knežević & A. Lemaitre, eds.

doi:10.1017/S1743921314007832

Tidal evolution in multiple planet systems: application to Kepler-62 and Kepler-186

Emeline Bolmont[1,2], Sean N. Raymond[1,2], Jérémy Leconte[3,4,5], Alexandre Correia[6,7] and Elisa Quintana[8,9]

[1]Univ. Bordeaux, Laboratoire d'Astrophysique de Bordeaux, UMR 5804, F-33270, Floirac, France

[2]CNRS, Laboratoire d'Astrophysique de Bordeaux, UMR 5804, F-33270, Floirac, France

[3]Canadian Institute for Theoretical Astrophysics, 60st St George Street, University of Toronto, Toronto, ON, M5S3H8, Canada

[4]Banting Fellow

[5]Center for Planetary Sciences, Department of Physical & Environmental Sciences, University of Toronto Scarborough, Toronto, ON, M1C 1A4

[6]Departamento de Fisica, I3N, Universidade de Aveiro, Campus de Santiago, 3810-193 Aveiro, Portugal

[7]ASD, IMCCE-CNRS UMR8028, Observatoire de Paris, UPMC, 77 Av. Denfert-Rochereau, 75014 Paris, France

[8]SETI Institute, 189 Bernardo Ave, Suite 100, Mountain View, CA 94043, USA

[9]NASA Ames Research Center, Moffett Field, CA 94035

Abstract. A large number of observed exoplanets are part of multiple planet systems. Most of these systems are sufficiently close-in to be tidally evolving. In such systems, there is a competition between the excitation caused by planet-planet interactions and tidal damping. Using as an example two multiple planet systems, which host planets in the surface liquid water habitable zone (HZ): Kepler-62 and Kepler-186, we show the importance and effect of both planetary and stellar tides on the dynamical evolution of planets and on the climate of the HZ planets.

Keywords. Planets and satellites: dynamical evolution and stability, Planet-star interactions, stars: individual (Kepler-62, Kepler-186), methods: N-body simulations.

---

1. Introduction

More than 1400 exoplanets have now been detected and about 20% of them are part of multiple planet systems (http://exoplanets.org/). Most of these planetary systems have close-in planets for which tides have an influence. In particular, tides can have an effect on the eccentricities of planets, and also on their rotation periods and their obliquities, which are important parameters for any climate studies (so-called Milankovitch cycles; Berger, 1988). Besides tides can influence the stability of multiple planet systems, by their effect on the eccentricities but also by their effect on precession rates.

We present here a few examples of tidal evolution of multiple planet systems including the study of the Kepler-62 (Borucki *et al.* 2013) and Kepler-186 (Quintana *et al.* 2014) systems. These recently discovered systems host 5 planets. Kepler-62 e and f are in the HZ of a 0.69 $M_\star$ star. Kepler-186 f is an Earth-sized planet in the HZ of a 0.42 $R_\star$ star. We study the possible evolution of these planets' orbits as well as their spin states.

2. Method

In order to study the dynamical evolution of these systems, we used a new code, which takes into account the evolution of the radius and spin of the central object (be it a star or brown dwarf evolving according to the evolution tracks of Chabrier *et al.* 2000). It also includes a rigorous treatment of tidal forces (Hut, 1981; Leconte *et al.* 2010), the effect of the rotation-induced flattening of planets and star (Correia *et al.* 2013), the correction due to general relativity (Kidder, 1995), and planet-planet gravitational interactions using a symplectic N-body algorithm (Chambers, 1999). This code will be introduced in Bolmont *et al.* 2015 (to be submitted) and has already been used in Bolmont *et al.* (2013, 2014) and Heller *et al.* (2014).

We used as initial orbital elements, stellar and planetary parameters the values given by Borucki *et al.* (2013) for Kepler-62 and Bolmont *et al.* (2014) for Kepler-186. In Bolmont *et al.* (2014), we derived two sets of planetary parameters: set A and set B both consistent with the data and self-consistent. These two sets were used here.

We derived the masses of the planets for different planetary compositions using Fortney *et al.* (2007). We chose for the Kepler-62 system an Earth-like composition for all planets. For Kepler-186, we tested compositions from 100% ice to 100% iron.

3. Results

3.1. *Stability*

We find that tides influence the stability of multiple planetary systems. This is especially clear in the case of Kepler-62 for which simulations without general relativity and tides lead to a destabilization in a few 10^7 yr. However simulations with general relativity and tides are stable during the 30 million years of the simulation. We also find that the system is very sensitive to the chosen planets' masses: a change of a few percent can stabilize significantly the system.

When planetary tides are taken into account, the stability of the system depends on the chosen planets' tidal dissipation. Instabilities tend to occur when the planets' dissipation are high. This is slightly counter-intuitive because increasing dissipation should contribute to decrease the eccentricity and stabilize the system. Changing dissipation changes the frequencies of the system, and some configurations happen to lead to a destabilization. This can be used as a way of reducing the size of the parameter space: dissipations leading to unstable configurations are unlikely. However, this could change with the tidal model used. Here we use the constant time lag model (Hut, 1981) and we plan to investigate how these features change using other tidal models.

A more thorough study of the stability of this system is needed, based on more longer simulations or different methods (such as Laskar, 1990, Correia *et al.* 2005, Couetdic *et al.* 2010), and should take into account all the effects considered here.

3.2. *Obliquity and rotation period evolution*

For Kepler-62, the rotation period of the three inner planets of the system evolves towards pseudo-synchronization in less than ten million years and their obliquities evolve towards small equilibrium values ($< 1°$). Given as the age of the system is estimated at 7 Gyr (Borucki *et al.* 2013), we expect that the three inner planets of the Kepler-62 system are now slowly rotating (their period is higher than 100 hr) and they have quasi null obliquities.

During the 30 million years of the simulation, the obliquities and rotation periods of the HZ planets Kepler-62e and f do not evolve significantly. So we performed longer

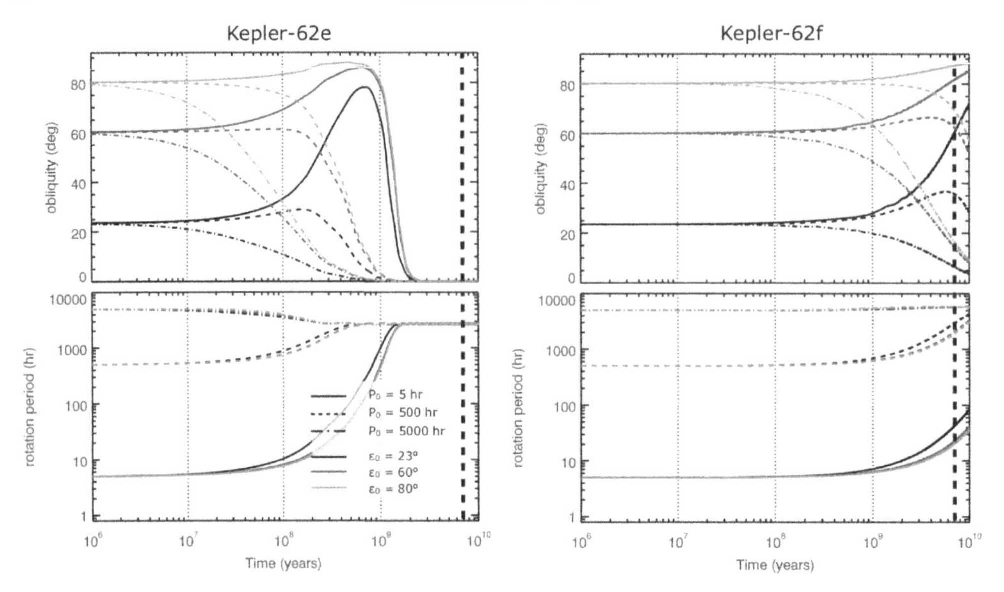

Figure 1. Evolution of the obliquity and the rotation period evolution for Kepler-62e and f for different initial obliquities (23°, 60° and 80°) and different initial rotation periods (5 hr, 500 hr and 5000 hr). The thick vertical black line corresponds to the estimated age of the system: 7 Gyr (Borucki *et al.* 2013).

simulations for these two outer planets. Assuming an Earth-like dissipation for the two planets, we found that Kepler-62e is likely to have reached pseudo-synchronization and have low obliquity (see Figure 1). The dissipation of the planets is not constrained and changing the dissipation would only shift the curves right (if the dissipation is lower) or left (if the dissipation is higher). As an Earth-like dissipation is actually probably a high dissipation value (Lambeck, 1977), it is likely that the curves should be shifted to the right. For Kepler-62f the timescales of evolution are higher and Figure 1 shows that the rotation period is still evolving towards pseudo-synchronization after 7 Gyr of evolution and that the obliquity can still be high.

Likewise, for Kepler-186, the four inner planets are located close-in to the star (orbital periods less than 25 days) and their rotation rates evolve in less than a million year. Given that the system is thought to be older than 4 Gyr (Quintana *et al.* 2014), we expect these four inner planets to be pseudo-synchronized and have low obliquities (Bolmont *et al.* 2014).

However, the tidal evolution timescales of the HZ planet Kepler-186f (orbital period of 130 days) are longer. Long term simulations showed us that giving the high uncertainty on the dissipation of this planet, it might still be tidally evolving or it might have reached the tidal equilibrium. If the dissipation of Kepler-186f is low or if the planetary system is young (~1 Gyr), the planet might have a high obliquity (up to 80°) and a relatively fast rotation (~30 hr). However, if the dissipation of Kepler-186f is high or the system is old (more than a few Gyr), the planet might have a low obliquity and a slow rotation (~3000 hr).

3.3. *An extra planet in the system Kepler-186*

In Bolmont *et al.* 2014, we show that formation simulations tend to form more than the 5 detected planets in a system like Kepler-186. We thus proposed the idea that there could be a sixth undetected planet in the system in the gap between planet e and f. However, this extra planet is not in a transit configuration because it would have been detected otherwise. We performed simulations to see if the dynamics of a system with an extra planet could be stable and consistent with the observations. We tested this hypothesis for different masses for the extra planet: from 0.1 $M_\oplus$ to $M_{\rm Jupiter}$, with an initial inclination of 2°.

Without an extra planet, Kepler-186f is dynamically isolated. Adding an extra planet in the system causes the inclination of the orbit of Kepler-186f to be excited. The more massive the planet, the higher the excitation of the inclination. If the mass of the extra planet is more than 1 $M_\oplus$, the inclination of Kepler-186f oscillates with values higher than the detectability limit ($i_{\rm lim}=\arctan(R_\star/a)$, where $R_\star$ is the radius of the star and a the semi-major axis of the planet). Thus, if an extra planet does exist, it should have a relatively low mass: $\lesssim 1\ M_\oplus$.

4. Conclusions

We showed here the importance of tides for the evolution of the planetary systems Kepler-62 and Kepler-186.

Taking into account tides increases the stability of the Kepler-62 system unless we assume the planets to be very dissipative.

We investigated the evolution of the spin states of Kepler-62e, Kepler-62f and Kepler-186f. The climate of an exoplanet depends strongly on its orbital elements and also its rotation. If it has a low obliquity and near synchronous rotation, the planet might have a cold trap (Kepler-62e). If it has a high obliquity and rapid rotation, it would have strong seasonal effects and Coriolis induced wind patterns (Kepler-62f). Kepler-186f could be in either configurations, the age of the system and the tidal dissipation are very poorly constrained.

Finally, we showed that there could be an extra planet in the Kepler-186 system between planets e and f. If such a planet existed, it would be a low mass planet $\lesssim 1\ M_\oplus$.

References

Berger, A. 1988, *Reviews of Geophysics*, 26, 624

Bolmont, E., Raymond, S. N., von Paris, P., *et al.* 2014, *ApJ*, 793, 3

Bolmont, E., Selsis, F., Raymond, S. N., Leconte, J., Hersant, F., Maurin, A.-S., Pericaud, J. 2013, *A&A*, 556, AA17.

Borucki, W. J., Agol, E., Fressin, F., *et al.* 2013, *Science*, 340, 587

Chabrier, G. & Baraffe, I. 2000, *ARAA*, 38, 337

Chambers, J. E. 1999, *MNRAS*, 304, 793

Correia, A. C. M., Udry, S., Mayor, M., *et al.* 2005, *A&A*, 440, 751

Correia, A. C. M. & Rodríguez, A. 2013, *ApJ*, 767, 128

Couetdic, J., Laskar, J., Correia, A. C. M., Mayor, M., & Udry, S. 2010, *A&A*, 519, A10

Fortney, J. J., Marley, M. S., & Barnes, J. W. 2007, *ApJ*, 659, 1661

Heller, R., Williams, D., Kipping, D., et al. 2014 *Astrobiology* 14, 798–835

Hut,P. 1981, *A&A*, 99,126

Kidder, L. E. 1995, *Phys. Rev. D*, 52, 821

Lambeck, K. 1977, *Royal Society of London Philosophical Transactions Series A*, 287, 545

Laskar, J. 1990, *Icarus*, 88, 266

Leconte, J., Chabrier, G., Baraffe, I., & Levrard, B. 2010, *A&A*, 516, A64+

Quintana, E. V., Barclay, T., Raymond, S. N., *et al.* 2014, *Science*, 344, 277

Complex Planetary Systems
Proceedings IAU Symposium No. 310, 2014
Z. Knežević & A. Lemaitre, eds.

doi:10.1017/S1743921314007844

Spin-orbit angle in compact planetary systems perturbed by an inclined companion. Application to the 55 Cancri system

Gwenaël Boué[1] and Daniel C. Fabrycky[2]

[1]Sorbonne Universités, UPMC Univ Paris 06, UMR 8028, ASD-IMCCE, Observatoire de Paris, F-75014 Paris, France
email: boue@imcce.fr

[2]Department of Astronomy and Astrophysics, University of Chicago, 5640 South Ellis Avenue, Chicago, IL 60637, USA
email: fabrycky@uchicago.edu

Abstract. The stellar spin orientation relative to the orbital planes of multiplanet systems are becoming accessible to observations. For example, 55 Cancri is a system composed of 5 planets orbiting a member of a stellar binary for which a projected obliquity of $72 \pm 12°$ relative to the orbit of the innermost planet has been reported (Bourrier & Hébrard 2014). This large obliquity has been attributed to the perturbation induced by the binary. Here we describe the secular evolution of similar systems and we discuss the case of the 55 Cancri system more deeply. We provide two different orbital configurations compatible with the currently available observations.

Keywords. celestial mechanics, methods: analytical, stars: planetary systems, stars: rotation, stars: binaries: general

---

1. Introduction

Consider a gravitational system composed of a central star, several planets on relatively tight orbits, and a perturbing body at a large semimajor axis. Any inclination of the outermost orbit relative to the inner ones would produce a long term precession of the planet orbital planes. Moreover, if the planets are close enough to each other, they all tilt in concert and their secular motion can be described as a solid rotation of the whole planet system (Innanen *et al.* 1997). In the following, this constraint is assumed to be fulfilled.

The goal of the present study is to determine the full range of spin-orbit angle that can be generated by the secular precession motion described above. By definition, the spin-orbit angle – also called the stellar obliquity or simply the obliquity – is the angle between the spin-axis of the central star and the normal of the planet plane. This specific problem has been analyzed extensively in Boué & Fabrycky (2014). Here, the method is applied to 55 Cancri, a 5 planet system in a stellar binary.

This work is motivated by a recent measurement of the projected spin-orbit angle of 55 Cancri performed by Bourrier & Hébrard (2014) using the Rossiter-McLaughlin effect. Their observations suggest a highly misaligned system with a projected obliquity $\lambda = 72 \pm 12°$. This measurement is, however, disputed by Lopez-Morales *et al.* (2014).

2. Mathematical description of the problem

The configuration of the system is uniquely defined by three unit vectors $\vec{s}$, $\vec{w}$, and $\vec{n}$ along the angular momentum of the stellar rotation, of the planet orbital motion, and

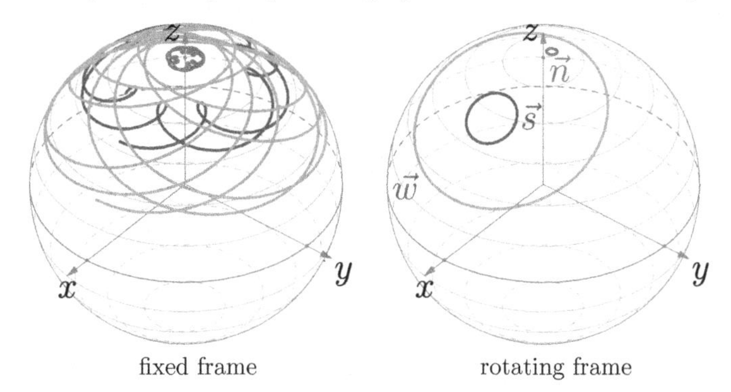

Figure 1. Typical solution of the equations (2.1) seen in a fixed reference frame (left) and in a rotating frame (right). The total angular momentum is aligned with the z-axis.

of the companion's orbit, respectively. Their motion is governed by the following secular equations (Boué & Fabrycky 2014)

$$\begin{aligned} \frac{d\vec{s}}{dt} &= -\nu_1(\vec{s}\cdot\vec{w})\vec{w}\times\vec{s}, \\ \frac{d\vec{w}}{dt} &= -\nu_2(\vec{s}\cdot\vec{w})\vec{s}\times\vec{w} - \nu_3(\vec{n}\cdot\vec{w})\vec{n}\times\vec{w}, \\ \frac{d\vec{n}}{dt} &= -\nu_4(\vec{n}\cdot\vec{w})\vec{w}\times\vec{n}. \end{aligned} \tag{2.1}$$

Thus, the evolution is fully characterized by four secular frequencies ν_1, ν_2, ν_3, and ν_4, only. These frequencies represent the speed at which each vector precesses around the others. The interaction between $\vec{s}$ and $\vec{w}$ is due to the oblateness of the central star induced by its rotation. A similar interaction should exist between $\vec{s}$ and $\vec{n}$, but given its weakness, it is neglected.

The system described by the differential equations (2.1) belongs to a class of integrable three vector problems (Boué & Laskar 2006, 2009). The general solution is a uniform rotation of the three vectors around the total angular momentum combined with periodic loops in the rotating frame (see figure 1).

3. Amplitude of spin-orbit oscillations

In the specific case where the perturber is a star, the binary's orbit possesses most of the total angular momentum. By consequence, its plane is practically invariant and the precession frequency ν_4 is negligible with respect to the other frequencies. Furthermore, if we discard the timescale of the evolution, the three remaining frequencies can be normalized by an arbitrary constant so that the effective dimension of the parameter space becomes equal to two. Using barycentric coordinates $(\nu_1/\nu_{\rm tot}, \nu_2/\nu_{\rm tot}, \nu_3/\nu_{\rm tot})$, where $\nu_{\rm tot} = \nu_1 + \nu_2 + \nu_3$, all systems with identical spin-orbit behavior but different timescales can be represented by a single point in a ternary diagram (Figure 2).

In this figure, close to the upper corner, as in the 55 Cancri case, the frequency ν_1 is the highest, thus the evolution is dominated by the precession of the central star around the planet plane. Near the bottom left vertex, the strongest motion is the precession of the planets with respect to the equator of the star at the frequency ν_2. Finally, in the vicinity of the bottom right corner where ν_3 dominates, the evolution is mainly characterized by the precession of the planets around the binary's orbital plane.

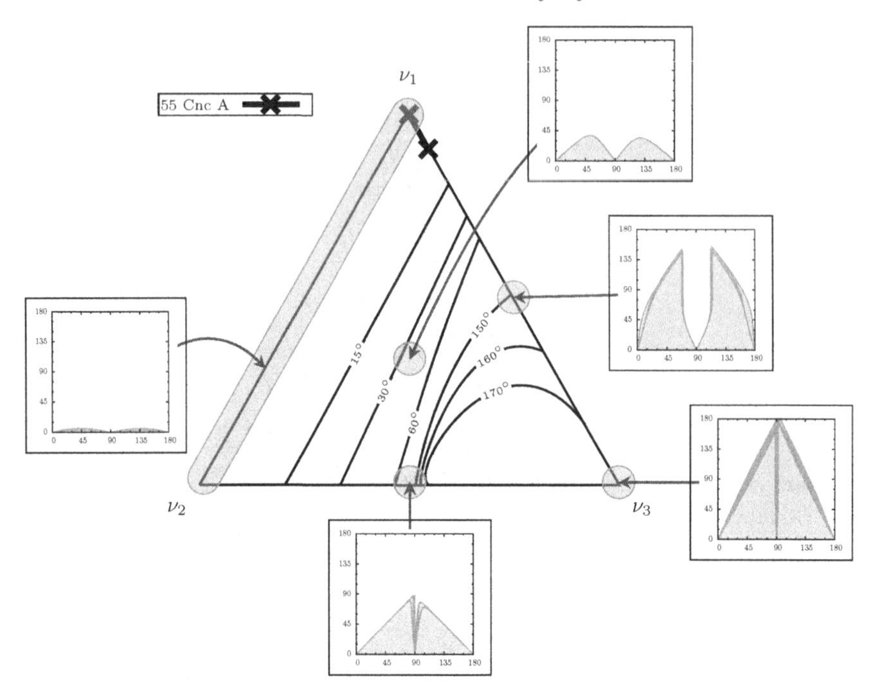

Figure 2. Ternary diagram with coordinates (ν_1, ν_2, ν_3). The system 55 Cancri is represented by a segment corresponding to the binary eccentricity ranging from 0 (top left end) to 0.95 (bottom right end). Any position in this diagram uniquely determines the behavior of the stellar obliquity. This value increases from a few degrees at the ν_1-ν_2 edge up to twice the initial inclination of the binary's orbit at the ν_3 vertex of the triangle. Five insets detail the obliquity's full amplitude (vertical axis, in degrees) as a function of the binary's initial inclination (hortizontal axis, in degrees), the maximum of which is shown as level curves across the diagram.

As a result, all systems in the neighborhood of the ν_1-ν_2 edge are lead by a strong coupling between the central star and the planets. Hence, if the spin-orbit angle is initially small, it remains small. Conversely, close to the ν_3 vertex the precession of the planets around the binary's orbit is too fast for the star to follow, thus the obliquity increases up to twice the initial inclination of the binary. The middle of the ν_1-ν_3 edge is particular because it corresponds to a secular spin-orbit resonance in the vicinity of a Cassini state. In this case, the maximal obliquity can exceed twice the initial inclination of the binary.

4. Discussion

Using the default values of Boué & Fabrycky (2014), the 55 Cancri system ends up in a region of the parameters space where no significant spin-orbit misalignment is expected (Figure 2). Yet, these parameters are the same as those considered by Kaib *et al.* (2011) who inferred a most probable projected obliquity of about 60°, in very good agreement with Bourrier & Hébrard (2014) observations. The apparent discrepancy among the two dynamical analyses is due to the spin-orbit coupling between the star and the planets which was not taken into account in Kaib *et al.* (2011). By consequence, ν_1 and ν_2 were forced to zero and the system was artificially shifted at the ν_3 corner in Figure 2.

In order to match the Rossiter-McLaughlin observations, it is necessary to enhance the coupling between the planets and the companion. One solution is to increase the eccentricity of the binary above $e' \gtrsim 0.987$ (Boué & Fabrycky 2014). With this value, the system reaches the Cassini regime (middle of the ν_1-ν_3 edge in Figure 2) and the obliquity can grow by a large amount.

Another solution is to assume that planet d is the actual perturber of the inner four planets. Indeed, planet d is well separated from the inner ones; its inclination with respect to the plane of the sky is estimated to be $i_d = 53 \pm 7°$ (McArthur *et al.* 2004) while planet e's inclination is close to $i_e = 90°$ (Winn *et al.* 2011), implying that these two planets are mutually inclined†; moreover, with this hypothesis the system falls in the pure orbital regime $(\nu_3 \sim \nu_4) \gg (\nu_1, \nu_2)$ compatible with large spin-orbit angles (Boué & Fabrycky 2014). Nevertheless, this scenario does not provide any hint to explain the origin of the required large inclination of planet d.

5. Conclusion

55 Cancri is a multiplanet system whose central star possesses a binary companion. This particular hierarchy can produce a misalignment of the stellar spin-axis with respect to the normal of the planets' plane. The secular evolution of such systems is uniquely determined by their position in a ternary diagram in which coordinates are three precession frequencies (ν_1, ν_2, ν_3). The 55 Cancri system is located in a region of the diagram with no excitation of the stellar obliquity in contradiction with recent Rossiter-McLaughlin measurements. Nevertheless, two different orbital configurations in agreement with the currently available observations are able to solve this issue. In the first one, the eccentricity of the stellar companion is set to a high value: $e' \gtrsim 0.987$. In the second, the outermost planet takes the place of the perturber.

References

Boué, G. & Fabrycky, D. C. 2014, *ApJ*, 789, 111

Boué, G. & Laskar, J. 2006, *Icarus*, 185, 312

Boué, G. & Laskar, J. 2009, *Icarus*, 201, 750

Bourrier, V. & Hébrard, G. 2014, *A&A*, 569, A65

Innanen, K. A., Zheng, J. Q., Mikkola, S., & Valtonen, M. J. 1997, *AJ*, 113, 1915

Kaib, N. A., Raymond, S. N., & Duncan, M. J. 2011, *ApJ*, 742, L24

Lopez-Morales, M., Triaud, A. H. M. J., Rodler, F., Dumusque, X., Buchhave, L. A., Harutyunyan, A., Hoyer, S., Alonso, R., Gillon, M., Kaib, N. A., Latham, D. W., Lovis, C., Pepe, F., Queloz, D., Raymond, S. N., Segransan, D., Waldmann, I. P., & Udry, S. 2014, *ApJ*, 792, L31

McArthur, B. E., Endl, M., Cochran, W. D., Benedict, G. F., Fischer, D. A., Marcy, G. W., Butler, R. P., Naef, D., Mayor, M., Queloz, D., Udry, S., & Harrison, T. E. 2004, *ApJ*, 614, L81

Nelson, B. E., Ford, E. B., Wright, J. T., Fischer, D. A., von Braun, K., Howard, A. W., Payne, M. J., & Dindar, S. 2014, *MNRAS*, 441, 442

Rein, H. 2012, *ArXiv*, e-print:1211.7121

Winn, J. N., Matthews, J. M., Dawson, R. I., Fabrycky, D., Holman, M. J., Kallinger, T., Kuschnig, R., Sasselov, D., Dragomir, D., Guenther, D. B., Moffat, A. F. J., Rowe, J. F., Rucinski, S., & Weiss, W. W. 2011, *ApJ*, 737, L18

† Nelson *et al.*. (2014) showed if the middle three planets are in the plane of the outermost, then Kozai cycles could destroy the innnermost planet; however, we suggest the middle three lie near the plane of the innermost.

Complex Planetary Systems
Proceedings IAU Symposium No. 310, 2014
Z. Knežević & A. Lemaitre, eds.

doi:10.1017/S1743921314007856

Pebble Delivery for Inside-Out Planet Formation

Xiao Hu[1] Jonathan C. Tan[1] and Sourav Chatterjee[2]

[1]Department of Astronomy, University of Florida, 32611, Gainesville, Florida, USA.
email: `ustcxhu@ufl.edu`, `jt@astro.ufl.edu`

[2]Center for Interdisciplinary Exploration and Research in Astrophysics (CIERA)
Department of Physics & Astronomy, Northwestern University, Evanston, IL 60208, USA.
email: `sourav.chatterjee@northwestern.edu`

Abstract. Inside-Out Planet Formation (IOPF; Chatterjee & Tan 2014, hereafter CT14) is a scenario for sequential *in situ* planet formation at the pressure traps of retreating dead zone inner boundaries (DZIBs) motivated to explain the many systems with tightly packed inner planets (STIPs) discovered by *Kepler*. The scenario involves build-up of a pebble-dominated protoplanetary ring, supplied by radial drift of pebbles from the outer disk. It may also involve further build-up of planetary masses to gap-opening scales via continued pebble accretion. Here we study radial drift & growth of pebbles delivered to the DZIB in fiducial IOPF disk models.

Keywords. planetary systems: protoplanetary disks, planetary systems: formation

---

1. Introduction

Kepler observations show STIPs are very common. The compact, well-aligned, but largely non-resonant architectures of these systems challenge formation scenarios involving migration of already-formed planets from the outer disk (e.g., Hands *et al.* 2014).

Chiang & Laughlin (2013) discussed aspects of *in situ* formation. Hansen & Murray (2012; 2013) studied STIP formation from a disk of protoplanets, requiring initial conditions with highly enriched solid surface densities above the minimum mass solar nebula.

CT14 proposed the IOPF scenario to link enrichment of solids in the inner disk by pebble drift with simultaneous and sequential formation of planets at the pressure maximum associated with the transition from a dead zone to a magneto-rotational instability (MRI)-dominated zone in the very inner disk, perhaps first set by the thermal ionization of alkali metals at ~ 1200 K. Here we calculate the radial drift timescale of different sizes of pebbles starting from various outer disk locations. We then couple this to a simple growth model for the pebbles. These are the first steps towards calculation of the global pebble supply rate to the DZIB, which will control the rate of IOPF.

2. Drag laws, radial drift velocity and pebble growth

Following Armitage (2010), we define the gas drag frictional timescale of a pebble of mass m_p moving at speed v_p relative to gas as $t_{\rm fric} = (m_p v_p)/F_D$. We consider four drag regimes (first is Epstein; others are Stokes regimes depending on Reynolds no., Re):

$$t_{\rm fric} = \begin{cases} \rho_p a_p/(\rho_g v_p) & \text{if } a_p < 9\lambda/4 \\ 2\rho_p a_p^2/(9\nu\rho) & \text{if } a_p > 9\lambda/4 \text{ and Re} < 1 \\ (\rho_p a_p/[9\rho_g v_p])(2av_p/\nu)^{0.6} & \text{if } 1 < \text{Re} < 800 \\ 8\rho_p a_p/(1.32\rho_g v_p) & \text{if Re} > 800, \end{cases} \tag{2.1}$$

where $a_p \equiv a_{p,1}$cm is pebble radius, $\rho_p \equiv \rho_{p,3}3$ g cm^{-3} is pebble density & ρ_g is gas density. $\nu \approx \lambda c_s$ is molecular viscosity, where λ is mean free path of gas molecules & c_s is sound speed. In terms of normalized frictional time, $\tau_{\rm fric} \equiv \Omega_K t_{\rm fric}$, where $\Omega_K = (Gm_*/r^3)^{1/2}$ is orbital angular frequency at radius $r \equiv r_{\rm AU}$ AU about star of mass $m_* \equiv m_{*,1} M_\odot$, radial drift speed is

$$v_{r,p} \simeq -k_P (c_s/v_K)^2 (\tau_{\rm fric} + \tau_{\rm fric}^{-1})^{-1} v_K, \tag{2.2}$$

where k_P describes the disk's pressure profile via $P = P_0(r/r_0)^{-k_P}$ (with fiducial value of 2.55) and v_K is the Keplerian velocity. We integrate $v_{r,p}$ to obtain the radial location of pebbles as a function of time $r_p(t)$. In the Epstein regime and the Stokes regime with Re < 1, $v_{r,p}$ is solved for analytically. When Re > 1, we solve numerically.

We also consider a simple pebble growth model. The pebble accumulates all small-grain dust material along its path within its physical cross section while in the Epstein regime. In the Stokes regimes we assume such accumulation is completely inefficient by deflection of the grains due to the pressure gradients of diverted gas streamlines.

3. Results and Discussion

Figure 1a shows examples of radial drift in a fiducial disk model with accretion rate $\dot{m} \equiv \dot{m}_{-9} 10^{-9} M_\odot {\rm yr}^{-1}$ and Shakura-Sunyaev viscosity parameter $\alpha \equiv \alpha_{-3} 10^{-3}$ (assumed constant outwards from the DZIB). Pebbles with radius 0.1 cm take about 10^5 yr to reach the inner region if starting from 10 AU. This timescale reduces to about 10^4 yr and 10^3 yr for 1 cm and 10 cm sized pebbles, respectively.

Figure 1b shows that varying the accretion rate over a range of a factor of 100 around the fiducial value has only modest effects on the drift timescales. Changing the value of α over a similar range has a somewhat larger effect. For most disk radii shown here we are in the Epstein regime with $\tau_{\rm fric} \ll 1$. In this case, $v_{r,p}$ is

$$v_{r,p} \simeq 1.92 a_{p,1} \rho_{p,3} \gamma_{1.4}^{8/5} \kappa_{10}^{2/5} \alpha_{-3}^{3/5} m_{*,1}^{-2/5} (f_r \dot{m}_{-9})^{-1/5} r_{\rm AU}^{1/5} \ {\rm m\,s^{-1}}, \tag{3.1}$$

where $\gamma \equiv 1.4\gamma_{1.4}$ is power-law exponent of the barotropic equation of state, $\kappa \equiv \kappa_{10} 10\,{\rm cm^2\,g^{-1}}$ is disk opacity, and $f_r \equiv 1 - \sqrt{r_*/r}$ (where r_* is stellar radius). So pebbles starting at 10 AU drift in faster in lower accretion rate disks and in more viscous disks. At ~100 AU this simple dependence on $\dot{m}$ breaks down as $\tau_{\rm fric} \gtrsim 1$.

Figure 1c shows the results for the pebble growth model, starting from 0.1 cm sizes. Such pebbles grow quickly, shortening delivery times from 100 AU to a few $\times 10^4$ yr.

Following CT14, the mass in solids (initially dust) inside disk radius r_1 is

$$M_s(< r_1) \approx 0.178 f_{s,-2} \gamma_{1.4}^{-4/5} \kappa_{10}^{-1/5} \alpha_{-3}^{-4/5} m_{*,1}^{1/5} \dot{m}_{-9}^{3/5} r_{1,\rm AU}^{7/5} \ M_\oplus, \tag{3.2}$$

(correcting minor typo in eq. 30 of CT14) where $f_s \equiv f_{s,-2} 0.01$ is the ratio of solid to gas mass. The gap-opening planetary mass at r_0 is

$$M_G \approx 5.67 \phi_{G,0.3} \gamma_{1.4}^{4/5} \kappa_{10}^{-1/5} \alpha_{-3}^{4/5} m_{*,1}^{3/10} (f_r \dot{m}_{-9})^{2/5} r_{0,\rm AU}^{1/10} \ M_\oplus, \tag{3.3}$$

based on fraction $\phi_G \equiv \phi_{G,0.3} 0.3$ of viscous-thermal mass (Zhu *et al.* 2013). If a fraction, $\epsilon_p \equiv \epsilon_{p,0.5} 0.5$, of disk solid mass forms pebbles that are accreted by the innermost forming planet with only minor gas accretion (as expected at $T \simeq 1200$ K DZIB conditions), then the required feeding radius of this first gap-opening mass planet is

$$r_1 \approx 19.4 \phi_{G,0.3}^{5/7} \gamma_{1.4}^{8/7} \kappa_{10}^{2/7} (f_{s,-2} \epsilon_{p,0.5})^{-5/7} \alpha_{-3}^{8/7} m_{*,1}^{1/14} \dot{m}_{-9}^{-1/7} r_{0,\rm AU}^{1/14} \ {\rm AU}. \tag{3.4}$$

With fiducial disk parameters and $r_0 = 0.1$ AU, we have $r_1 = 16.5$ AU. The drift time

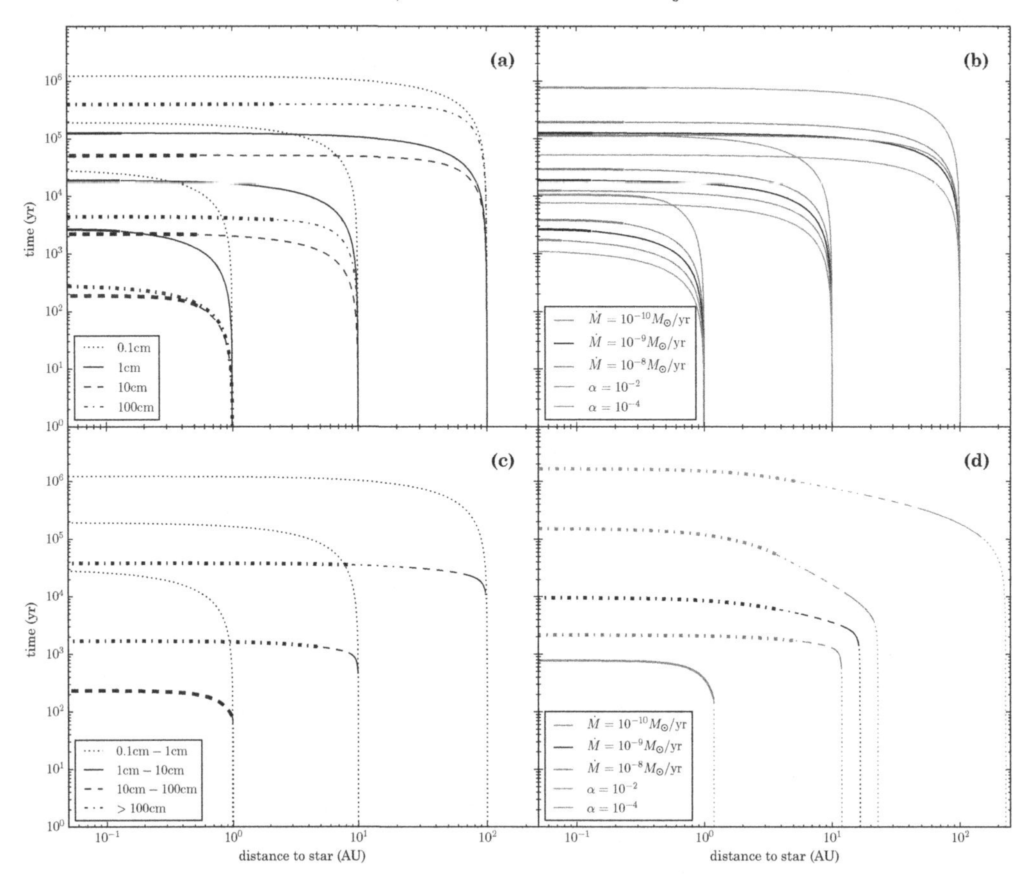

Figure 1. **a**: Radial drift timescale of pebbles with different fixed radial sizes starting from various disk radii in the fiducial disk model ($\dot{m} = 10^{-9}\ M_{\odot}\ \mathrm{yr}^{-1}$, $\alpha = 10^{-3}$). Thicker lines show the Stokes drag regimes. **b**: Radial drift time scale of 1 cm fixed-sized pebbles in disk models with different accretion rates and α viscosities, varied around the fiducial model. **c**: Comparisons between fixed-size 0.1 cm pebbles and the growth model with same initial size. The different line styles indicate growth in pebble radius. **d**: Growth model drift times for initially 0.1 cm pebbles starting from the outer radius of the feeding zone r_1 for gap-opening mass planets (see text).

for a 0.1 cm pebble of constant size from this distance is 2.88×10^5 yr. The same pebble growing by sweeping up dust drifts in after 1.15×10^4 yr with final size of 233 cm. This result is not very sensitive to the choice of inititial pebble size: a growing pebble starting with 0.01 cm radius has a drift time of 1.33×10^4 yr and a very similar final size. Note, if ϵ_p is much smaller than our fiducial value, e.g., due to interception of pebbles by a population of outer disk planetesimals (Guillot *et al.* 2014), then this would increase the radius of the required feeding zone and thus lengthen the drift timescales.

These drift times thus set lower limits for the timescale of first, innermost planet formation in the IOPF model. Figure 1d shows these times for various $\dot{m}$ & α. They are shorter for disks that are denser, i.e., due to higher $\dot{m}$ or lower α, given $M_G(\dot{m}, \alpha)$ and that the feeding zone is then smaller and pebble growth more efficient. These formation times may also be lower limits if pebble formation by dust grain coagulation (e.g., Birnstiel *et al.* 2012) is the rate limiting step, to be investigated in future work. Still, Fig. 1d shows IOPF requires $\alpha \lesssim 10^{-3}$, i.e., dead zone conditions, given observed disk lifetimes ~ 1 Myr. Variation in dead zone properties, e.g., from different disk midplane ionization rates by cosmic rays or radionuclide decay, could thus lead to a variety of planetary system formation mechanisms, perhaps helping to explain STIP vs. Solar System formation.

References

Armitage, P. J. 2010, Astrophysics of Planet Formation, Cambridge University Press, 2010
Birnstiel, T., Klahr, H., & Ercolano, B. 2012, *A&A*, 539, 148
Chatterjee, S. & Tan, J. C. 2014, *ApJ*, 780, 53
Chiang, E. & Laughlin, G. 2013, *MNRAS*, 431, 3444
Guillot, T, Ida, S. & Ormel, C. W. 2014, *arXiv*: 1409.7328
Hands, T. O., Alexander, R. D., & Dehnen, W. 2014, *arXiv*: 1409.0532
Hansen, B. M. S. & Murray, N. 2012, *ApJ*, 751, 158
Hansen, B. M. S. & Murray, N. 2013, *ApJ*, 775, 53
Zhu, Z., Stone, J. M., & Rafikov, R. R. 2013 *ApJ*, 768, 143

Complex Planetary Systems
Proceedings IAU Symposium No. 310, 2014
Z. Knežević & A. Lemaitre, eds.

doi:10.1017/S1743921314007868

Modeling resonant trojan motions in planetary systems

Christos Efthymiopoulos[1] **and Rocío I. Páez**[2]

[1]Research Center for Astronomy and Applied Mathematics, Academy of Athens, Greece
[2]Dip. di Matematica, Universitá di Roma "Tor Vergata", Italy
emails: cefthim@academyofathens.gr, paez@mat.uniroma2.it

Abstract. We consider the dynamics of a small trojan companion of a hypothetical giant exoplanet under the secular perturbations of additional planets. By a suitable choice of action-angle variables, the problem is amenable to the study of the slow modulation, induced by secular perturbations, to the dynamics of an otherwise called 'basic' Hamiltonian model of two degrees of freedom (planar case). We present this Hamiltonian decomposition, which implies that the slow chaotic diffusion at resonances is best described by the paradigm of modulational diffusion.

Keywords. celestial mechanics, methods: analytical, chaotic diffusion

1. Introduction

Despite extensive search, no pairs of co-orbital exoplanets have been discovered so far. Some reasons for the unlikeliness of the co-orbital configuration are discussed in Giuppone *et al.* (2012), Haghighipour (2013), Dobrovolskis (2013) and Pierens and Raymond (2014). Dynamical obstructions appear in the formation process as well as during the migration and/or capture of the planets into resonance. Besides these constraints, however, there is also the question of the *long-term stability* of co-orbital motions. This means the stability of the orbits over timescales comparable to the age of the hosting system.

In a recent work (Páez and Efthymiopoulos 2014) we initiated a study of the long-term stability in a hypothetical configuration in which a small (considered massless) planet moves around the Lagrangian points of a giant primary. Numerical simulations have shown that up to Earth-sized trojan planets can appear close to gaseous giants (Beaugé *et al.* 2007, Lyra *et al.* 2009). This dynamical system is a case of the elliptic restricted three body problem (ERTBP), or, with additional planets, the 'restricted multi-body problem' (RMBP). Alternative applications of the RMBP encompass co-orbital satellites of a planet, asteroids, and artificial trojan objects in a Sun-planet or planet-moon system.

In an accompanying poster (Páez and Efthymiopoulos, this volume) we outline one of our so-far obtained numerical results, referring to the diffusion timescales in the case of initial conditions taken close to some so-called *secondary resonances* within the co-orbital domain. Several authors (e.g. Érdi *et al.* 2007, 2009, Schwarz *et al.* 2007) have stressed the importance of secondary resonances in the problem of long-term stability. Related numerical works, applied to Jupiter's trojan asteroids, are Marzari *et al.* (2003), Robutel and Gabern (2006), Robutel and Bodossian (2009). Our own numerical work compares maps of the resonant structure, as depicted in a suitably defined domain of action variables (i.e. proper elements), with maps of the stability times for initial conditions within the resonance web. We found evidence of a tight correlation between the two maps (see Páez and Efthymiopoulos, this volume).

In the sequel we briefly discuss how our Hamiltonian formulation in action-angle variables is introduced in the framework of the RMBP, as well as the consequences this formulation leads to regarding the dynamical characterization of the problem.

2. Summary of the Hamiltonian formulation

A summary of our formulation is the following: assuming all perturbing planets far from mean motion resonances, by a suitable sequence of canonical transformations we arrive (in the planar case) at expressing the Hamiltonian of the RMBP as:

$$H = H_b(J_s, \phi_s, Y_f, \phi_f, Y_p; e_0) + H_{sec}(J_s, \phi_s, Y_f, \phi_f, Y_p, \phi, P_1, \phi_1, ..., P_S, \phi_S). \qquad (2.1)$$

i) The pairs (Y_f, ϕ_f), (J_s, ϕ_s), (Y_p, ϕ) are action-angle conjugate variables corresponding to the 'short-period', 'synodic' and 'secular' motions of the trojan body respectively. The short-period terms correspond physically to epicyclic oscillations. The synodic oscillations describe the 'long period' librations around the Lagrangian points L4 or L5. The action variable J_s determines the value of the 'proper libration' (see Milani (1993), or Beaugé and Roig (2001) for the definition of trojan proper elements). The action Y_p labels the 'proper eccentricity'. The angle ϕ measures phase oscillations around an angle β (see below) which expresses the relative difference between the arguments of perihelia of the trojan body and the giant primary. We note that an analysis omitted here allows to see that the form of the Hamiltonian (2.1) implies that the oscillations of β are bounded. Finally, the pairs (ϕ_i, P_i), $i = 1, ..., s$ are action-angle variables for the oscillations of the eccentricity vectors of the S additional planets.

ii) We call the first term H_b in (2.1) the 'basic model'. The angle ϕ is ignorable in H_b, implying that Y_p is a constant of motion under the dynamics of H_b alone. The parameter e_0 is the mean modulus of the eccentricity vector of the giant primary. Thus, H_b represents a system of two degrees of freedom, wherein both e_0 and Y_p act as parameters, i.e. the 'forced' (e_0) and 'proper' ($e_p = \sqrt{-2Y_p}$) eccentricity.

iii) The term H_{sec} contains only trigonometric terms depending on the slowly varying angles ϕ,ϕ_i, $i = 1, ..., s$. Hence, H_{sec} introduces only secular perturbations to the dynamics under H_b. In particular, H_{sec} causes a slow pulsation of the chaotic separatrix-like layers at the borders of the resonances arising under H_b. As shown in Páez and Efthymiopoulos 2014, this phenomenon is best described by the paradigm of 'modulational diffusion' (Chirikov *et al.* 1985).

iv) The form of the function H_b is identical in the ERTBP and the RMBP, setting $e_0 = e'$ and $\beta = \omega$ in the former, where e' is the (constant) eccentricity of the primary, and $\omega' = 0$ its pericentric position. This formal equivalence implies that the qualitative features of the diffusion along resonances, as they appear in the plane of the action variables J_s, Y_p, are similar in the RMPP and the ERTBP. Examples of the latter are studied in Páez and Efthymiopoulos (2014).

We now summarize the derivation of the Hamiltonian (2.1). We assume that, far from mean-motion resonances, the time evolution of the eccentricity vectors of all massive bodies can be approximated by quasi-periodic formulae

$$e' \exp i\omega' = e'_0 \exp i(\omega'_0 + g't) + \sum_{k=1}^{s} A_k \exp i(\omega'_{k0} + g_k t)$$

$$e_j \exp i\omega_j = B_{j0} \exp i(\omega_{0j} + g't) + \sum_{k=1}^{s} B_{kj} \exp i(\omega'_{kj} + g_k t) \qquad (2.2)$$

setting, without loss of generality, $\omega'_0 = 0$. The constants g', and g_j, $j = 1, \ldots s$ are secular frequencies associated with the primary and the S planets respectively. Also, we assume that the condition $e_0' > \sum_{k=1}^{S} A_k$ holds for the giant primary, implying an average constant rate of precession of its perihelion with frequency g'. One has $e' = e_0' + F$, $\omega' = \phi' + G$, where $\phi' = g't$ and F and G are of first order in the amplitudes A_k, $k = 1, ..., s$. Averaged over the mean longitudes $\lambda_1, \ldots, \lambda_S$ the Hamiltonian reads

$$H = -\frac{1}{2(1+x)^2} + I_3 + g'I' + \sum_{j=1}^{S} g_j I_j - \mu R(\lambda, \omega, x, y, \lambda', \phi'; e_0') - \mu R_2 - \sum_{j=1}^{S} \mu_j \mathcal{R}_j \quad (2.3)$$

where: i) $x = \sqrt{a} - 1$, $y = \sqrt{a}\left(\sqrt{1-e^2} - 1\right)$ are Delaunay action variables, (a, e) being the major semi-axis and eccentricity of the trojan body (in units in which $a' = 1$ for the primary), and (λ, ω) the mean longitude and argument of the perihelion. The variables I_3, I', I_j, $j = 1, \ldots, S$ are dummy actions congugate to the angles λ', $\phi' = g't$ and $\phi_j = g_j t$. ii) R is has the same form as the disturbing function in the ERTBP with the substitution $e_0 \to e'$, $\phi' \to \omega'$, with μ equal to the primary's mass parameter (all functions and variables are considered in the heliocentric frame). iii) R_2, expressing the indirect effects of the S additional planets, comes from replacing $e' = e_0' + F(\phi', \phi_j)$, $\omega' = \phi' + G(\phi', \phi_j)$ in the disturbing function of the ERTBP and Taylor-expanding around e_0' and ϕ', assuming F and G small quantities. $\mathcal{R}$ is of degree one or higher in the mass parameters μ_j, $j = 1, \ldots, S$. iv) Finally, $\mathcal{R}_j$ are the (averaged over mean longitudes) direct terms of the S additional planets.

The canonical transformation $\tau = \lambda - \lambda'$, $\beta = \omega - \phi'$, $J_3 = I_3 + x$, $P' = I' + y$ allows to re-express the hamiltonian in terms of the resonant angle τ and the relative argument of pericenter difference β. The Hamiltonian can be recast as $H = \langle H \rangle + H_1$, where

$$\langle H \rangle = -\frac{1}{2(1+x)^2} - x + J_3 - g'y - \mu \langle R \rangle(\tau, \beta, x, y; e_0')$$

$$H_1 = g'P' + \sum_{j=1}^{S} g_j I_j - \mu \tilde{R}(\tau, \beta, x, y, \lambda', \phi'; e_0')$$

$$- \sum_{j=1}^{S} \mu_j R_j(x, y, \beta, \phi', \phi_1, ..., \phi_s) - \mu \mathcal{R}_2(x, y, \tau, \beta, \phi', \phi_1, ..., \phi_s)$$

with $\langle R \rangle = \frac{1}{2\pi} \int_0^{2\pi} R d\lambda'$, $\tilde{R} = R - \langle R \rangle$. The Hamiltonian $\langle H \rangle$ allows to determine the forced equilibrium by the solution to the system of equations $\partial \langle H \rangle / \partial x = \partial \langle H \rangle / \partial y = \partial \langle H \rangle / \partial \tau = \partial \langle H \rangle / \partial \beta = 0$. One finds that

$$(\tau_0, \beta_0, x_0, y_0) = \left(\pi/3, \pi/3, 0, \sqrt{1 - e_0'^2} - 1\right) + O(g'). \quad (2.4)$$

Note that the forced equilibrium represents a relative configuration, i.e., the eccentricity vector of the trojan body has the same modulus e_0 and a constant *relative* angle with respect to the *mean* eccentricity vector of the primary. This result follows also by a careful inspection of the formulae provided in Morais (2001).

Expanding around the forced equilibrium, we introduce new variables

$$v = x - x_0, \quad u = \tau - \tau_0, \quad Y = -(W^2 + V^2)/2, \quad \phi = \arctan(V/W) \quad (2.5)$$

$$V = \sqrt{-2y} \sin \beta - \sqrt{-2y_0} \sin \beta_0, \qquad W = \sqrt{-2y} \cos \beta - \sqrt{-2y_0} \cos \beta_0.$$

The variables (v, u) describe the motion in the synodic plane, while the action variable Y measures the distance of an orbit from the forced equilibrium position in the secular plane $(\sqrt{-2y}\cos\beta, \sqrt{-2y}\sin\beta)$. Finally, we introduce the canonical transformations $Y_p = Y + J_3$, $\phi_f = \lambda' - \phi$, and

$$J_s = \frac{1}{2\pi}\int_C (v - v_0)d(u - u_0) \tag{2.6}$$

where the integration is over a closed invariant curve C around (u_0, v_0), with conjugate angle ϕ_s. Substituting these transformations yields the form (2.1) of the Hamiltonian.

The study of the basic model allows to identify the most important secondary resonances, which are commensurabilities between the fast and synodic frequencies $\omega_f = \dot{\phi}_f$, $\omega_s = \dot{\phi}_s$. The fast frequency is related to the secular frequency $g = \dot{\phi}$ by $\omega_f = 1 - g$, in units in which the mean motion of the giant primary is equal to 1. The general form of a resonance is

$$m_f\omega_f + m_s\omega_s + mg + m'g' + m_1 g_1 + \ldots + m_S g_S = 0 \tag{2.7}$$

with m_f, m_s, m, m', m_j (with $j = 1, \ldots, S$) integers. The resonances of the basic model exist in the complete hierarchy of problems, from the planar circular restricted three body problem ($s = 0$, $g' = 0$, $e_0' = 0$) up to the complete multi-body problem. For the mass parameters of giant exoplanets the most important resonances are of the form $\omega_f - n\omega_s = 0$, with n in the range $4 \leqslant n \leqslant 12$ for typical mass parameters of the gaseous primary. In the frequency space (ω_f, ω_s, g), these resonances define planes normal to the plane (ω_f, ω_s) which intersect each other along the g–axis. All other resonances with $|m| + |m'| + |m_1| + \ldots + |m_S| > 0$ intersect transversally one or more planes of the main resonances. We refer to such resonances as 'transverse' if $|m_f| + |n| > 0$, or 'secular' if $|m_f| + |n| = 0$. In Efthymiopoulos and Páez (2014), we show that the diffusion along transverse or secular resonances is of the Arnold type, hence very slow. On the other hand, there are transverse resonances which accumulate to multiplets around the main ones, thus producing a faster (modulational) diffusion.

Acknowledgments

R. I. P. was supported by the Astronet-II Training Network (PITN-GA-2011-289240). C. E. was supported by the Research Committee of the Academy of Athens (Grant 200/815) and by an IAU Symposium Grant.

References

Beaugé, C. & Roig, F. 2001, *Icarus* 153, 391–415.

Chirikov, B. V., Lieberman, M. A., Shepelyansky, D. L., & Vivaldi, F. M. 1985, *Physica D* 14, 289–304.

Dobrovolskis, A. 2013, *Icarus* 226, 1636–1641.

Érdi B., Nagy I, Sándor Zs. & Süli, A., Fröhlich G. 2007, *MNRAS* 381, 33–40.

Érdi, B., Forgács-Dajka, E., Nagy, I., & Rajnai, R. 2009, *Cel. Mech. Dyn. Astron.* 104, 145–158

Giuppone, C. A., Benítez-Llambay, P., & Beaugé, C. 2012, *MNRAS* 421, 356–368.

Haghighipour, N., Capen, S., & Hinse, T. 2013, *Cel. Mech. Dyn. Astron.* 117, 75–89.

Lyra, W., Johansen, A., Klahr, H., & Piskunov, N. 2009, *A&A* 493, 1125–1139.

Marzari, F., Tricarino, P. & Scholl, H. 2003, *MNRAS* 345, 1091–1100.

Milani, A. 1993, *Cel. Mech. Dyn. Astron.* 57, 59–94.

Morais, M. H. M. 2001, *A&A* 369, 677–689.

Páez , R.I. & Efthymiopoulos, C. 2014, *Cel. Mech. Dyn. Astron.* (in press), arXiV 1410.1407.

Pierens, A. & Raymond, S. N. 2014, *MNRAS* 442, 2296–2303.

Robutel, P. & Gabern, F. 2006, *MNRAS* 372, 1463–1482.

Robutel, P. & Bodossian, J. 2009, *MNRAS* 399, 69–87.

Schwarz, R., Dvorak, R., Süli, A., & Érdi B. 2007 *A&A* 474, 1023–11029.

Complex Planetary Systems
Proceedings IAU Symposium No. 310, 2014
Z. Knežević & A. Lemaitre, eds.

doi:10.1017/S174392131400787X

On the relativistic Lagrange-Laplace secular dynamics for extrasolar systems

M. Sansottera[1], L. Grassi[1], and A. Giorgilli[1]

[1]Dipartimento di Matematica,
Università degli Studi di Milano,
20133 — Milano, Italia
email: `marco.sansottera@unimi.it`
`lorenzo.grassi@studenti.unimi.it`
`antonio.giorgilli@unimi.it`

Abstract. We study the secular dynamics of extrasolar planetary systems by extending the Lagrange-Laplace theory to high order and by including the relativistic effects. We investigate the long-term evolution of the planetary eccentricities via normal form and we find an excellent agreement with direct numerical integrations. Finally we set up a simple analytic criterion that allows to evaluate the impact of the relativistic effects in the long-time evolution.

Keywords. celestial mechanics, methods: analytical, secular dynamics, relativistic effects, extrasolar planetary systems, normal forms

1. Introduction

The study of the secular evolution of planetary systems is a long standing and challenging problem. The discoveries of hundreds of extrasolar planetary systems raised many interesting problems concerning their long-term evolution. In the present paper, we study the secular dynamics of two non-resonant coplanar planets in an extrasolar system. We extend the Lagrange-Laplace theory to high order and include the main relativistic effects.

The study of extrasolar system raised two particularly relevant problems, namely: (i) most exoplanets have highly eccentric orbits, in contrast with the almost circular orbits of the Solar System; (ii) there are many giant planets orbiting at a low distance from the central star, with periods of a few months or even a few days. In the latter case relativistic effects could have a significant impact and should be taken into account. The General Theory of Relativity, despite having been widely used in astrophysics, is not commonly adopted in the study of planetary system dynamics.

The generalization of the Lagrange-Laplace secular theory to high order in the eccentricities has been exploited so as to obtain an analytic model that gives an accurate description of the behavior of planetary systems, up to surprisingly high eccentricities (see, Libert & Henrard (2005, 2006)). The results appear to be quite good for systems which are not close to a mean-motion resonance. In Libert & Sansottera (2013) the secular theory has also been extended to order two in the masses, by using a first-order approximation of an elliptic lower dimensional torus in place of the usual circular approximation. In particular this allows to deal with systems close to a mean-motion resonance. The relevance of the relativistic corrections and tidal effects on the long-term evolution of extrasolar planetary systems has been studied, e.g., in Adams & Laughlin (2006) and Migaszewski & Goździewski (2008).

On the other hand, the application of Kolmogorov and Nekhoroshev theorems, allowed to make substantial progress for the problem of stability of the Solar System. Indeed, in recent years, the estimates for the applicability of both theorems to realistic models of some part of the Solar System have been improved by some authors (e.g., Robutel (1995), Celletti & Chierchia (2005), Locatelli & Giorgilli (2007), Giorgilli *et al.* (2009, 2014) and Sansottera *et al.* (2011, 2013)).

In the present paper we exploit the idea of extending the Lagrange-Laplace theory, already used in the above-cited papers, to the case of high eccentricities. The technical tool is the construction of a suitable normal form which allows us to investigate the long-time evolution of the planetary eccentricities. In this contribution we neglect the tidal effects, although we know that for many system they can be relevant. We decided to just consider the relativistic correction in order to keep the discussion at a simple level and to show that the extension of the Lagrange-Laplace theory, including relativistic effects, produces accurate results. We plan to further investigate the problem in a forthcoming work.

2. Classical expansion of the Hamiltonian

We consider a system of three coplanar point bodies, mutually interacting according to Newton's gravitational law: a central star P_0 of mass m_0 and two planets P_1 and P_2 of mass m_1 and m_2 and semi-major axis a_1 and a_2, respectively.

We refer to Libert & Sansottera (2013) for a detailed exposition concerning the expansion of the Hamiltonian in the Poincaré canonical variables, that reads

$$H(\boldsymbol{\Lambda},\boldsymbol{\lambda},\boldsymbol{\xi},\boldsymbol{\eta}) = H_0(\boldsymbol{\Lambda}) + \varepsilon H_1(\boldsymbol{\Lambda},\boldsymbol{\lambda},\boldsymbol{\xi},\boldsymbol{\eta}), \tag{2.1}$$

where H_0 is the Keplerian part and εH_1 the perturbation due to the mutual attraction between the planets. Using the standard notation, we will refer to $(\boldsymbol{\Lambda},\boldsymbol{\lambda})$ as the *fast variables* and to $(\boldsymbol{\xi},\boldsymbol{\eta})$ as the *secular variables*.

3. Relativistic corrections

Starting from the Hamiltonian of the Newton model, we add the relativistic corrections due to the mutual interaction between the star and each of the two planets. That is, we consider the correction included in the relativistic Hamiltonian of the problem of two-body in heliocentric coordinates $(\mathbf{r},\mathbf{p})$. The relativistic Hamiltonian takes the form

$$H = H_0 + \varepsilon H_1 + \frac{1}{c^2}H_2, \tag{3.1}$$

with H_0 and εH_1 as in the Newtonian model, while $\frac{1}{c^2}H_2$ is

$$\frac{1}{c^2}H_2 = \frac{1}{c^2}\sum_{i=1}^{2}\left[-\frac{\gamma_{1,i}}{\mu_i^3}(\mathbf{P}_i\cdot\mathbf{P}_i)^2 - \frac{\gamma_{2,i}}{\mu_i}\frac{\mathbf{P}_i\cdot\mathbf{P}_i}{\parallel \mathbf{r}_i \parallel} - \frac{\gamma_{3,i}}{\mu_i}\frac{(\mathbf{r}_i\cdot\mathbf{P}_i)^2}{\parallel \mathbf{r}_i \parallel^3} + \gamma_{4,i}\mu_i\frac{1}{\parallel \mathbf{r}_i \parallel^2}\right], \tag{3.2}$$

with

$$\begin{aligned}
&\mu_i = \frac{m_0 m_i}{m_0+m_i}, \qquad \beta_i = \mathcal{G}(m_0+m_i), \qquad \upsilon_i = \frac{m_0 m_i}{(m_0+m_i)^2},\\
&\gamma_{1,i} = \frac{1-3\upsilon_i}{8}, \qquad \gamma_{2,i} = \frac{\beta_i(3+\upsilon_i)}{2}, \qquad \gamma_{3,i} = \frac{\beta_i\upsilon_i}{2}, \qquad \gamma_{4,i} = \frac{\beta_i^2}{2},
\end{aligned} \tag{3.3}$$

and $\mathbf{P}_i = \mathbf{p}_i + \frac{\mu_i}{m_0}\mathbf{p}_{3-i} + \mathcal{O}(c^{-2})$ for $i = 1, 2$.

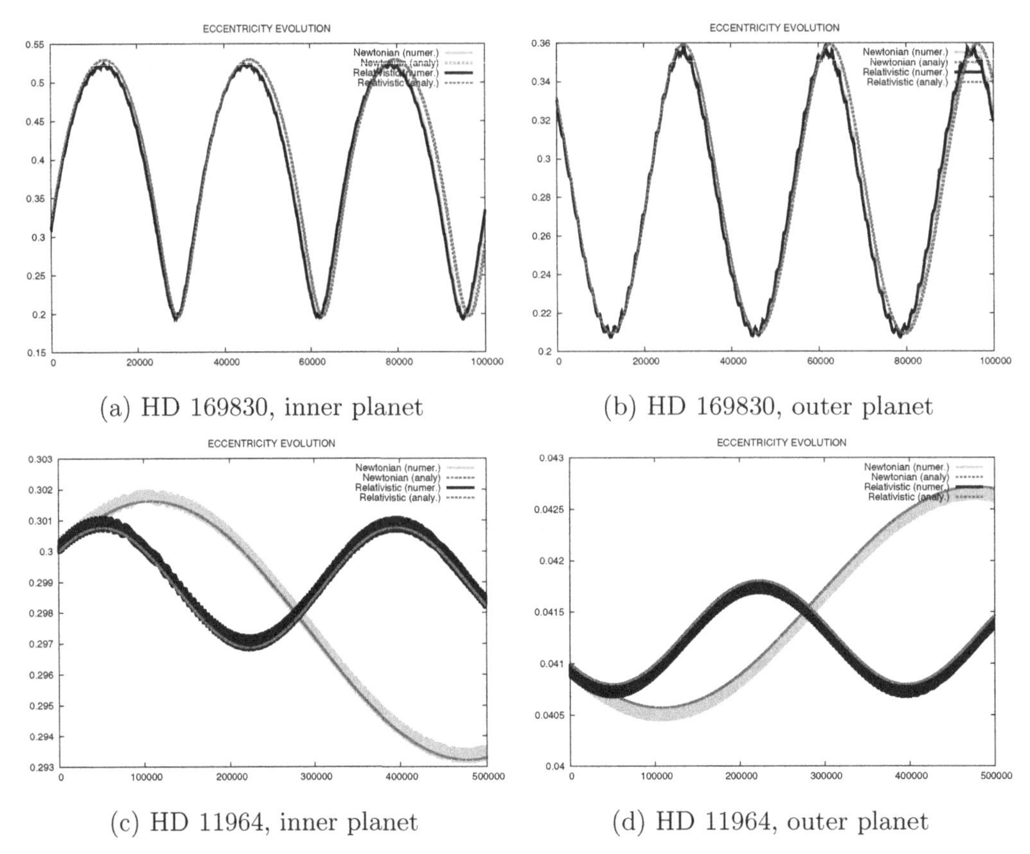

(a) HD 169830, inner planet

(b) HD 169830, outer planet

(c) HD 11964, inner planet

(d) HD 11964, outer planet

Figure 1. Long-term evolution of the eccentricities for the HD 169830 and HD 11964 planetary systems. Comparison of the results obtained via direct numerical integration (green-black) against normal form (blue-red), for the Newtonian approximation (green-blue) and the model including the relativistic effects (black-red).

4. Long-term evolution

As we are interested in the long-term dynamics, we remove the dependency of the Hamiltonian from the fast angles. The classical approach consists in replacing the Hamiltonian with its average, the so-called approximation at order one in the masses. We replace this procedure by a Kolmogorov-like step, that allows us to include in the secular model the effects of the main near-resonances effects (see Libert & Sansottera (2013) for a detailed exposition). This is the secular Hamiltonian at order two in the masses.

After averaging, the secular Hamiltonian has two degrees of freedom and its quadratic part differs from the one considered in the Lagrange-Laplace theory by relativistic corrections and contributions of order two in the masses, which however are small.

As in Libert & Sansottera (2013), we introduce the action-angle variables via normal form. In the normalized coordinates, the equations of motion take a simple form that can be analytically integrated. We validate the results by comparing the analytic integration with the direct numerical integration of the full three-body system.

In Figure 1a–1b we report the results for the HD 169830 system. In this case the relativistic effects are negligible and the Newtonian approximation allows to accurately describe the long-term evolution for a time interval of 10^5 years. Instead, for the HD 11964, the relativistic corrections play a major role, as it is clearly shown in Figure 1c–1d. In this case the calculations cover 5×10^5 years. In all cases, the evolutions via normal and via numerical integration are in excellent agreement. We emphasize that the use of normal form provides us with a natural criterion for deciding whether or not the relativistic corrections are relevant. Indeed the difference is seen in the precession frequencies: if the relativistic corrections are relevant then so is the difference, as the figures clearly show.

5. Relevance of the relativistic corrections

In order to evaluate the impact of the relativistic corrections, we look at the quadratic parts of the secular Hamiltonians, namely

$$H_q^{(\mathrm{New})}(\boldsymbol{\eta},\boldsymbol{\xi}) = \boldsymbol{\eta}\cdot A\boldsymbol{\eta} + \boldsymbol{\xi}\cdot A\boldsymbol{\xi} \quad \text{and} \quad H_q^{(\mathrm{Rel})}(\boldsymbol{\eta},\boldsymbol{\xi}) = \boldsymbol{\eta}\cdot B\boldsymbol{\eta} + \boldsymbol{\xi}\cdot B\boldsymbol{\xi},$$

where A and B are real symmetric 2×2 with

$$B = A - \frac{3}{2}\frac{\mathcal{G}^{3/2}}{c^2}\begin{bmatrix} \frac{(m_0+m_1)^{3/2}}{a_1^{5/2}} & 0 \\ 0 & \frac{(m_0+m_2)^{3/2}}{a_2^{5/2}} \end{bmatrix}.$$

Clearly, the relativistic effects are more important if

$$A_{ii} \sim -\frac{3}{2}\frac{\mathcal{G}^{3/2}}{c^2}\frac{(m_0+m_i)^{3/2}}{a_i^{5/2}}, \qquad \text{i.e. if} \quad \Pi_i \equiv \frac{4\mathcal{G}a_2^3 m_0(m_0+m_i)}{c^2 a_i^2 a_1^2 m_{3-i}} \sim 1. \tag{5.1}$$

In the following table we report the dimensionless quantities Π_i for the extrasolar systems considered above.

HD 169830	Π_1: 0.0021779	Π_2: 0.0001547
HD 11964	Π_1: 0.9651708	Π_2: 0.0399271

We observed that for the majority of the extrasolar systems taken into consideration, the relevance of relativistic corrections may be inferred from the difference between the matrices. This provides us with a rough criterion based on the first order approximation. Normal form provides a more refined criterion.

References

F. C. Adams & G. Laughlin, 2006, *ApJ*, 649, 992
A. Celletti & L. Chierchia, 2007, *Mem. Amer. Math. Soc.*, 187, 1
A. Giorgilli, U. Locatelli & M. Sansottera, 2009, *Celes. Mech. Dyn. Astr.*, 104, 159
A. Giorgilli, U. Locatelli & M. Sansottera, 2014, *Celes. Mech. Dyn. Astr.*, 119, 397
A. Giorgilli & M, Sansottera, 2011, *Workshop Series of the Asociacion Argentina de Astronomia*, 3, 147.
A.-S. Libert & J. Henrard, 2005, *Celes. Mech. Dyn. Astr.*, 93, 187
A.-S. Libert & J. Henrard, 2006, *Icarus*, 183, 186
A.-S. Libert & M. Sansottera, 2013, *Celes. Mech. Dyn. Astr.*, 117, 149
U. Locatelli & A. Giorgilli, 2007, *DCDS-B*, 7, 377
C. Migaszewski & K. Goździewski, 2008, *MNRAS*, 3927, (1): 2
P. Robutel, 1995, *Celes. Mech. Dyn. Astr.*, 62, 219
M. Sansottera, U. Locatelli & A. Giorgilli, 2013, *Math. Comput. Simulat.*, 88, 1
M. Sansottera, U. Locatelli & A. Giorgilli, 2011, *Celes. Mech. Dyn. Astr.*, 111, 337

Complex Planetary Systems
Proceedings IAU Symposium No. 310, 2014
Z. Knežević & A. Lemaitre, eds.

doi:10.1017/S1743921314007881

Orbital fitting of Fomalhaut b and subsequent interaction with the dust belt

Hervé Beust[1], Virginie Faramaz[1] and Jean-Charles Augereau[1]
[1]Univ. Grenoble Alpes, CNRS, IPAG, B. P. 53, F-38041 Grenoble, France
email: Herve.Beust@obs.ujf-grenoble.fr

Abstract. Fomalhaut harbours a moderately eccentric dust belt with a planet candidate (Fom b) imaged near its inner edge. MCMC-based orbital determination of Fom b shows that the orbit is highly eccentric ($e \simeq 0.9$), nearly apsidally aligned with the belt. We study the secular interaction between the planet and the dust ring. We show that only if it is a small mass object, Fom b can perturb the belt without destroying it. But Fom b's perturbing action inevitably drives the belt to high eccentricity and apsidal misalignment. This behaviour is due to the planet's high eccentricity.

This dynamical outcome contradicts both observations and orbital determination. We conclude that another, more massive and less eccentric planet (Fom c) is required to dynamically control the belt. We show that Fom b is likely to have been formerly trapped in mean-motion resonance with Fom c and that subsequent eccentricity increase caused it to cross Fom c's orbit and to jump on its present day orbit via a scattering event.

Keywords. Stars: Fomalhaut – Planetary systems – Planet-disk interactions – Methods: numerical – Celestial mechanics

---

1. Introduction

Fomalhaut A (αPsa) is a 440 Myr old (Mamajek (2012)) A3V star, located at 7.7 pc. As revealed by HST, Fomalhaut A is surrounded by an eccentric dust ring ($e = 0.11 \pm 0.01$) with a sharp inner edge at 133 AU and extending up to 158 AU (Kalas *et al.* (2005)). Afterwards, a companion near the inner edge of the belt, Fomalhaut b (herafter Fom b) was directly imaged (Kalas *et al.* (2008)). The nature of Fom b is controversial (Kalas *et al.* (2008), Marengo *et al.* (2009), Janson *et al.* (2012)). It is viewed today as a low mass planet surrounded by a population of dust (Kennedy & Wyatt (2011), Kenyon *et al.* (2014)). Further observations of this body led to the detection of its orbital motion (Graham *et al.* (2013), Beust *et al.* (2014)).

The purpose of this paper is to summarize the orbital fit process and the most recent subsequent dynamical studies (Beust *et al.* (2014), Faramaz *et al.* (2014)).

2. Orbital fit

Four astrometric points of Fom b are available between 2004 and 2012 (Kalas *et al.* (2013)). We use a Markov-Chain Monte Carlo (MCMC) approach to fit the orbit, assuming $d = 7.7$ pc and $M = 1.92\,M_\odot$ for the distance and the mass of Fomalhaut. The result of the fit is described in Beust *et al.* (2014). We display in Fig. 1 the posterior distribution of the eccentricity. The semi-major axis (not shown here) peaks as expected around $\sim$ 110–120 AU, but surprisingly, the eccentricity is very high, peaking above 0.9. The distributions of the inclination, longitude of ascending node and argument of periastron (not shown here) also favour coplanarity and apsidal alignment with the disk.

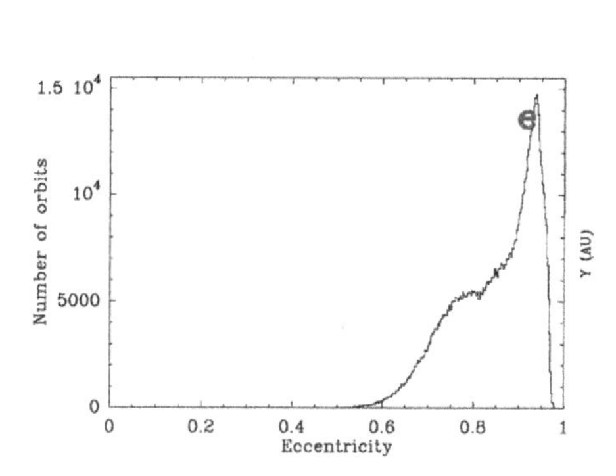

Figure 1. Resulting MCMC posterior distribution of the eccentricity of Fom b's orbit.

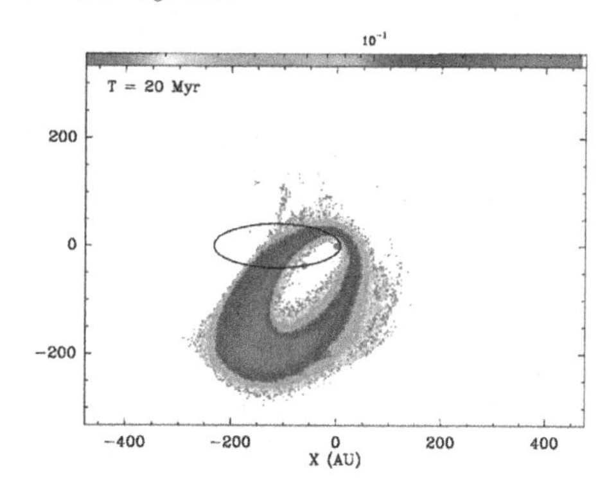

Figure 2. Example of result of the numerical investigation : upper view of the particle disk as perturbed by a super-Earth sized Fom b, after 20 Myr of evolution.

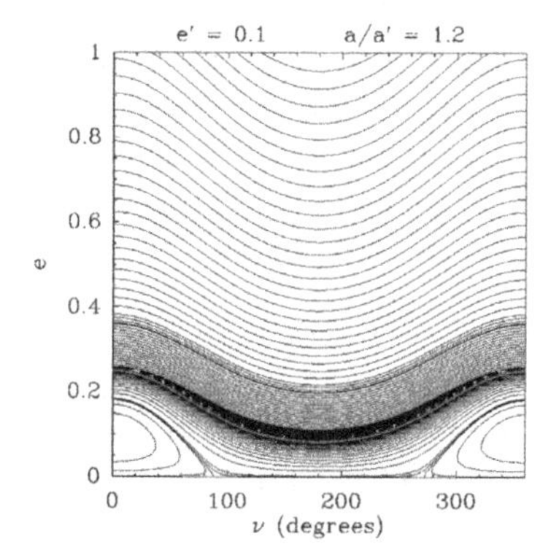

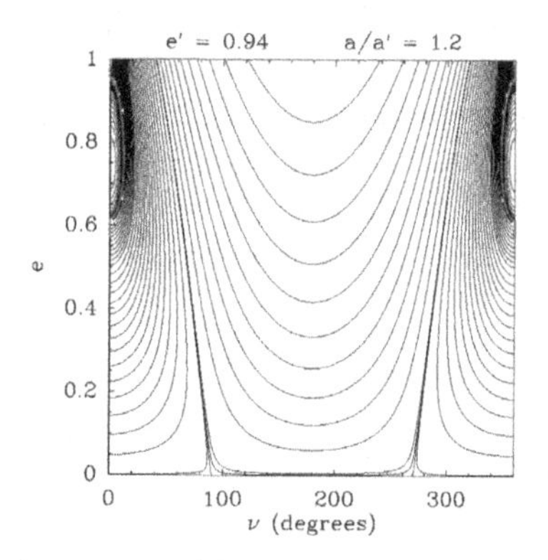

Figure 3. Phase portraits in (ν, e) space of secular averaged Hamiltonian for different values of the perturber's eccentricity e' and a fixed semi-major axis ratio $a/a' = 1.2$. The red curves separate regions where the orbits actually cross from regions where they do not.

3. Dynamical study

The conclusion of the orbital fit is that Fom b very probably has high eccentricity, is nearly coplanar with the dust belt and apsidally aligned. It is thus expected to be disk crossing. The potential effect of this planet on the dust belt needs to be investigated.

We present a numerical study of the Fomalhaut system, taking an initial ring of 10^5 massless particles between 110 AU and 170 AU, adding a planet orbiting on an orbit corresponding to our best fit: $a = 120$ AU and $e = 0.94$. We ran several N-body simulations changing Fom b's mass from sub-Earth to Jupiter size.

Our integrations first showed that with a Jovian-sized Fom b, the disk is quickly destroyed (see Beust *et al.* (2014) for details), thus excluding this mass range. For lighter planets the disk does not erode so much, but Fig. 2 shows that it is stirred up to a highly eccentric shape misaligned by $\sim 70°$ with respect to Fom b's periastron. This obviously does not match the observation.

This behaviour difference can be understood via a semi-analytical study. In Fig. 3, we display phase portraits of the secular doubly averaged interaction Hamiltonian for a test particle orbiting a star, and perturbed by a coplanar planet, assuming a semi-major axis 1.2 times that of the planet. The phase protrait is done in (ν, e) space, where ν is the longitude of periastron of the particle with respect to that of the planet. The left plot is done assuming that the planet's eccentricity is $e' = 0.1$, while the right one is done for $e' = 0.94$. In the $e' = 0.1$ case, we see that particles starting at low eccentricity are expected to remain so, having their maximum eccentricity at $\nu = 0$. In the $e' = 0.94$ case, all particles are nearly subject to evolve up to $e \simeq 1$, and most of the eccentricity growth is done around $\nu = 70°$. This explains our simulation results.

4. The need for another planet

The perturbations triggered on the disk by Fom b do not match the observation. We thus have two conclusions. First, Fom b is very likely to be a low mass planet (~Earth or super-Earth sized), otherwise the dust belt would not survive more than a few Myrs. Second, Fom b is not responsible for the shaping of the dust ring. Subsequently, there must be another, more massive planet shepherding the dust ring. Chiang *et al.* (2009) and Kalas *et al.* (2013) also came to the same conclusion. Given its high eccentricity, Fom b's orbit will inevitably cross that of that additional planet (hereafter called Fom c), which raises the issue of its dynamical stability.

We conclude that Fom b is presently on a moderately unstable orbit. We suggest that Fom b could have resided initially closer to the star, and it would have been put more or less recently on its present orbit by a scattering event. The likelihood of this scenario is related to the evolution and survival timescales of the transient configuration, as compared to Fomalhaut's age. This requires a dedicated study.

5. Dynamical history of Fomalhaut b

We investigate the past history of Fom b, assuming that it formed inside Fom c's orbit and was furthermore scattered by a close encounter. Considering the constraints given by Chiang *et al.* (2009), we first chose a mass of $\mathrm{m_c} = 3\,\mathrm{M_{Jup}}$ for Fom c, a semi-major axis $\mathrm{a_c} = 108\,\mathrm{AU}$, and an eccentricity $\mathrm{e_c} = 0.1$.

We must guess where Fom b could have initially resided before undergoing a significant eccentricity increase to cross Fom c's chaotic zone. Mean-motion resonances (MMRs) are good dynamical routes to achieve this. It has been indeed demonstrated (Yoshikawa (1989), Beust & Morbidelli (1996)) that some inner MMRs can trigger important eccentricity increases provided the perturbing planet is moderately eccentric (~ 0.1).

To test this scenario, we ran several simulations over 500 Myr with different initial sets of particles chosen close the major MMRs. In each case, we monitor the number of particles that are put on Fom b-like orbits, and how long they survive there. We define Fom b-like orbits as orbits with semi-major axis $a > 80\,\mathrm{AU}$ and eccentricity $e > 0.7$. The

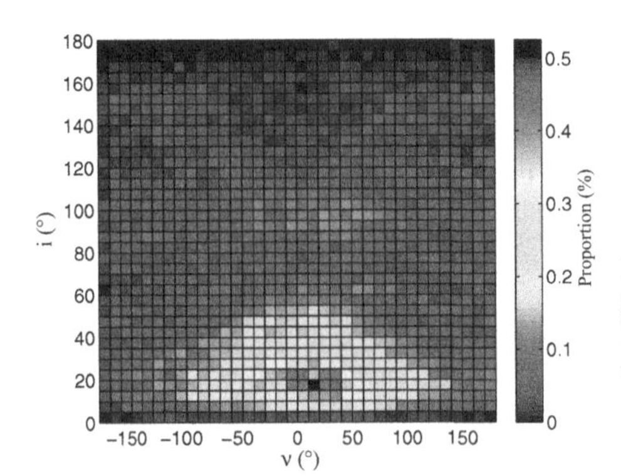

Figure 4. Distribution in inclination i and longitude of periastron ν with respect to that of Fom c of Fom b-like orbits produced in the case of the and 5:2 MMR

3:1, 7:3, 5:2 and 2:1 MMRs all appear to be sources of Fom b like orbits, the most efficient one being the 5:2 (Faramaz *et al.* (2014)) with 3.8% of the initial particles moving to Fom b-like state. It takes approximately 2 Myrs for particles to reach this state, but their survival time in a Fom b-like state is also a few Myrs, which is clearly too short. If we assume a Saturn-sized Fom c, then the number of particles becoming Fom b-like decreases down to ~0.1%, but it takes now more than 100 Myrs to get there and they are able to stay there for several 10^7 yrs, which makes them good candidates.

Figure 4 shows the distribution of 5:2 MMR originating Fom b-like particles in (ν, i) space (ν having the same definition ans above, and i accounting for the inclination). It reveals that the particles concentrate around $\nu = 0$ and $i = 0$, i.e., coplanar and apsidally aligned orbits, which is in agreement with the observation ! This behaviour is however not due to the encounters, but rather to post-encounter evolution. Considering the near conservation of the Tisserand parameter in encounters, we can show (Faramaz *et al.* (2014)) that particles initially trapped in inner MMRs hardly reach the $e = 0.7$ threshold characteristic for Fom b-like orbits after encounter, and even less easily $e = 0.9$.

Once a particle has been scattered, it is no longer trapped in a MMR, but it keeps evolving secularly as perturbed by Fom c. This evolution can be viewed in the left plot of Fig. 3, which shows the non-resonant phase-space diagram of particles perturbed by a $e' = 0.1$ planet. One must consider that after the encounter, a Fom b progenitor appears somewhere in this diagram around $e \simeq 0.6$ and starts evolving following one of the curves. If it appears around $\nu = 0$ (i.e., apsidally aligned), it naturally evolves towards lower eccentricity at $\nu = 180°$. It never becomes Fom b-like. Conversely, if it appears around $\nu = 180°$, its secular evolution inevitably drives it to higher eccentricity, i.e., Fom-b-like state, when reaching $\nu = 0$. Post-encounter evolution can thus drive particles towards higher eccentricity, but naturally selecting apsidally aligned (and also coplanar) orbits. This scenario explains all characteristics of Fom b. Faramaz *et al.* (2014) show examples of particles out of our simulation following this 3-step dynamical route : Resonant eccentricity increase, then scattering by close encounter, and finally post-encounter evolution that drives it to apsidal alignment with the current eccentricity.

Further work will investigate the effect of this scenario on the dust belt and aid the search for the hypothetical Fom c exoplanet.

References

Beust, H. & Morbidelli, A. 1996, *Icarus* 120, 358
Beust, H., Augereau, J.-C., & Bonsor, A., *et al.* 2014, *A&A* 561, A43
Chiang, E. I., Kite, E., Kalas, P., Graham, J. R.,& Clampin, M. 2009, *ApJ* 693, 734
Faramaz, V., Beust, H., Augereau, J.-C., *et al.* 2014, *A&A* in press, ArXiv:1409.6868
Graham, J. R., Fitzgerald, M. P., & Kalas, P., Clampin, M. 2013, *AAS* 221, 324.03
Janson, M., Carson, J. C., Lafrenière, D., *et al.* 2012, *ApJ* 747, 116
Kalas, P. & Graham, J. R., Chiang, E. *et al.* 2005, *Nature* 435, 1067
Kalas, P. & Graham, J. R., Chiang, E. *et al.* 2008, *Science* 322, 1345
Kalas, P., Graham, J. R., Fitzgerald, M. P., Clampin, M. 2013, *ApJ* 775, 56
Kennedy, G. M. & Wyatt, M. C. 2011, *MNRAS* 412, 2137
Kenyon, S. J., Currie, T., & Bromley, B. C. 2014, *ApJ* 786, 70
Mamajek, E. E. 2012, *ApJ* 754, L20
Marengo, M., Stapelfeldt, K., Werner, M. W., *et al.* 2009, *ApJ* 700, 1647
Yoshikawa, M. 1989, *A&A* 213, 436

Complex Planetary Systems
Proceedings IAU Symposium No. 310, 2014
Z. Knežević & A. Lemaitre, eds.

doi:10.1017/S1743921314007893

1/1 resonant periodic orbits in three dimensional planetary systems

Kyriaki I. Antoniadou[1], George Voyatzis[2] and Harry Varvoglis[3]
Sect. of Astrophysics, Astronomy and Mechanics, Dept. of Physics,
Aristotle University of Thessaloniki, 54124, Greece
email: [1] kyant@auth.gr, [2] voyatzis@auth.gr, [3] varvogli@physics.auth.gr

Abstract. We study the dynamics of a two-planet system, which evolves being in a 1/1 mean motion resonance (co-orbital motion) with non-zero mutual inclination. In particular, we examine the existence of bifurcations of periodic orbits from the planar to the spatial case. We find that such bifurcations exist only for planetary mass ratios $\rho = \frac{m_2}{m_1} < 0.0205$. For ρ in the interval $0 < \rho < 0.0205$, we compute the generated families of spatial periodic orbits and their linear stability. These spatial families form bridges, which start and end at the same planar family. Along them the mutual planetary inclination varies. We construct maps of dynamical stability and show the existence of regions of regular orbits in phase space.

Keywords. celestial mechanics, co-orbital motion, spatial periodic orbits, planetary systems

---

Planar co-orbital motion for a planetary system has been studied analytically (e.g. see Robutel and Pousse (2013)) or numerically (e.g. Hadjidemetriou *et al.* (2009), Hadjidemetriou and Voyatzis (2011)). In the last two papers, the main families of periodic orbits have been determined and classified as *planetary* or *satellite*. Herein, we utilize the spatial general TBP in a rotating frame of reference (Antoniadou and Voyatzis (2014)) and study co-orbital motion in space.

Firstly, we examine the vertical stability (Antoniadou and Voyatzis (2014)) of all families of planar stable periodic orbits mentioned above. We find that all periodic co-planar orbits are vertically stable, which means that, if we assume small deviations from the co-planar motion, the long-term stability of the orbits will be preserved with only small anti-phase oscillations in the planetary inclinations (i.e. when i_1 increases, i_2 decreases and vice-versa, due to the invariance of the total angular momentum). However, the families of planar symmetric planetary orbits, which are projected on the eccentricity plane in Fig. 1a, show segments of vertical instability, if the planetary mass ratio is $\rho = m_2/m_1 \leqslant 0.02049$ (or $\rho \geqslant 48.804$ if $m_1 < m_2$). At the edges of vertically unstable segments we have vertical critical orbits (v.c.o) (see Table 1 and (+) symbols in Fig. 1a).

Families of spatial xz-symmetric periodic orbits bifurcate from the v.c.o. and form bridges which connect the couples of v.c.o. In Fig. 1b, these families are presented on the $3D$ space $e_1 - e_2 - \Delta i$, where Δi denotes the mutual inclination. The spatial periodic orbits correspond to almost Keplerian orbits with $\omega_2 - \omega_1 = 0$, $\Omega_2 - \Omega_1 = \pi$, where ω_i and Ω_i denote the arguments of pericenter and the longitudes of ascending node, respectively. Also, at $t = 0$ one planet starts at periastron and the other at apoastron.

Table 1. Eccentricity values (e_1, e_2) of v.c.o. in 1/1 resonance.

ρ $(1/\rho)$	0.01 (100)	0.018 (55.55)	0.02 (50)	0.0205 (48.8)
v.c.o. 1	(0.015, 0.713)	(0.033, 0.731)	(0.045, 0.744)	(0.047, 0.746)
v.c.o. 2	(0.085, 0.782)	(0.066, 0.766)	(0.053, 0.752)	(0.051, 0.750)

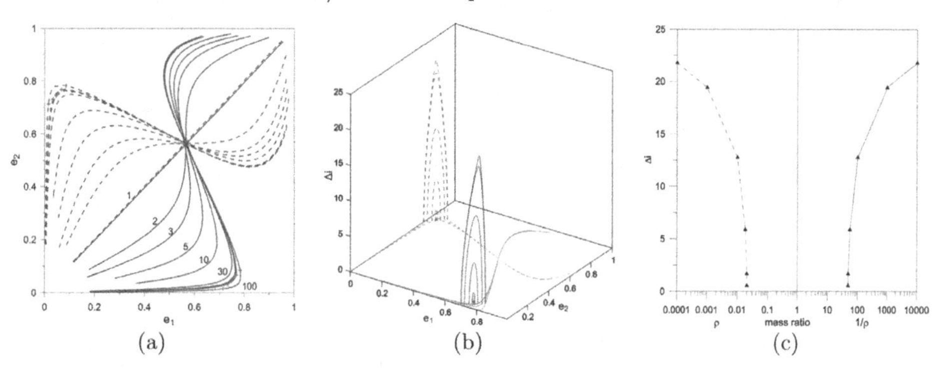

Figure 1. **a** Planar and **b** spatial families of symmetric periodic orbits in 1/1 resonance. **c** The maximum mutual inclination along families as a function of ρ.

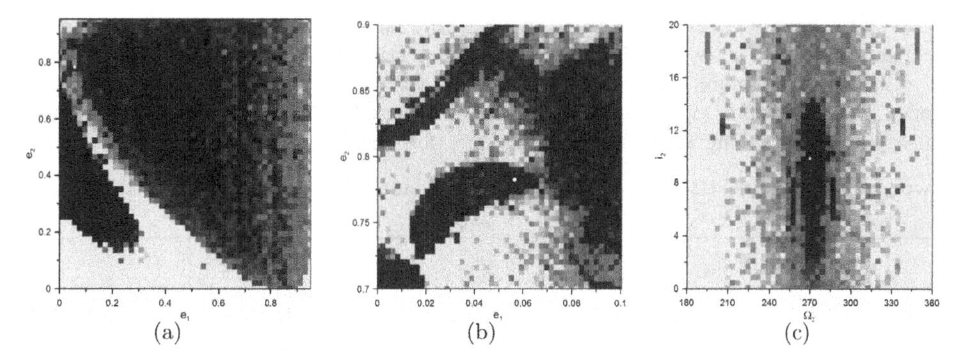

Figure 2. Maps of dynamical stability for $\rho = m_2/m_1 = 0.01$.

Along the family Δi reaches a maximum value, which depends on ρ and is presented in Fig. 1c. All spatial periodic orbits computed are linearly stable.

We construct maps of dynamical stability assuming plane grids of initial conditions and computing a FLI for chaos detection (Voyatzis (2008)). On these maps, dark colour indicates regular orbits, while regions of pale colour indicate strongly irregular orbits (Fig. 2). Apart from the initial conditions that correspond to the particular grid, the other ones are those of the periodic orbit with $\Delta i \approx 10°$. The map of panel (a) indicates the existence of stable orbits in a large eccentricity domain (but for the particular phase configuration). Panel (b) is a magnification of the previous map around the periodic orbit, which is indicated by a white dot, and seems to define an isolated domain (island) of regular motion. In panel (c), it is clearly shown that stability is restricted for $\Omega_2 \approx 270°$ and consequently, for $\Omega_2 - \Omega_1 \approx 180°$.

Acknowledgments: This research has been co-financed by the European Union (European Social Fund - ESF) and Greek national funds through the Operational Program "Education and Lifelong Learning" of the National Strategic Reference Framework (NSRF) - Research Funding Program: Thales. Investing in knowledge society through the European Social Fund.

References

Antoniadou, K. I. & Voyatzis, G. 2014, *Ap&SS*, 349, 657
Hadjidemetriou, J. D., Psychoyos, D., & Voyatzis, G. 2009, *Cel. Mech. Dyn. Astron.*, 104, 23
Hadjidemetriou, J. D. & Voyatzis, G. 2011, *Cel. Mech. Dyn. Astron.*, 111, 179
Robutel, P. & Pousse, A. 2013, *Cel. Mech. Dyn. Astron.*, 117, 17
Voyatzis, G. 2008, *ApJ*, 675, 802

Complex Planetary Systems
Proceedings IAU Symposium No. 310, 2014
Z. Knežević & A. Lemaitre, eds.

doi:10.1017/S174392131400790X

PlanetPack software tool for exoplanets detection: coming new features

Roman V. Baluev[1,2]

[1]Central Astronomical Observatory at Pulkovo of Russian Academy of Sciences, Pulkovskoje shosse 65, St Petersburg 196140, Russia

[2]Sobolev Astronomical Institute, St Petersburg State University, Universitetskij prospekt 28, Petrodvorets, St Petersburg 198504, Russia
email: `r.baluev@spbu.ru`

Abstract. We briefly overview the new features of PlanetPack2, the forthcoming update of PlanetPack, which is a software tool for exoplanets detection and characterization from Doppler radial velocity data. Among other things, this major update brings parallelized computing, new advanced models of the Doppler noise, handling of the so-called Keplerian periodogram, and routines for transits fitting and transit timing variation analysis.

Keywords. stars: planetary systems - methods: data analysis - methods: statistical

1. Introduction

PlanetPack is a software tool that facilitates the detection and characterization of exoplanets from the radial velocity (RV) data, as well as basic tasks of long-term dynamical simulations in exoplanetary systems. The detailed description of the numeric algorithms implemented in PlanetPack is given in the paper (Baluev 2013), coming with its initial 1.0 release. After that several updates of the package were released, offering a lot of bug fixes, minor improvements, as well as moderate expansions of the functionality. As of this writing, the current downloadable version of PlanetPack is 1.8.1. The current source code, as well as the technical manual, can be downloaded at `http://sourceforge.net/projects/planetpack`.

Here we pre-announce the first major update of the package, PlanetPack 2.0, which should be released in the near future. In addition to numerous bug fixes, this update includes a reorganization of the large parts of its architecture, and several new major algorithms. Now we briefly describe the main changes.

2. PlanetPack2: transits fitting and other new features

The following new features of the PlanetPack 2.0 release deserve noticing:

(*a*) Multithreading and parallelized computing, increasing the performance of some computationally heavy algorithms. This was achieved by migrating to the new ANSI standard of the C++ language, C++11.

(*b*) Several new models of the Doppler noise can be selected by the user, including e.g. the regularized model from (Baluev 2014). This regularized model often helps to suppress the non-linearity of the RV curve fit.

(*c*) The optimized computation algorithm of the so-called Keplerian periodogram (Cumming 2004), equipped with an efficient analytic method of calculating its significance levels (Baluev 2014, in prep.).

(*d*) Fitting exoplanetary transit lightcurves is now implemented in PlanetPack. This algorithm can fit just a single transit lightcurve, as well as a series of transits for the same star to generate the transit timing variation (TTV) data. These TTV data can be further analysed as well in order to e.g. reveal possible periodic variations indicating the presence of additional (non-transiting) planets in the system. The transit lightcurve model is based on the stellar limb darkening model by (Abubekerov & Gostev 2013). Also, the transit fitting can be performed taking into account the red (correlated) noise in the photometry data.

3. Plans for future work

Some results of the PlanetPack TTV analysis of the photometric data from the Exoplanet Transit Database, `http://var2.astro.cz/ETD/`, will be soon presented in a separate work. Concerning the evolution of the PlanetPack code, we plan to further develop the transit and TTV analysis module and to better integrate it with the Doppler analysis block. We expect that in a rather near future PlanetPack should be able to solve such complicated tasks as the simultaneous fitting of the RV, transit, and TTV data for the same star. This integration should also take into account subtle intervenue between the Doppler and photometry measurements like the Rositter-McLaughlin effect.

The work was supported by the President of Russia grant for young scientists (MK-733.2014.2), by the Russian Foundation for Basic Research (project 14-02-92615 KO_a), and by the programme of the Presidium of Russian Academy of Sciences "Non-stationary phenomena in the objects of the Universe".

References

Abubekerov, M. K. & Gostev, N. Y. 2013, *MNRAS*, 432, 2216
Baluev, R. V. 2013, *Astron. & Comput.*, 2, 18
Baluev, R. V. 2014, *MNRAS*, submitted, arXiv:1407.8482
Cumming, A. 2004 *MNRAS*, 354, 1165

Complex Planetary Systems
Proceedings IAU Symposium No. 310, 2014
Z. Knežević & A. Lemaitre, eds.

doi:10.1017/S1743921314007911

Impact flux of asteroids and water transport to the habitable zone in binary star systems

D. Bancelin[1,3], E. Pilat-Lohinger[2,1], S. Eggl[3] and R. Dvorak[1]

[1]Institute of Astrophysics, University of Vienna, Austria
email: david.bancelin@univie.ac.at

[2]Institute of Physics, University of Graz, Austria

[3]IMCCE, Paris Observatory, France

Abstract. By now, observations of exoplanets have found more than 50 binary star systems hosting 71 planets. We expect these numbers to increase as more than 70% of the main sequence stars in the solar neighborhood are members of binary or multiple systems. The planetary motion in such systems depends strongly on both the parameters of the stellar system (stellar separation and eccentricity) and the architecture of the planetary system (number of planets and their orbital behaviour). In case a terrestrial planet moves in the so-called habitable zone (HZ) of its host star, the habitability of this planet depends on many parameters. A crucial factor is certainly the amount of water. We investigate in this work the transport of water from beyond the snow-line to the HZ in a binary star system and compare it to a single star system.

Keywords. celestial mechanics, stars: binaries, stars: planetary systems

1. Introduction

Water is the main ingredient defining a habitable planet. Therefore, the main question we would like to answer in our study is if a dry or almost dry planet can be fed with water by a bombardement of wet small bodies in binary systems. First simulations of planetary formation in such systems show the stochastic behaviour of the water to mass ratio of planetary embryos (Haghighipour & Raymond 2007) . After the planetary formation, a remnent disc of small bodies can be found around the main star. It mainly contains asteroids and comets whose initial water distribution depends on their relative position to the snow-line. In our study, we mainly tackle the question of the amount of water available in the HZ. To this purpose, we consider various binary star systems where both stars are G types. We analyse the dynamics of a ring containing 100 asteroids (with maximum mass equal to the mass of Ceres) initially placed beyond the snow-line (2.7 au for a primary G star) moving under the gravitational perturbation of the binary stars and a Jupiter at 5.2 au. The systems were studied numerically for 10 Myr. To make this study statistical, each disk is cloned 100 times. Thus the statistics is made with 10000 asteroids.

2. Results

Figure 1 (left panel) shows the statistics on the asteroids dynamics. The perturbations of both Jupiter and the binary companion will drastically influence the asteroids in the ring and may increase their eccentricities and evolve beyond their initial semi-major axis borders. This behaviour is highlighted by the increasing value of the probability for an asteroid to cross the HZ as a function of the periapsis distance. They are called habitable zone crossers (HZc). As a consequence, the ring will be depopulated because the dynamics

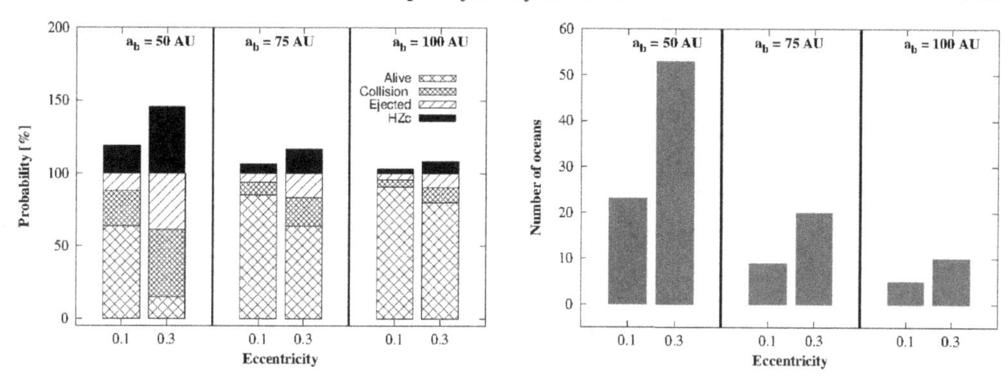

Figure 1. Left panel: Statistics on the asteroid's dynamics as a function of the binary's eccentricity and semi-major axis a_b. Each pattern refers to the probability for an asteroid initially in the ring to collide with the stars or Jupiter, be ejected or still be present in the system after 10 Myr. HZc refers to the probability for an asteroid to cross the HZ. Right panel: Quantity of water (in ocean units) transported to the HZ assuming also a water loss due to ice sublimation.

will induce ejections of asteroids and collisions with Jupiter and the stars. Therefore, the probability for an asteroid initially in the ring to stay in the system ("alive") after 10 Myr will decrease with the periapsis distance of the secondary. Assuming an initial water content of 10% for each asteroid, we derived the water to mass fraction when it first enters the HZ. This corresponds to the maximum value of transported water by this asteroid. We also assume a water loss process due to ice sublimation caused by the radiation of both stars. Indeed, we estimated up to 14% contribution of the secondary star during the sublimation process. The right panel of Fig. 1 shows the total amount of water transported (in ocean units) by the HZc among the 10000 asteroids of all rings. As we can see, a binary companion on an eccentric orbit helps for a more efficient transport. Indeed, the ring is highly perturbed because the presence of Jupiter will induce secular perturbations. Only a few oceans can be transported for the most distant secondary stars because of the small flux of incoming asteroids in the HZ. This means that the timescale to drastically increase the transport of water must be beyond 10 Myr. This test shows how efficient a system can be to rapidly transport water to the HZ. However, we estimated that the maximum duration for the last HZc to bring water to the HZ is less than 0.25 Myr. This suggest a transport in more than one step as some systems still have more than 50% of asteroids in the ring (i.e. they have never crossed the HZ). Finally, we estimate that the water transport is ~4–17 less efficient if the system is not a binary.

Acknowledgments

DB and EPL acknowledge the support of the FWF project S11608-N16 and S11603-N16 (NFN-subprojects) and the Vienna Scientific Cluster (VSC) for the computational ressources.

References

Haghighipour, N. & Raymond, S. N. 2007, *ApJ*, 666:436–446

Complex Planetary Systems
Proceedings IAU Symposium No. 310, 2014
Z. Knežević & A. Lemaitre, eds.

doi:10.1017/S1743921314007923

Eccentricity estimates in hierarchical triple systems

Nikolaos Georgakarakos[1] and Siegfried Eggl[2]

[1]Higher Technological Educational Institute of Central Macedonia, Serres, Greece
email: georgakarakos@hotmail.com

[2]IMCCE, Observatoire de Paris, Paris, France
email: siegfried.eggl@imcce.fr

Abstract. Perturbation theory in the three body problem has greatly advanced our ability to understand and model a variety of systems ranging from artificial satellites to stars and from extrasolar planets to asteroid-Jupiter interactions. In a series of papers, we developed an analytical technique for estimating the orbital eccentricity of the inner binary in hierarchical triple systems. The method combined the secular theory with calculations of short period terms. The derivation of the short term component was based on an expansion of the rate of change of the Runge-Lenz vector by using first order perturbation theory, while canonical perturbation theory was used to investigate the secular evolution of the system. In the present work we extend the calculation to the orbit of the outer binary. At the same time, we provide an improved version for some previous results. A post-Newtonian correction is included in our model. Our analytical estimates are compared with numerical and analytical results on the subject and applications to stellar triples and extrasolar planets are discussed.

Keywords. (Stars:) binaries: general, Celestial mechanics, Planets and satellites: dynamical evolution and stability.

---

1. Introduction

In a series of papers (Georgakarakos 2002, 2003, 2004, 2006, 2009), we focused on the evolution of the inner eccentricity of a hierarchical triple system. In the present work, we derive analytical expressions for the outer eccentricity of a coplanar hierarchical triple system, with a particular interest in the case where the outer body is of planetary mass.

2. Theory

The derivation of the short period terms follows the same basic technique described in the papers for the inner eccentricity (differentiation and then integration of the Runge-Lenz vector). However, in the expressions for the two Jacobian vectors, instead of the true anomaly, we use the eccentric anomaly: $\mathbf{r} = a_1(\cos E_1 - e_1, \sqrt{1-e_1^2}\sin E_1)$ and $\mathbf{R} = a_2(\cos l_2, \sin l_2)$, where the indices 1 and 2 refer to the inner and outer orbit respectively, $\mathbf{r}$ is the relative position vector of the inner orbit, $\mathbf{R}$ is the relative position vector of the outer orbit, a, e, E and l denote semi-major axis, eccentricity, eccentric anomaly and mean anomaly respectively. Eventually, retaining the two dominant terms in our expressions, we get:

$$e_{21sh}(t) = \frac{m_0 m_1}{(m_0+m_1)^{\frac{4}{3}} M^{\frac{2}{3}}} \frac{1}{X^{\frac{4}{3}}} \left[\frac{3}{4}\cos l_2 + e_1^2\left(\frac{33}{16}\cos l_2 + \frac{35}{16}\cos 3l_2\right) + \frac{P_1(t)}{X}\right] + C_{e_{21sh}} \tag{2.1}$$

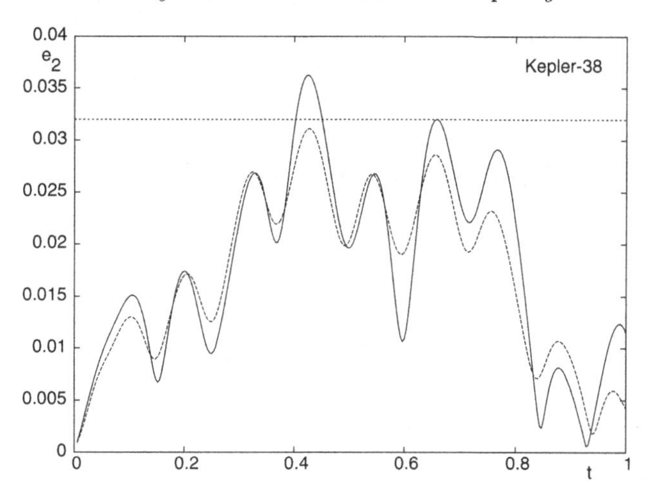

Figure 1. Short term eccentricity evolution of the circumbinary planet in the Kepler-38 system (continues line: numerical solution of the full equations of motion, dashed line: our analytical estimate). The planet was started with $e_2 = 0$ and the straight line shows its current eccentricity of 0.032. The time is given in planetary periods.

$$\begin{aligned} e_{22sh}(t) = & \frac{m_0 m_1}{(m_0+m_1)^{\frac{4}{3}} M^{\frac{2}{3}}} \frac{1}{X^{\frac{4}{3}}} \left[\frac{3}{4} \sin l_2 + e_1^2 \left(\frac{3}{16} \sin l_2 + \frac{35}{16} \sin 3l_2 \right) \right. \\ & \left. + \frac{P_2(t)}{X} \right] + C_{e_{22sh}}, \end{aligned} \tag{2.2}$$

where m_0 and m_1 are the masses of the inner binary, M is the total mass of the system, X is the period ratio, $P_1(t)$ and $P_2(t)$ are lengthy expressions containing terms with frequencies of the order of the inner mean motion and $C_{e_{21sh}}$, $C_{e_{22sh}}$ are constants of integration.

3. Applications

Fig.1 is an application of the above equations to the Kepler-38 system.

4. Outlook

The next step is to obtain expressions for the secular evolution of the system (including a post-Newtonian correction), derive estimates for the maximum and averaged eccentricity, check the range of their validity by means of numerical simulations and finally use the estimates for the determination of planetary habitable zones in stellar binaries (e.g. Eggl *et al.* 2012).

References

Eggl, S., Pilat-Lohinger, E., Georgakarakos, N., Gyergyovits, M., & Funk, B. 2012, *ApJ*, 752, 74
Georgakarakos, N. 2002, *MNRAS*, 337, 559
Georgakarakos, N. 2003, *MNRAS*, 345, 340
Georgakarakos, N. 2004, *Celes. Mech. Dyn. Astr.*, 89, 63
Georgakarakos, N. 2006, *MNRAS*, 366, 566
Georgakarakos, N. 2009, *MNRAS*, 392, 1253

Complex Planetary Systems
Proceedings IAU Symposium No. 310, 2014
Z. Knežević & A. Lemaitre, eds.

doi:10.1017/S1743921314007935

Understanding the assembly of *Kepler's* tightly-packed planetary systems

Thomas O. Hands, Richard D. Alexander and Walter Dehnen

Department of Physics & Astronomy,
University of Leicester, University Road,
Leicester, LE1 7RH, UK
email: tom.hands@le.ac.uk

Abstract. The Kepler mission has recently discovered a number of exoplanetary systems, such as Kepler 11, in which ensembles of several planets are found in very closely packed orbits. These systems present a challenge for traditional formation and migration scenarios. We present a dynamical study of the evolution of these systems using an N-body approach, incorporating both smooth and stochastic migration forces and a variety of initial conditions, in order to assess the feasibility of assembling such systems via traditional, disc-driven migration.

Keywords. planets and satellites: individual (Kepler-11; Kepler-32; Kepler-80), planets and satellites: dynamical evolution and stability, planets and satellites: formation

---

1. Motivation

An intriguing result of the *Kepler* mission is the relative abundance of a new class of closely-packed planetary systems. These systems contain 5 or 6 planets, each separated from its neighbours by fractions of an AU. The prototype for this class of systems is Kepler-11 (Lissauer *et al.* 2011), a 6-planet system in which the 5 innermost planets all orbit at smaller radii than Mercury. All of the planets appear to have low (< 0.05) eccentricities (Lissauer *et al.* 2013). Kepler-32 (Swift *et al.* 2013) and Kepler-80 (Ragozzine & Kepler Team 2012) each contain 5 planets and are even more tightly-packed than Kepler-11. Kepler-11 displays no obvious resonant behaviour, but there are possible 3 and 4 body interlocking mean-motion resonances in Kepler-32 and -80 respectively.

It is theoretically challenging to explain how such systems might form and remain dynamically cold. If the planets underwent convergent migration, one would expect that the vast majority of adjacent pairs would be trapped in mean-motion resonances with high forced eccentricities, which is not the case. This has led several authors to consider *in situ* formation models (e.g., Hansen & Murray 2013), while others have considered groups of embryos colliding and growing as they migrate inwards (e.g., Terquem & Papaloizou 2007). We investigate the possibility of assembling these systems from planets formed further out in a disc, which then migrate to their observed locations. A full account of the method and results can be found in Hands, Alexander & Dehnen (2014).

2. Method

We follow the evolution of these systems using an N-body integrator, allowing us to explore a wider parameter space than would be possible with hydrodynamical simulations. We use parametrized forces to migrate the planets and damp their eccentricities, with the time-scale of both being proportional to the inverse of the planetary mass analogous to Type I migration. We follow the method of Rein & Papaloizou (2009), adding

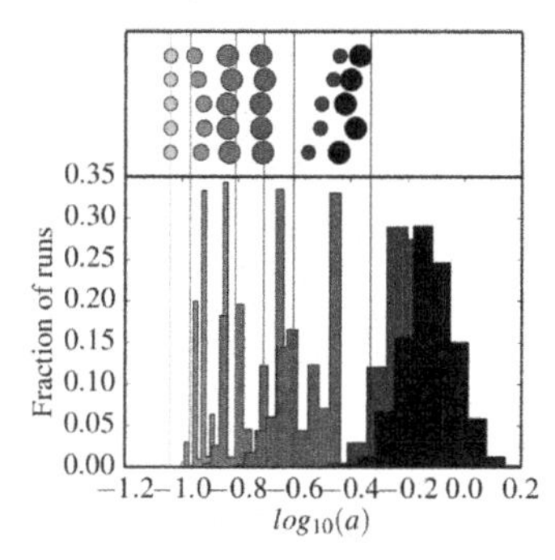

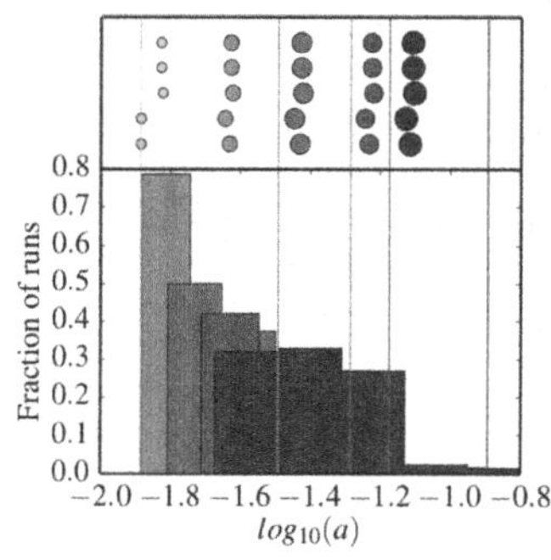

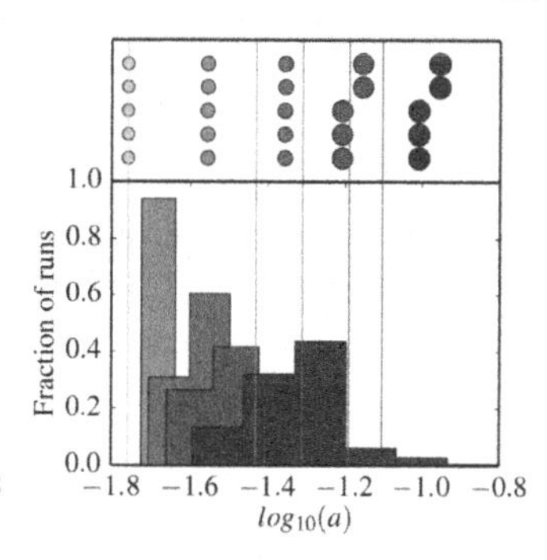

Figure 1. Histograms showing distribution of each planet in each system across all successful runs. Each shade represents a specific planet, with darker shades representing planets that are further from the star. Solid lines show the actual positions of the planets in each system, while the circles plotted above show planetary positions in some representative "best-fit" models, fitted by semi-major axis. Left to right: Kepler-11, Kepler-32, Kepler-80. Note that there is no histogram for the innermost planet in each system since this is always halted in the same place.

a stochastic component to the forces acting on the planets to simulate the effect of disc turbulence. The strength of the forces can be controlled via free parameters, and we perform 10,000 simulations with each system in order to sample the vast 3-D parameter space. We begin each simulation with random initial orbital separations and phases, and consider a simulation successful if the innermost planet reaches its observed orbital radius with the other planets in their observed order and with no collisions or ejections.

3. Results & Discussion

Figure 1 shows that systems analogous to Kepler-11, -32 and -80 are easily reproduced by our model. We find a preference for weaker turbulence and short eccentricity damping time-scales, with no particular preference in migration time-scale across the range of $10^{3.5}$–10^{6} yr that we explore. The lack of mean-motion resonances is difficult to reproduce.

This proof-of-concept model shows that forming tightly-packed planetary systems via simultaneous, disc-driven migration of fully-formed planets is possible for sensible disc models. Further work is required to understand why Type I migration should stop at the current locations of the planets, and the paucity of resonances in the observed systems.

References

Hands, T. O., Alexander, R. D., & Dehnen, W. 2014, *MNRAS*, in press, arXiv:1409.0532

Lissauer, J. J., Fabrycky, D. C., Ford, E. B., Borucki, W. J., Fressin, F., Marcy, G. W., Orosz, J. A., Rowe, J. F., Torres, G., Welsh, W. F., Batalha, N. M., Bryson, S. T., Buchhave, L. A., Caldwell, D. A., Carter, J. A., Charbonneau, D., Christiansen, J. L., Cochran, W. D., Desert, J.-M., Dunham, E. W., Fanelli, M. N., Fortney, J. J., Gautier, T. N., III, Geary, J. C., Gilliland, R. L., Haas, M. R., Hall, J. R., Holman, M. J., Koch, D. G., Latham, D. W., Lopez, E., McCauliff, S., Miller, N., Morehead, R. C., Quintana, E. V., Ragozzine, D., Sasselov, D., Short, D. R., & Steffen, J. H. 2011, *Nature*, 470, 53

Rein, H. & Papaloizou, J. C. B. 2009, *A&A*, 497, 595

Lissauer, J. J., Jontof-Hutter, D., Rowe, J. F., Fabrycky, D. C., Lopez, E. D., Agol, E., Marcy, G. W., Deck, K. M., Fischer, D. A., Fortney, J. J., Howell, S. B., Isaacson, H., Jenkins, J. M., Kolbl, R., Sasselov, D., Short, D. R., & Welsh, W. F. 2013, *ApJ*, 770, 15

Swift, J. J., Johnson, J. A., Morton, T. D., Crepp, J. R., Montet, B. T., Fabrycky, D. C., & Muirhead, P. S. 2013, *ApJ*, 764, 14

Ragozzine, D. & Kepler Team 2012, *AAS/Division for Planetary Sciences Meeting Abstracts*, 44, # 200.04

Terquem, C. & Papaloizou, J. C. B. 2007, *ApJ*, 654, 1110

Hansen, B. M. S. & Murray, N. 2013, *ApJ*, 775, 17

Complex Planetary Systems
Proceedings IAU Symposium No. 310, 2014
Z. Knežević & A. Lemaitre, eds.

doi:10.1017/S1743921314007947

Empirically Derived Dynamical Models for the 55 Cancri and GJ 876 Planetary Systems

Benjamin E. Nelson[1], **Eric B. Ford**[1], **Jason T. Wright**[1], **and Debra A. Fischer**[2]

[1]Dept. of Astronomy & Astrophysics, Pennsylvania State University
Center for Exoplanets and Habitable Worlds
525 Davey Laboratory, University Park, Pennsylvania, USA, 16803
email: benelson@psu.edu

[2]Dept. of Astronomy, Yale University
New Haven, CT 06520 USA

Abstract. In this paper we present empirically derived dynamical models for the 55 Cancri and GJ 876 planetary systems.

Keywords. planets and satellites: individual (55 Cancri; GJ 876), radial velocity

---

55 Cancri and GJ 876 are two landmark radial velocity (RV) systems with 4+ planets. These systems display strong planet-planet interactions over their ~10-20 year observing baseline. Characterizing the orbits and masses of such planets can be difficult for a few reasons: covariance among model parameters, an oddly shaped posterior probability distribution, and the large dimensionality of the problem (30+ parameters). Standard parameter estimation algorithms such as MCMC can be insufficient without some clever techniques. Even with a sophisticated algorithm, it becomes computationally tedious to perform an N-body integration for every model evaluation.

We apply our Radial velocity Using N-body Differential evolution Markov chain Monte Carlo code (RUN DMC; Nelson *et al.* (2014a)) to these two systems. The n-body integrations have been parallelized using SWARM-NG, a CUDA library to integrate an ensemble of few-body systems on GPUs. We perform subsequent long-term dynamical integrations using the MERCURY symplectic integrator (Chambers 1999).

55 Cancri: 55 Cancri is a Sun-like star hosting five known planets with a wide array of orbital separations and masses. We compile 1,418 unbinned RV measurements taken with the Lick Hamilton Spectrograph [Fischer *et al.* (2008)], HIRES [Fischer *et al.* (2008); Nelson *et al.* (2014b)], HET HRS, and HJST TS1 [Endl *et al.* (2012)]. In our orbital model, the outer four planets were assigned coplanar orbits and a separate inclination was given to the transiting inner-most planet *e*. Integrating the orbital evolution of 10,000 initial conditions drawn from our posterior samples for 10^5 years, we find the mutual inclination between planet *e* and the outer four planets must lie below 60° or above 120°. The cause of these instabilities is planet *e* undergoing Kozai-like perturbations, which pump its eccentricity such that the periastron of its orbit crossed the stellar surface. Accounting for relativistic precession in *e*'s orbit causes the unstable domain in mutual inclination to widen slightly.

Planets *b* and *c* orbit near a 3:1 period commensurability. Over the timespan of the observations, we can detect the planets' orbits deviating from Keplerian behavior. We modeled the secular evolution of the 55 Cancri planets based on 10,000 sample systems. We find 36-45% of our model systems have *b* and *c* apsidally aligned about 180° and undergo low amplitude libration ($51° \pm^{6°}_{10°}$). Other cases showed short-term perturbations in the libration of the secular angle, circulation, and nodding. We find the arguments associated with the 3:1 mean motion resonance are circulating over 99% of the time, so we conclude that these planets are not in resonance.

GJ 876: GJ 876 is an M dwarf hosting four known planets, where the outer three (c, b, and e) are orbiting in a 1:2:4 resonant chain [Rivera *et al.* (2010)]. We compile 367 measurements from Keck HIRES [Rivera *et al.* (2010)], HARPS, and the ELODIE and CORALIE spectrographs [Correia *et al.* (2010)]. In our orbital model, we allowed every planet in the system to have its own inclination and longitude of ascending node.

The RVs meaningfully constrain the three-dimensional orbital architecture of all four planets. We see the mutual inclination between planets c and b (i_{cb}) has an upper limit of $\sim$7° and i_{be} has an upper limit of $\sim$30°. We investigate the secular evolution of these models and find that higher mutual inclination in any planet pair typically leads to an instability. By demanding orbital stability, we infer that i_{cb} has an upper limit of $\sim$4° and i_{be} has an upper limit of $\sim$10°.

We can investigate the dynamical nature of this system by looking at the libration amplitude distribution of various angles associated with mean-motion resonances. Figure 1 shows these distributions for a coplanar (dashed) and full three-dimensional (solid) model. For at least one resonant argument associated with each planet pair, we find low or medium libration amplitudes, all of which are higher for the three-dimensional case than the coplanar case. Even so, these low amplitudes are strongly indicative of past disk migration of planets c, b, and e.

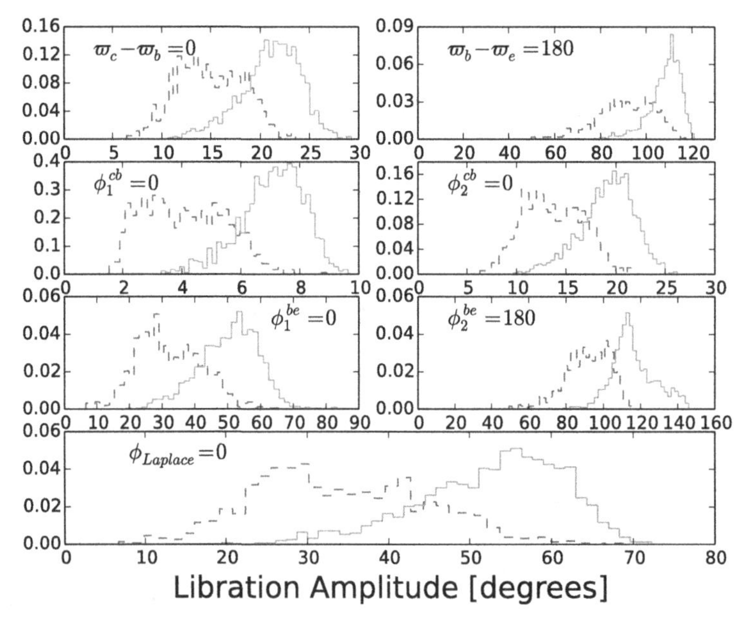

Figure 1. Libration amplitude distributions for various resonant arguments for a coplanar (dashed) and three-dimensional (solid) orbital model for GJ 876. Each panel shows an angle and what value it is librating about. **First row, left:** The apsidal alignment of c and b is librating about 0°. **First row, right:** The apsidal alignment of b and e is circulating. **Second row:** The two 2:1 MMR arguments for c and b, both of which are librating about 0°. **Third row:** The two 2:1 MMR arguments for b and e. The left panel is librating about 0° and the the right panel is circulating. **Bottom row:** The Laplace argument is librating about 0°.

References

Chambers, J. 1999, *MNRAS*, 304, 793
Correia, A. *et al.* 2010, *A&A*, 511, 21
Endl, M. *et al.* 2012, *ApJ*, 759, 19
Fischer, D. *et al.* 2008, *ApJ*, 675, 790
Nelson, B., Ford, E., & Payne, M. 2014, *ApJS*, 210, 11
Nelson, B. *et al.* 2014b, *MNRAS*, 441, 442
Rivera, E. *et al.* 2010, *ApJ*, 719, 890

Complex Planetary Systems
Proceedings IAU Symposium No. 310, 2014
Z. Knežević & A. Lemaitre, eds.

doi:10.1017/S1743921314007959

Modeling Trojan dynamics: diffusion mechanisms through resonances

Rocío I. Páez[1] **and Christos Efthymiopoulos**[2]

[1]Dip. di Matematica, Universitá di Roma "Tor Vergata", Italy
[2]Research Center for Astronomy and Applied Mathematics, Academy of Athens, Greece
emails: paez@mat.uniroma2.it, cefthim@academyofathens.gr

Abstract. In the framework of the ERTBP, we study an example of the influence of secondary resonances over the long term stability of Trojan motions. By the integration of ensembles of orbits, we find various types of chaotic diffusion, slow and fast. We show that the distribution of escape times is bi-modular, corresponding to two populations of short and long escape times. The objects with long escape times produce a power-law tail in the distribution.

Keywords. Celestial mechanics, methods:numerical.

---

1. Resonances

We study an example of mass parameter $\mu = 0.0041$ and eccentricity $e' = 0.02$ of the primary in the framework of the ERTBP. Following Páez & Efthymiopoulos 2014 (hereafter, P&E14), we describe Trojan orbits in terms of modified Delaunay variables given by

$$x = \sqrt{a} - 1, \quad y = \sqrt{a}\left(\sqrt{1-e^2} - 1\right), \quad \Delta u = \lambda - \frac{\pi}{3} - u_0, \quad \omega,$$

where a, e, λ and ω are the major semi-axis, eccentricity, mean longitude, and argument of the perihelion of the Trojan body, and u_0 is such that $\Delta u = 0$ for the 1:1 short period orbit at L_4.

In this problem, the secondary resonances (see P&E14) are of the form $m_f\omega_f + m_s\omega_s + m_g\omega_g = 0$, involving the fast frequency ω_f, the synodic frequency ω_s and the secular frequency ω_g of the Trojan body. Resonances are denoted below as $[m_f{:}m_s{:}m_g]$. The most important resonances, called the 'main' secondary resonances, correspond to the condition $\omega_f - n\omega_s = 0$ ([1:$-n$:0]). For $\mu = 0.0041$, this corresponds to [1:-6:0].

2. Diffusion and stability

Numerical experiments show that, for $e' > 0$, at least two different mechanisms of diffusion are present. Along non-overlapping resonances, a slow (and practically undetectable) Arnold-like diffusion (Arnold, 1964) takes place. On the other hand, for initial conditions along partly overlapping resonances, due to the phenomenon of pulsating separatrices (P&E14), we observe a faster 'modulational' diffusion (Chirikov *et al.*, 1985) leading to relatively fast escapes.

In order to distinguish which parts of the resonant web provide each behavior, we integrate 3600 initial conditions with $0.33 \leqslant \Delta u \leqslant 0.93$ and $0 \leqslant e_p \leqslant 0.06$, where Δu (libration angle) and e_p (proper eccentricity) are proper elements (see Efthymiopoulos and Páez, this volume). We visualize the resonance web by color maps of the Fast Lyapunov Indicator FLI (Froeschlé *et al.*, 2000) of the orbits. The resonances are identified by Frequency Analysis (Laskar, 1990). We integrate all orbits up to 5 different integration times along 10^7 periods of the primaries. After each integration, the initial conditions are categorized as ***Regular*** (if $\Psi(t) < \log_{10}(\frac{N}{10})$, where Ψ denotes the FLI value and N

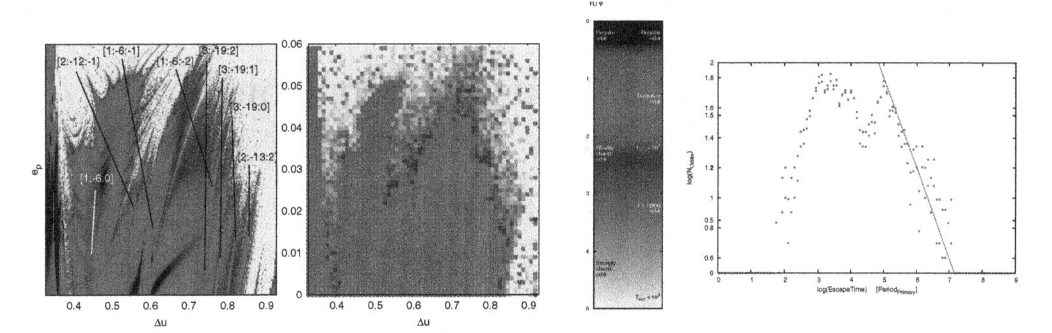

Figure 1. Left: FLI map for initial conditions described in the text where various secondary resonances are distinguished in the space of proper elements $(\Delta u, e_p)$. Middle: Color distribution of escaping times for the same initial conditions (color scale indicated). Right: distribution of the escaping times of the orbits.

is the total number of integration periods), ***Escaping*** (if the orbit undergoes a sudden jump in the numerical energy error greater than 10^{-3}) or ***Transition*** (non Regular nor Escaping).

N. of periods	Regular Orb	Transition Orb	Escaping Orb
10^3	1220 (33.8%)	2027 (56.3%)	353 (09.9%)
10^4	1263 (35.0%)	1388 (38.5%)	946 (26.5%)
10^5	1296 (36.0%)	966 (26.8%)	1338 (37.2%)
10^6	1299 (36.1%)	699 (19.4%)	1602 (44.5%)
10^7	1309 (36.3%)	603 (16.8%)	1688 (46.9%)

After 10^7 periods, 46.9% of the orbits have escaped. However, a significant portion (16.8%) still remain trapped, despite having a high FLI value. Figure 1 resumes the results. The histogram in the right panel shows two distinct timescales. The first peak (10^3 periods), corresponds to fast escapes, and the second (10^5 periods), to slow escapes. When we compare the FLI map (left) with the color distribution of the escaping times (middle), we find that the majority of fast escaping orbits lay within the chaotic sea surrounding the secondary resonances. The thin chaotic layers delimiting the resonances provide both slowly escaping orbits and *transition* orbits (*sticky* set of initial conditions that do not escape after 10^7 periods). For escaping orbits, beyond $t \sim 10^5$ periods, the distribution of the escape times is given by $P(t_{esc}) \propto t_{esc}^{-\alpha}$, $\alpha \approx 0.8$, while the sticky orbits exhibit features of 'stable chaos' (Milani & Nobili, 1992), since their Lyapunov times are much shorter than 10^7 periods.

Acknowledgments

R. I. P. was supported by the Astronet-II Training Network (PITN-GA-2011-289240). C. E. was supported by the Research Committee of the Academy of Athens (Grant 200/815).

References

Arnold, V. I., 1964, *Sov.Math.Dokr.*, 5, 581

Chirikov, B. V., Lieberman, M. A., Shepelyansky, D. L., & Vivaldi, F. M., 1985, *Phys. D*, 14, 289

Froeschlé, C., Guzzo, M., & Lega, E., 2000, *Science*, 289, 2108

Laskar, J., 1990, *Icarus*, 88, 266

Milani, A. & Nobili, A., 1992, *Nature*, 357, 569

Páez, R. I. & Efthymiopoulos, C., 2014, *Cel. Mech. Dyn. Astron.*, in press. ArXiv:1410.1407

Complex Planetary Systems
Proceedings IAU Symposium No. 310, 2014
Z. Knežević & A. Lemaitre, eds.

doi:10.1017/S1743921314007960

Diagrams of stability of circumbinary planetary systems

Elena Popova

Pulkovo Observatory of the Russian Academy of Sciences
Pulkovskoye ave. 65, Saint Petersburg 196140, Russia
email: m02pea@gmail.com

Abstract. The stability diagrams in the "pericentric distance — eccentricity" plane of initial data are built and analyzed for Kepler-38, Kepler-47, and Kepler-64 (PH1). This completes a survey of stability of the known up to now circumbinary planetary systems, initiated by Popova & Shevchenko (2013), where the analysis was performed for Kepler-16, 34, and 35. In the diagrams, the planets appear to be "embedded" in the fractal chaos border; however, I make an attempt to measure the "distance" to the chaos border in a physically consistent way. The obtained distances are compared to those given by the widely used numerical-experimental criterion by Holman & Wiegert (1999), who employed smooth polynomial approximations to describe the border. I identify the resonance cells, hosting the planets.

Keywords. Celestial mechanics, planetary systems, methods: numerical.

The systems Kepler-16, 34, and 35 were considered by Popova & Shevchenko (2013): stability diagrams were constructed in the plane of initial parameters "pericentric distance — eccentricity", which showed that all three planets are situated in resonance cells bounded by unstable resonances. Here we consider the planetary dynamics in the systems Kepler-38, 47, and 64. The parameters of the systems were determined by Orosz *et al.* (2012a), Orosz *et al.* (2012b), Kostov *et al.* (2013).

To explore the stability problem for the planetary motion, we use two stability criteria, following Popova & Shevchenko (2013). The first one is the "escape-collision" criterion and the second one is based on the value of the maximum Lyapunov characteristic exponent (maximum LCE). The computations are performed using the algorithms and codes by von Bremen *et al.* (1997), Shevchenko & Kouprianov (2002), Kouprianov & Shevchenko (2003), Kouprianov & Shevchenko (2005). The employed statistical method for separation of regular and chaotic orbits was proposed by Melnikov & Shevchenko (1998), Shevchenko & Melnikov (2003).

Holman & Wiegert (1999) obtained an empirical formula for the critical semimajor axis (separating the chaotic and regular domains) in function of the binary mass ratio and the binary eccentricity, for zero eccentricity planetary orbits. In Fig. 1(d) the Holman–Wiegert $a_{\rm cr}$ values are given in column 2, whereas our numerical estimates $a'_{\rm cr}$ (the main border at zero eccentricity) in column 3.

The computed stability diagrams for Kepler-38, 47, and 64 are given in Fig. 1. The chaotic domains revealed by the maximum LCE criterion are shown in black, and those revealed by the "escape-collision" criterion are shown in gray. Triangles are the nominal planet positions.

Our basic conclusions are as following. (*i*) The planets turn out to be situated in resonance cells between the "teeth" of instability corresponding to the resonances between the orbital periods of the planet and binary (5/1 and 6/1 for Kepler-38b, 6/1 and 7/1 for

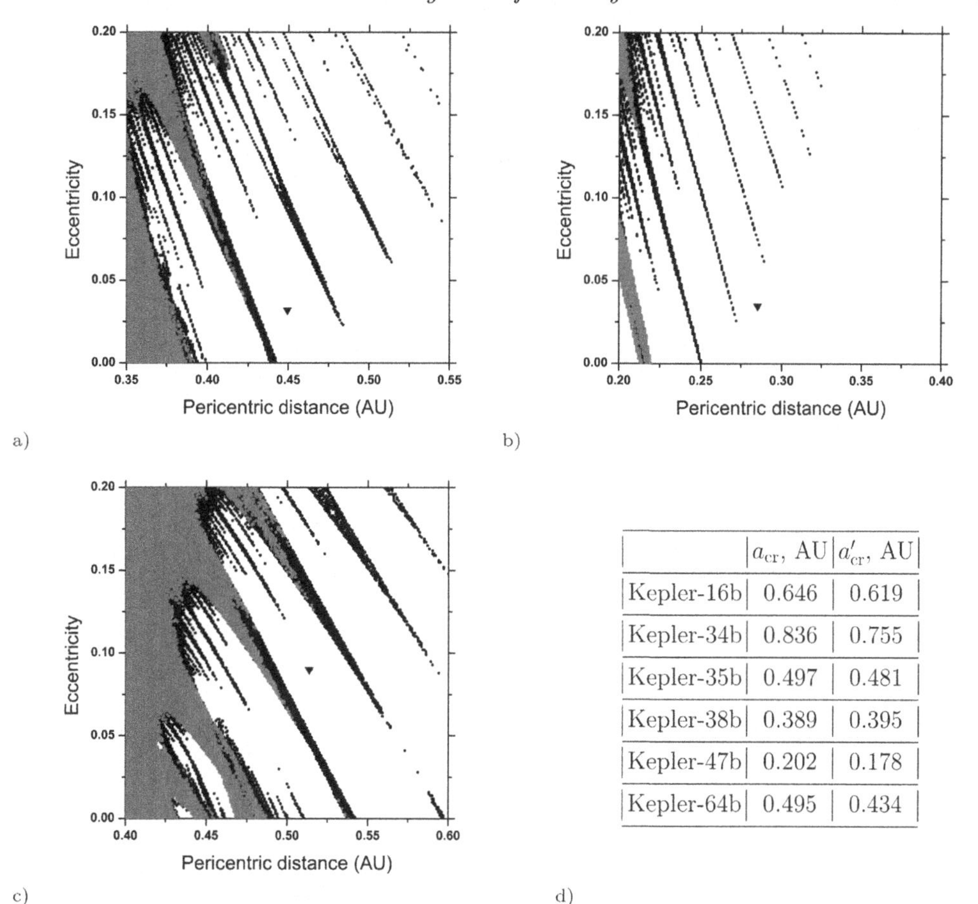

	$a_{\rm cr}$, AU	$a'_{\rm cr}$, AU
Kepler-16b	0.646	0.619
Kepler-34b	0.836	0.755
Kepler-35b	0.497	0.481
Kepler-38b	0.389	0.395
Kepler-47b	0.202	0.178
Kepler-64b	0.495	0.434

Figure 1. The stability diagrams for (a) Kepler-38, (b) 47, and (c) 64. (d) The critical semimajor axis values from Holman–Wiegert (a_{cr}) and from stability diagrams (a'_{cr}).

Kepler-47b and 64b). Thus all the circumbinary planets are "embedded" in the fractal chaos border in the stability diagrams. (*ii*) The semimajor axis critical values, given by the Holman–Wiegert empirical criterion, may differ significantly from real ones (directly obtained from the stability diagrams). (*iii*) The measured distances between the planet locations and the nearest unstable resonant "teeth" do not exceed 6% of the planet semimajor axes. (*iv*) The representative values of the Lyapunov time in the chaotic domains for the studied planets are very small: 1–1.5 years.

I am grateful to Ivan Shevchenko for useful remarks and advices. This work was supported in part by the Russian Foundation for Basic Research (project No. 14-02-00464).

References

Popova, E. & Shevchenko, I. 2013, *ApJ*, 769, 152
Holman, M. & Wiegert, P. 1999, *AJ*, 117, 621
Orosz, J., Welsh, W., Carter, J. *et al.* 2012, *ApJ*, 758, 82
Orosz, J., Welsh, W., Carter, J. *et al.* 2012, *Science*, 337, 1511
Kostov, V., McCullough, P., Hinse, T. *et al.* 2013, *ApJ*, 770, 52
von Bremen, H., Udwadia, F., & Proskurowski, W. 1997, *Physica D*, 101, 1

Shevchenko, I. & Kouprianov, V. 2002, *A&A*, 394, 663
Kouprianov, V. & Shevchenko, I. 2003, *A&A*, 410, 749
Kouprianov, V. & Shevchenko, I. 2005, *Icarus*, 176, 224
Melnikov, A. & Shevchenko, I. 1998, *Sol. Sys. Res.*, 32, 480
Melnikov, A. & Shevchenko, I. 2003, *JETP Lett.*, 77, 642

Complex Planetary Systems
Proceedings IAU Symposium No. 310, 2014
Z. Knežević & A. Lemaitre, eds.

doi:10.1017/S1743921314007972

Transit observations with the three San Pedro Mártir telescopes

D. Ricci[1], M. Reyes-Ruiz[1], R. Michel[1], C. Ayala-Loera[1], G. Ramón-Fox[1], L. Fox Machado[1], S. Navarro-Meza[1], S. Brown Sevilla[2] and S. Curiel[3]

[1]IA-UNAM, Km 103 Carr. Tijuana Ensenada, B.C, Mexico.
email: indy@astrosen.unam.mx

[2]Facultad de Ciencias Fsico-Matemticas, BUAP, Av. San Claudio y 18 Sur Puebla, Mexico

[3]IA-UNAM, Cir. Exterior S/N, Ciudad Universitaria, Coyoacn, D. F., Mexico

Abstract. Exoplanetary transit observations were carried out for the first time with all the three telescopes at the San Pedro Mártir National Astronomical Observatory in Baja California, Mexico.

We present preliminary results on WASP-39 and WASP-43, two Hot Jupiters known for the presence of a highly-inflated radius. Using the defocused photometry technique, we observed these systems, achieving photometric precision of ±3–5mmag peak-to-valley. The preliminary fit of their lightcurves shows physical and orbital parameters consistent with published results.

Keywords. techniques: photometric, exoplanets

1. Introduction

In the framework of a campaign for the observation of extrasolar planets, in the first semester of 2014 we successfully obtained light curves of 15 transiting systems, some of which present peculiar features.

For the first time the three San Pedro Mártir telescopes at the Observatorio Astronomico Nacional (Baja California, Mexico) were involved in this kind of research: the traditional 84cm telescope, the robotic 1.5m, and the traditional 2.12m provided with active optics on the primary mirror. This represents a unique opportunity in the northern hemisphere for the follow-up of dedicated surveys such as SuperWasp and HAT, but also for alert of forthcoming programs such as TAOS-II (see Lehner *et al.* (2013)). The selected objects are of the Hot Jupiter-type, and are characterized in the literature with clues of a strong interaction with their parent star such as a highly inflated radius. Here, we focus on WASP-39 and WASP-43. In the case of WASP-39, the presence of an accretion disk around the parent star is also indicated in the literature We present a preliminary analysis of these two systems, which have been extensively observed during our ongoing survey.

2. Observations, data reduction and analysis

The observations were carried on in a wide range of photometric bands using the defocused photometry method (Southworth *et al.* (2014)). Concerning WASP-39, we observed the same transit in the Johnson U filter with the 2.12m telescope, and in the I filter with the 84cm telescope, while an additional transit was observed in the R filter with the 84cm telescope. Concerning WASP-43, five transits were observed with the 84cm telescope: one in the Johnson V filter, two in the R filter, and two in the I filter. An

Table 1. Result of the fit of the orbital and physical parameters of WASP-39 and WASP-43.

Parameter	filter		WASP-39	WASP-43
Orbital inclination		i [deg]	87.78 ± 0.43	81.92 ± 0.54
Scaled Semi-major Axis		a/R_*	11.3242 ± 0.4237	4.815 ± 0.114
	U	R_p/R_*	0.1462 ± 0.0116	
	V	R_p/R_*		0.1615 ± 0.0041
Planet / Star radius ratio	R	R_p/R_*	0.1424 ± 0.0023	0.1599 ± 0.0025
	I	R_p/R_*	0.1424 ± 0.0023	0.1653 ± 0.0054
	i	R_p/R_*		0.1738 ± 0.0033

additional observation was carried out the same night as the later R observation, but in the SDSS i filter used with the 1.5m telescope. We then reduced the data using the aperture photometry technique with the IDL pipeline `defot` (Southworth *et al.* (2014)), and we compared the results with standard IRAF procedures developed by our team, finding consistent results. As reference star for photometry, we used for each target the field star closest in terms of flux and color index. The shape of the defocused PSFs allowed to calculate the FWHM of the target and of the reference stars. This information was used to dynamically adjust the aperture during the reduction with the IRAF pipeline. We used an aperture diameter of 2.5 times the previously calculated FWHM on each image frame. The light curves were corrected for slight trends that we attribute to differential extinction between target star and reference star. The value of HJD reported in the `fits` header was converted to BJD. We used the IDL software TAP by Gazak *et al.* (2011), a Monte Carlo method using the Mandel & Agol(2002) model, to fit the light curves, in order to obtain physical and orbital parameters. Each system was fitted using a set of fixed parameters and varying other parameters after providing initial guesses. We fixed the period, the limb darkening, the eccentricity and the argument of periastron ω. We varied the inclination, the scaled semi-major axis, and the mid-transit time, fitting together the curves for each of the observed systems. Finally, we varied the star/planet radius, fitting separate values for each filter. In the fit parameters, the eccentricity and ω were fixed to 0, and the limb darkening was calculated using the Exofast (see Eastman *et al.* (2012)) online tool. The other fixed values and the initial guesses were taken from the most recently published results: Faedi *et al.* (2011) for WASP-39, and Chen *et al.* (2014) for WASP-43.

3. Results

We obtain a photometric accuracy (peak-to-valley) of 3mmag with the 2.12m telescope, and of ±5mmag with the 84cm and 1.5m telescopes. Our results are shown in Table 1, and they are in agreement with previous studies. In particular, we find for WASP-43 a slight dependence of the planet/star radius on the photometric band. Although the variation is within the error bars, we are planning additional observations in order to investigate this effect in more detail. Moreover, we are ready to apply a genetic algorithm developed by Cantó *et al.* (2009) to our data.

References

Cantó, J., Curiel, S., & Martínez-Gómez, E. 2009, *A&A*, 501, 1259
Chen, G., van Boekel, R., Wang, H., *et al.* 2014, *A&A*,, 563, A40

Eastman, J., Gaudi, B. S., & Agol, E. 2012, *Astrophysics Source Code Library*, 7001
Faedi, F., Barros, S. C. C., Anderson, D. R., *et al.* , 2011, *A&A*, 531, A40
Gazak, J. Z., Johnson, J. A., Tonry, J., *et al.* , 2011, *Astrophysics Source Code Library*, 6014
Lehner, M., Wang, S., Ho, P., *et al.* 2013, *AAS/DPS Meeting Abstracts*, 45, #414.08
Mandel, K. & Agol, E. 2002, *Astrophysical Journal Letters*, 580, L171
Southworth, J., Hinse, T. C., Burgdorf, M., *et al.* 2014, *MNRAS*, 444, 776

Complex Planetary Systems
Proceedings IAU Symposium No. 310, 2014
Z. Knežević & A. Lemaitre, eds.

doi:10.1017/S1743921314007984

Dynamic study of possible host stars for extrasolar planetary systems

N. A. Shakht, L. G. Romanenko, D. L. Gorshanov and O. O. Vasilkova

Central (Pulkovo) Observatory RAS,
St- Petersburg 196140, Russia
email: shakht@gao.spb.ru

Abstract. We present the stellar systems which consist of double and multiple stars with distances $3.5 \div 25$ pc from the Sun, belonging to spectral classes F, G, K, M, having masses from 0.3 to 1.5 solar mass and can, in principle, possess planetary systems. On the basis of observations with Pulkovo 65 cm refractor the relative positions of double stars, the parameters of motion, the orbits and also the ephemeris for the nearest epochs have been computed.

Keywords. stars: binaries, stars: planetary systems, habitable zones

1. Introduction

The astrometric positional observations of single, double and multiple stars, which are located near the Sun, are carrying out in Pulkovo observatory many years. Observations are made by means of 26-inch refractor (D = 65 cm; F =10413 mm) since 1956. We have long-term series to $40 \div 50$ years of photographic observations with yearly error about 10 mas. The telescope functions in an automatic mode. Since 2007 year the CCD camera is used. The results of our observations of about 300 stars are collected in the Catalog of Relative Positions of Double Stars (Kiselev *et al.* 2014). Owing to the physical properties (a spectral class, masses, the distance to the Sun and the duration of positional observations, which allow to obtain precise parameters of movement) these stars could become objects of observations for programs exoplanets searches. It is known, that some projects are aimed at searches of planets by means of astrometric methods. These programs include the research of stars of F, G, K and M spectral classes (see, for instance, CAPS program by Boss *et al.* 2014 and also future project NEAT with expected microseconds precision by Malbet *et al.* 2014).

2. Results of observations

We investigate wide double stars with separation $\geqslant 4''$ and with $m < 13$ mag. For computing orbits we use Apparent Motion Parameters method, see the description in Kiselev & Romanenko (1996) and Kiselev *et al.* (2014). This method allows to calculate an orbit and dynamical mass of a double star on the basis of observations of a short arc if the parallax and the relative radial velocity are known. The orbits for about 50 pairs have been determined. We have the close pairs with semi-axis major a from 33 to 200 AU and periods $P < 2000$ years and more wide pairs with periods up to several tens of thousands of years. In the table 1 we give a short list of stars having Pulkovo orbits, their ADS and WDS numbers and spectral classes for two components. In the table 2 we give the new orbits of three stars, the distance of a star from the Sun - D and the minimum

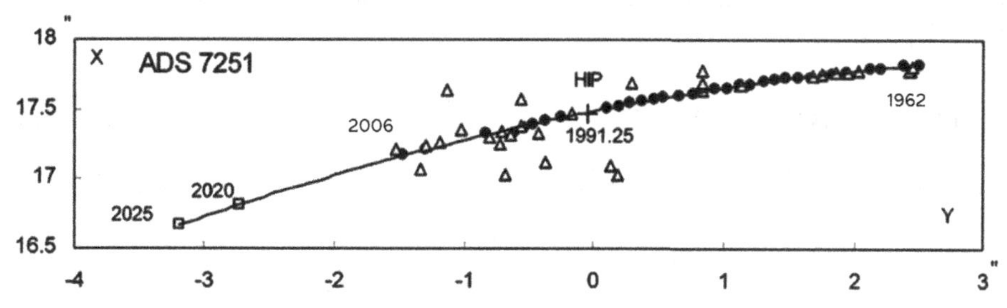

Figure 1. ADS 7251, the solid circles - Pulkovo observations in 1962-2006, the triangles - the positions from WDS catalog, the cross - Hipparcos 1991.25 and the squares - the ephemeris for 2020 and 2025 years. X=$\rho \sin \theta$, Y=$\rho \cos \theta$.

Table 1. List of the selected stars.

ADS	WDS	Sp	ADS	WDS	Sp	ADS	WDS	Sp	ADS	WDS	Sp	ADS	WD	Sp
48	00057 +4549	K6V M0V	8002	10596 +2527	K0 -	8861	13195 +3508	M1V M3V	10288	16579 +4722	K0V K0V	11632	18428 +5938	M4 M4
2427	03162 +5810	M2V M2V	7251	09144 +5241	M0V M0V	9031	13491 +2659	K4V K6V	10329	17033 +5935	K4V M0	12169	19121 +4051	G5V G5V
2757	03470 +4126	K1V K2V	8100	11152 +7329	K5 M0	9090	14025 +4620	M3 M3	10386	17101 +5429	K8V K8V	12815 16Cyg	19418 +5032	G2V G2V
5983	07202 +2159	F0IV K6V	8250	11387 +4507	G0V -	9696	15292 +8027	K0V K0V	10759	17419 +7209	F5IV F8V	14636 61Cyg	21069 +3845	K5V K7V

Table 2. Orbits of binary stars and astrometric signal due to possible planets.

ADS	a, au	P, yr	e	i°	ω°	Ω°	T_p, yr	D, pc	a_p, au	$A_{\oplus}$, μas	A_{Jup}, mas
7251 A	136.9	1528.0	0.08	141.0	210.4	216.6	1882.8	6.1	0.25	0.20	0.06
B									0.24	0.20	0.06
11632 A	93.4	1094.1	0.42	106.7	319.3	145.1	1835.2	3.5	0.22	0.52	0.16
B									0.14	0.36	0.11
14636 A	81.8	674.3	0.50	132.5	156.3	177.1	1728.5	3.5	0.35	0.44	0.14
B									0.26	0.39	0.12

Note: a_p is a semi-major axis of a planet's orbit, which we proposed equal to the inner edge of habitable zone. HZ was computed according to Selsis *et al.* 2007.

of astrometric signal from planets located near to the inner edge of HZ with the Earth's ($A_{\oplus}$) and Jupiter's (A_{Jup}) masses.

We hope that our astrometric observations can be useful for comprehensive study of stars with possible planetary systems.

References

Boss, A. P., Weinberger, A. G., Anglada-Escudé, G.,Thompson, I. B., & Brahm, R. 2014, *Formation, Detection and Characterization of Extrasolar Habitable Planets* Proc. IAU Symp. No 293 , 2012, ed. N. Haghighipoor, p.183

Kiselev, A. A., Kiyaeva, O. V., Izmailov, I. S., Romanenko, L. G., Kalinichenko, O. A., Vasilkova, O. O., Vasil'eva, T. A., Shakht, N. A., Gorshanov, D. L., & Roschina, E. A. 2014, *Astron.Rep.* 58, 78

Kiselev, A. A. & Romanenko, L. G.1996, *Astron.Rep.* 40, 795

Malbet, F.,Crouzier, A., Renaud, G. *et al.* 2014, *Formation, Detection and Characterization of Extrasolar Habitable Planets* Proc. IAU Symp. No 293 , 2012, N. Haghighipoor ed., p.448

Selsis, F., Kasting, J. F., Levrard, B. *et al.* 2007, *A&A* 476, 1373

Complex Planetary Systems
Proceedings IAU Symposium No. 310, 2014
Z. Knežević & A. Lemaitre, eds.

doi:10.1017/S1743921314007996

Symmetric Four-mass Schubart-like Systems

Winston L. Sweatman

Institute of Natural and Mathematical Sciences,
Massey University, Albany, Auckland, New Zealand
email: w.sweatman@massey.ac.nz

Abstract. The general four-body problem can be simplified by considering the special case where the system contains two pairs of identical masses and is symmetrical. The simple models that occur may aid our understanding of the general problem. Systems that arise from Schubart-like interplay orbits are an important feature of the dynamics.

Keywords. celestial mechanics, symmetrical four-body problem, binary-binary interactions, Caledonian four-body problem, Schubart-like interplay orbits

1. Introduction

Although somewhat simpler, a four-body system with symmetry may provide some insight into understanding the general four-body problem (Sweatman 2015). The Caledonian four-body problem, Fig. 1, is a four-body system which has a rotational symmetry about its centre of mass (Roy & Steves 1998, 2001; Steves & Roy 1998). The symmetry requires masses m_1 and m_3 to be equal and images of one another on opposite sides of the centre of mass. Likewise, masses m_2 and m_4 are equal and images. If motion is strictly one-dimensional, the Caledonian four-body problem becomes the symmetrical collinear four-body problem (Sweatman 2002, 2006; Sekiguchi & Tanikawa 2004). For this the masses remain in the order m_1, m_2, m_4, m_3. The inevitable collisions between mass pairs are resolved with an elastic bounce, the limit of a planar encounter as eccentricity tends to unity. The system's state is given by the positions and momenta of m_1 and m_2: x_1 and x_2, and $w_1 = 2m_1\dot{x}_1$ and w_2, respectively, with Hamiltonian

$$H = \frac{w_1^2}{4m_1} + \frac{w_2^2}{4m_2} - \frac{m_1^2}{2x_1} - \frac{m_2^2}{2x_2} - \frac{2m_1m_2}{x_1 + x_2} - \frac{2m_1m_2}{x_1 - x_2}. \tag{1.1}$$

We regularise equations of motion using the square-root of inter-mass distances, conjugate momenta and rescaled time (Sweatman 2002, 2006). The Caledonian four-body problem can be handled similarly with Levi-Civita regularisation (Sivasankaran *et al.* 2010).

As for three bodies a key dynamical feature of the one-dimensional four-body problem is the periodic orbit analogous to Schubart's orbit (Schubart 1956). In this orbit, there are alternately interactions between masses m_2 and m_4 at the centre, and simultaneous interactions, between m_1 and m_2 and between m_4 and m_3, on the outside.

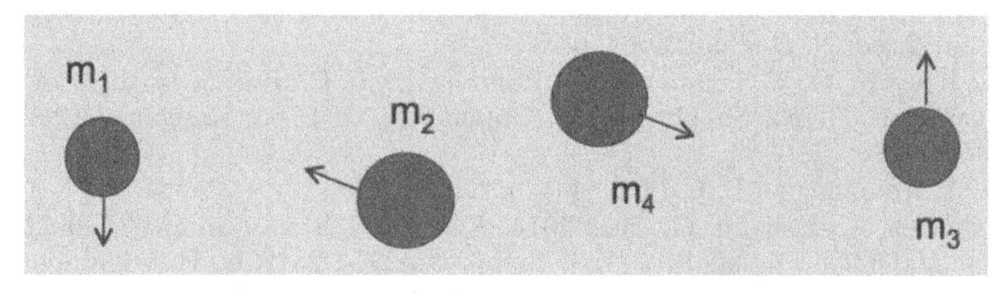

Figure 1. The Caledonian four-body problem.

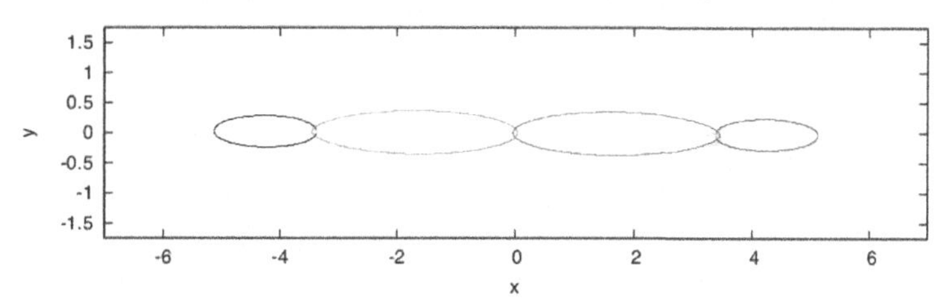

Figure 2. A planar orbit near the equal mass Schubart-like orbit. Approximate initial positions are ± 3.45413106, ± 3.38409367, with transverse velocities ∓ 3.79430129, ± 3.70734831.

2. Families of Schubart-like Interplay Orbits

The collinear equal-mass Schubart-like orbit is found by progressive refinement of a Poincaré section (Sweatman 2002). It is collinearly stable but spatially unstable.

From the equal-mass Schubart-like orbit, families of related periodic orbits can be found using approaches similar to those of three-body investigations (Schubart 1956; Hénon 1976, 1977; Mikkola & Hietarinta 1991; Hietarinta & Mikkola 1993). In one-dimension, a family of Schubart-like interplay orbits arises from different mass ratios (Sweatman 2006). Some of these orbits are spatially stable. The orbits are found by applying differential correction. Beginning with the known equal-mass solution, the results for previous mass ratios are used to generate approximate initial conditions for the current mass ratio. The orbit is integrated through one period and the difference between initial and final points is used to produce improved initial conditions. The beginning and end of the period were taken to be times at which the inter-body distances were at a fixed ratio.

Hénon (1976) generated a family of periodic orbits in the plane from Schubart's orbit. Ongoing investigation suggests a similar family of planar orbits for the corresponding four-body case. Fig. 2 shows one such orbit which is relatively close to the equal-mass Schubart-like interplay orbit. Each outer mass orbits the closer inner mass. The inner masses alternately pass around one another, and the corresponding outer mass. Again, differential correction is used to find this periodic orbit. The search started with collinear masses close to the positions of simultaneous double collision for the Schubart-like orbit. Velocities perpendicular to the line of the bodies are determined to provide a small fixed non-zero angular momentum. The beginning and end of the period are taken to be maxima of the distance from m_2 to m_4. The caption has approximate initial conditions.

References

Hénon, M. 1976, *Cel. Mech. Dyn. Astron.*, 13, 267

Hénon, M. 1977, *Cel. Mech. Dyn. Astron.*, 15, 243

Hietarinta, J. & Mikkola, S. 1993, *Chaos*, 3,183

Mikkola, S. & Hietarinta, J. 1991, *Cel. Mech. Dyn. Astron.*, 51, 379

Roy, A. E. & Steves, B. A. 1998, *Planet. Space Sci.*, 46, 1475

Roy, A. E. & Steves, B. A. 2001, *Cel. Mech. Dyn. Astron.*, 78, 299

Schubart, J. 1956, *Astron. Nachr.*, 283, 17

Sekiguchi, M. & Tanikawa, K. 2004, *PASJ*, 56, 235

Sivasankaran, A., Steves, B. A., & Sweatman, W. L. 2010, *Cel. Mech. Dyn. Astron.*, 107, 157

Steves, B. A. & Roy, A. E. 1998, *Planet. Space Sci.*, 46, 1465

Sweatman, W. L. 2002, *Cel. Mech. Dyn. Astron.*, 82, 179

Sweatman, W. L. 2006, *Cel. Mech. Dyn. Astron.*, 94, 37

Sweatman, W. L. 2015, *Symmetric four-body problems*, in: M. Cojocaru, I. S. Kotsireas, R. N. Makarov, R. Melnik & H. Shodiev (eds.), *Interdisciplinary Topics in Applied Mathematics, Modeling, and Computational Science*, Springer Proceedings in Mathematics and Statistics, Volume 117, in press

Complex Planetary Systems
Proceedings IAU Symposium No. 310, 2014
Z. Knežević & A. Lemaitre, eds.

doi:10.1017/S174392131400800X

Small asteroid system evolution

Seth A. Jacobson[1,2]

[1]Laboratoire Lagrange, Observatoire de la Côte d'Azur,
Boulevard de l'Observatoire, CS 34229, F 06304 Nice Cedex 4, France
[2]Bayerisches Geoinstitut, Universtät Bayreuth,
D 95440 Bayreuth, Germany
email: seth.jacobson@oca.eu

Abstract. Recently, the discovery of small unbound asteroid systems called asteroid pairs have revolutionized the study of small asteroid systems. Observations with radar, photometric and direct imaging techniques have discovered that multiple asteroid systems can be divided clearly into a handful of different morphologies. Simultaneously, new theoretical advances have demonstrated that solar radiation dictates the evolution of small asteroids with strong implications for asteroid internal structure. We review our current understanding of how small asteroid systems evolve and point to the future.

Keywords. minor planets, dynamics

---

1. Asteroid system morphologies

Asteroid systems cannot be represented by a single distribution of characteristics, instead they are apportioned among a number of distinct morphologies defined by size, multiplicity, spin states and shapes. Asteroids are divided into small and large asteroids according to whether or not rotational acceleration from the YORP effect controls their spin states over other factors. We focus on small asteroids with YORP-driven evolution.

1.1. *Size-determined morphology*

The YORP effect is radiative torque due to asymmetrically emitted thermal photons (Rubincam 2000). The timescale for an asteroid of radius R to rotational accelerate from rest to the critical disruption spin rate is (Scheeres 2007b):

$$\tau_{\mathrm{YORP}} = \frac{2\pi\rho\omega_d R^2}{\bar{Y} H_\odot} \tag{1.1}$$

where $H_\odot = F_\odot / (a_\odot^2 \sqrt{1-e_\odot^2})$ is a term containing heliocentric orbit factors, $F_\odot \approx 10^{22}$ g cm s^{-2} is the solar radiation constant, $a_\odot$ and $e_\odot$ are the semi-major axis and eccentricity of the asteroid system's heliocentric orbit, and the critical disruption spin rate $\omega_d = \sqrt{4\pi\rho G/3}$ is the spin rate at which the centrifugal accelerations match gravitational accelerations for a massless test particle resting on the surface of an asteroid of density ρ (i.e. the spin rate that casts a test particle from the surface into orbit), G is the gravitational constant, and, lastly, $\bar{Y}$ is the effective YORP coefficient, which takes into account small changes in obliquity and shape over the rotational acceleration of the asteroid. Furthermore, it is the sum of the normal YORP components (Rubincam 2000, Scheeres 2007b) and the tangential YORP components (Golubov & Krugly 2012). Typical values for the effective YORP coefficient are $\bar{Y} \sim 0.01$ (Taylor *et al.* 2007, Kaasalainen *et al.* 2007), and so characteristic timescales are $\tau_{\mathrm{NEA}} \sim 6.7\left(R^2/1 \text{ km}\right)$ My for the near-Earth asteroid (NEA) population and $\tau_{\mathrm{MBA}} \sim 41.6\left(R^2/1 \text{ km}\right)$ My for the main belt asteroid (MBA) population. These timescales are shown in Figure 1. Amongst the NEAs, the

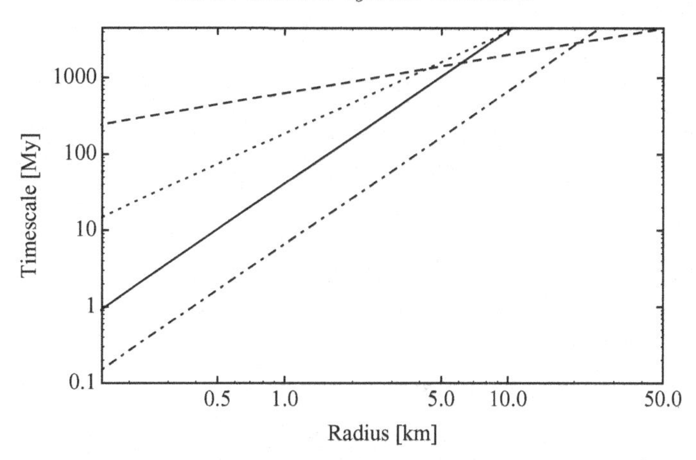

Figure 1. Important small asteroid evolution timescales. Each line indicates the time in millions of years at which an asteroid of a radius R given standard parameter choices will undergo the following processes: rotational acceleration to critical disruption spin rate ω_d from collisions in the main belt ($\tau_{\rm rot}$, dashed line), catastrophic disruption due to collisions in the main belt ($\tau_{\rm disr}$, dotted line), rotational acceleration to critical disruption spin rate ω_d from the YORP effect in the main belt ($\tau_{\rm MBA}$, solid line), rotational acceleration to critical disruption spin rate ω_d from the YORP effect in the NEAs ($\tau_{\rm NEA}$, dot-dashed line).

YORP effect is more important than collisions (Bottke *et al.* 1994) and planetary flybys (Rossi *et al.* 2009), because there are no NEAs significantly larger than a few tens of kilometers and those events are comparatively rare. Thus, the NEA population contains only small asteroids, however the YORP effect timescale can exceed the dynamical lifetime of a NEA, so observed NEAs have not necessarily undergone significant YORP effect evolution in NEA space.

The MBA population does contain both small and large asteroids. Collisions catastrophically disrupt asteroids of radius R on a timescale $\tau_{\rm disr} = 633\,(R/1\ {\rm km})^{1/2}$ My and deliver enough angular momentum to rotationally accelerate the asteroid significantly on a timescale $\tau_{\rm rot} = 188\,(R/1\ {\rm km})^{3/4}$ My (Farinella *et al.* 1998). These timescales are shown in Figure 1, and it is clear that the division between small, YORP-driven asteroids and large asteroids is when the asteroid radius $R \sim 6$ km. This is confirmed by more sophisticated modeling (Jacobson *et al.* 2014a). Furthermore, a division amongst the large asteroids at $R \sim 50$ km is apparent between those small enough to be significantly evolved by collisions during the current dynamical epoch of the main belt (e.g. ~ 4 Gy for a Nice model explanation of the late heavy bombardment, Marchi *et al.* 2013) and those even larger whose spin states on average reflect an earlier epoch, possibly primordial.

1.2. *Multiplicity-determined morphology*

Multiplicity is usually taken to refer to the number of bound members within an asteroid system, but it has been known since the beginning of the 20th century that unbound asteroids can be linked together. Unbound asteroid systems include the well known asteroid families formed from the debris from collisions and the recently discovered asteroid pairs (Vokrouhlický & Nesvorný 2008, Vokrouhlický & Nesvorný 2009, Pravec & Vokrouhlický 2009), which is a misnomer since asteroid 'pairs' of more than two components exist (Vokrouhlický 2009, Pravec *et al.* 2013). Color photometry and spectroscopic analysis have confirmed that asteroid pairs are compositionally related as well (Moskovitz 2012, Duddy *et al.* 2012, Polishook *et al.* 2014a). Unlike asteroid families, which are

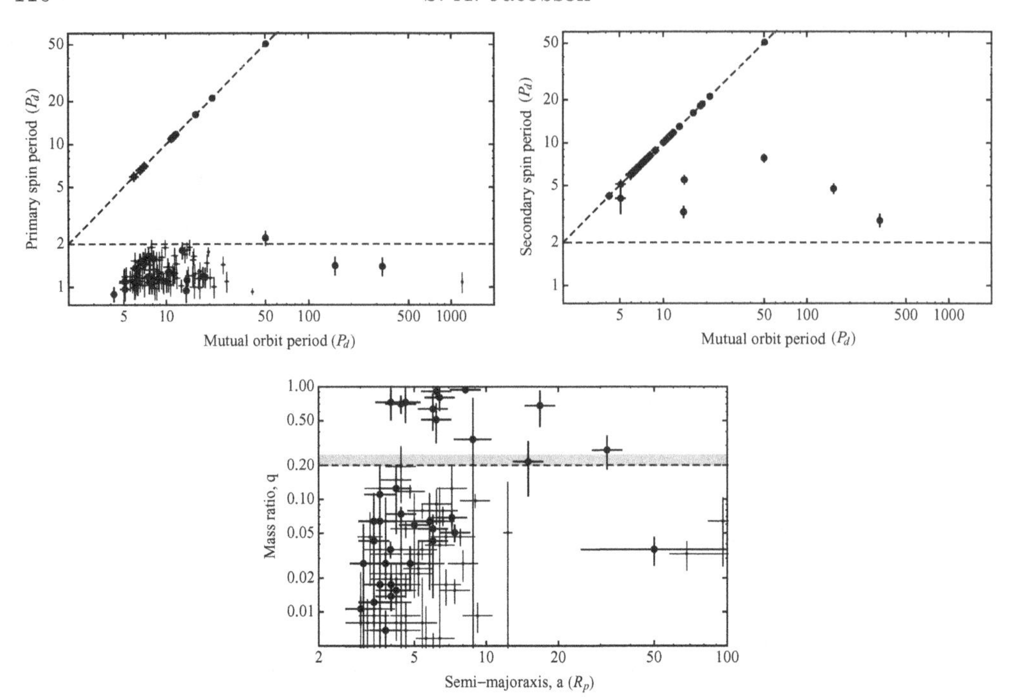

Figure 2. Observational data for the small binary asteroid dataset from the September 2013 release of the Binary Asteroid Parameters dataset compiled according to Pravec & Harris 2007 and updated by Petr Pravec and colleagues. Each binary is shown as a point with 3-σ uncertainties shown (some uncertainties are very small). Small black points represent incompletely characterized binaries, i.e. no measured secondary spin period. Large points represent well-characterized binaries. The upper left graph shows the primary spin period P_p measured in units of the critical disruption period $P_d = \sqrt{3\pi/\rho G}$ as a function of the mutual orbit period P_o measured in the same units. Any systems on the diagonal dashed line have a synchronicity between the primary spin periods and orbit periods; these are the doubly synchronous binaries. The horizontal dashed line at 2 P_d is the approximate rapid rotator cutoff. The upper right graph shows the secondary spin period P_s as a function of the mutual orbit period P_o measured in critical disruption periods P_d. Synchronous secondaries and the rapid rotator cutoff are again indicated by dashed lines; both doubly and singly synchronous binaries lie along the synchronous dashed line while both tight and wide asynchronous binaries are below it but above the rapid rotator dashed line. The bottom plot shows the mass ratio q of each binary pair as a function of the semi-major axis a measured in primary radii R_p. The dashed line indicates the special mass ratio of $q \approx 0.2$ (Scheeres 2007a) and the gray region above it indicates some uncertainty up to $q \sim 0.25$ due to variations in shape (Taylor & Margot 2014); only doubly synchronous binaries exist with mass ratios greater than this cutoff.

confirmed by examining the statistical significance of their number densities against the background asteroid population, asteroid pairs are confirmed through direct backwards numerical integration. This limits the timescale for which they can be discovered to ~2 My because the uncertainty of the asteroid's position and velocity is chaotically lost due to perturbations from the planets and other asteroids (Vokrouhlický & Nesvorný 2008, Pravec & Vokrouhlický 2009). For discovered asteroid pairs, these dynamical ages are also likely the surface ages of the asteroid components and so they are a powerful tool to explore space weathering and other surface processes (Polishook *et al.* 2014a). Unbound systems share a common ancestor, but are no longer interacting—a possible exception to this may be a feedback effect in large asteroid families, e.g. the increase in local number density increases the local collision rates.

Bound multiple asteroid systems are observed to contain two or three members, but there is no theoretical limit. Amongst the small multiple asteroid systems, there are distinct patterns related to the orbit period, spin periods and mass ratio as shown in Figure 2. Binary asteroid systems have a larger primary and a smaller secondary, and their mass ratio is the mass of the secondary divided by the mass of there primary. The most common observed small binary system is the singly synchronous, so named because the spin period of the secondary is synchronous with the orbit period. These binaries also have low mass ratios, which is defined as $q < 0.2$, tight semi-major axes, which is defined as $a \lesssim 8\ R_p$, and a primary with rapid rotation, which is defined as a period within a factor of two of the critical disruption period $P_d = 2\pi/\omega_d = \sqrt{3\pi/\rho G}$. The tight asynchronous binary systems also have rapidly rotating primaries and tight semi-major axes, but the secondary period is neither rapidly rotating nor synchronous with the orbit period. Wide asynchronous binary systems are very similar but naturally have larger semi-major axes. This distinction between tight and wide asynchronous binaries comes from comparing and contrasting with the singly synchronous binaries, respectively. All three morphologies are low mass ratio with the exception of a few wide asynchronous binaries which skirt the $q \sim 0.2$ cut-off. This is unlike the doubly synchronous binary asteroids, which all have high mass ratios $q > 0.2$, and take their names from the fact that all three periods in the system are synchronous. As shown in Figure 2, most of the non-classified binary systems are consistent with the singly synchronous or tight asynchronous morphologies.

There are also a few confirmed small asteroid triple systems. All have rapidly rotating primaries and low mass ratios between the smaller two members and the primary. Some of the inner ternary periods are confirmed synchronous. All of the outer ternary periods are asynchronous. Both (3749) Balam and (8306) Shoko are 'paired' to other asteroids 2009 BR_{60} and 2011 SR_{158}, respectively (Vokrouhlický 2009, Pravec *et al.* 2013).

1.3. *Spin-determined morphology*

The distribution of spin periods amongst the asteroid population is a strong function of size as shown in Figure 3. Very small asteroids with radii less than 0.15 km are found at very high rotation rates well in excess of the rapid rotator population. It is difficult to observe the rotation of very small asteroids if the period is slow, since they are only observable for very short epochs. Small asteroids with radii between 0.15 km and 6 km, asteroids larger than this are not significantly torqued by the YORP effect (Figure 1), have three distinguishing characteristics due to the YORP effect: rapid rotating single asteroids are common (Pravec & Harris 2000), binary asteroids are common and mostly amongst the rapid rotator population (Pravec & Harris 2007), and tumbling asteroids are common (Pravec *et al.* 2005). Amongst large asteroids with radii between 6 and 50 km, there is a dearth of rapid rotators, binary asteroids, and tumbling asteroids. It is difficult to assess whether this absence is real or an artifact, but since so many binaries have been discovered amongst larger and small systems, it is likely real unless there is a pathologic preference for a binary morphology that is difficult to detect such as extremely low mass ratios or wide semi-major axes. The dearth of tumblers is not as extreme as the dearth of rapid rotators. The very large asteroids with radii greater than 50 km do have binaries. The primaries are not rapid rotators but are amongst the fast rotators of this population.

While this work focuses on small asteroid evolution, it is worth noting that large multiple asteroids systems do appear very consistent with the formation of satellites in large impacts (Durda *et al.* 2004). The very large primary asteroids with rotation periods near a few P_d are consistent with the smashed target satellite formation mechanism

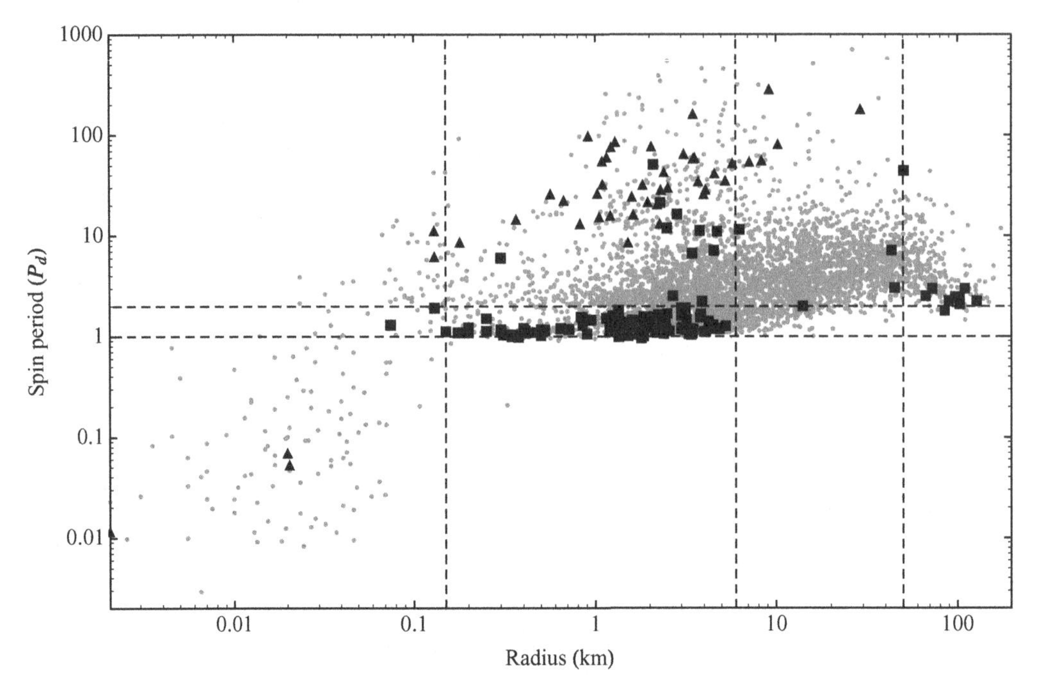

Figure 3. The spin period of the NEAs and MBAs as a function of radius from the March 2013 release of the Asteroid Lightcurve Database compiled according to Warner *et al.* 2009 and updated by Brian Warner and colleagues. Single asteroids (or at least no detected companion) are shown as small gray points with the exception of the tumbling asteroids which are black triangles. The primary spin periods and radii of multiple component asteroid systems are shown as large black squares. The spin period is measured in critical disruption periods $P_d = \sqrt{3\pi/\rho G}$ assuming a bulk density $\rho = 2$ g cm^{-3}. The horizontal dashed lines indicate spin periods of 1 and 2 P_d, which brackets the rapid rotating population. The vertical dashed lines indicate radii of 6 and 50 km, which are identified as special in the text and Figure 1.

(SMATS) (Durda *et al.* 2004) and a few wide asynchronous binaries (317 Roxanne and 1717 Arlon) are consistent with the escaping ejected binaries (EEB) (Durda *et al.* 2010, Jacobson *et al.* 2014b). These very large multiple asteroid system likely date to prior to the current collision environment (>4 Gy) from timescale arguments (See Figure 1). The dearth of binaries in the large asteroid morphology can be explained by considering that the YORP effect is not significant and that given the typically high velocities of collisions in the main belt, impacts on asteroids in this size range are too violent to create binaries and too frequent for systems created prior to the current collision environment to survive (<4 Gy).

1.4. *Shape-determined morphology*

Lastly, it is worth noting that asteroids are not drawn from just a spectrum of oblate to prolate potato shapes, but also represent bi-modal, necked or contact binary shape distributions and diamond-shape distributions. Bi-modal and necked shapes range in appearance from two distinct asteroids resting on one another to only vaguely having a 'head' and 'body'. Collectively, they are often referred to as contact binaries. Diamond-shaped asteroids are often the primaries of binary asteroids, but single asteroids can also have this shape. This shape is characterized by an equatorial bulge, nearly oblate, shallow slopes near the equator and steep slopes near the poles.

2. Rotational fission hypothesis

Margot *et al.* (2002) rejected formation of small binary asteroids from a sub-catastrophic impact, capture from a three-body interaction in the NEAs or MBAs, or capture after a catastrophic impact. None of these mechanisms systematically produce rapidly rotating primaries or the generally tight orbits observed amongst the binary populations. They concluded that rotational fission could explain these observations.

The rotational fission hypothesis posits that a parent asteroid can be torqued to a rotation rate so great that the centrifugal accelerations overpower the gravitational accelerations holding a rubble pile, i.e. strengthless, asteroid together (Weidenschilling 1980). This rotational fission hypothesis is consistent with the most common singly synchronous binaries having super-critical amounts of angular momentum (Pravec & Harris 2007). Planetary flyby-induced rotational fission was found to be insufficient to explain the existence of main belt and even near-Earth binaries (Walsh & Richardson 2008), but the YORP effect can rotationally accelerate asteroids to the critical disruption period, where rotational fission occurs (Scheeres 2007a, Walsh *et al.* 2008, Sánchez & Scheeres 2012). YORP-induced rotational fission is naturally limited to only sizes where the YORP effect controls the spin state, so any binaries larger than $R \sim 6$ km are unlikely to be formed from this mechanism. Large binaries do not have rapidly rotating primaries or tight semi-major axes, and are consistent with the SMATS formation mechanism. It is likely that very small asteroids have cohesive or physical strength, and so in these cases the centrifugal accelerations must overcome these additional forces in order for the asteroid to fission (Pravec & Harris 2000, Sánchez & Scheeres 2014). The rotational energy at the time of the rotational fission event determines the possible orbit and spin configuration immediately after the disruption (Scheeres 2007a). If the asteroid breaks into two massive components, then the spin period necessary for disruption is much lower. If the mass ratio between the two components is nearly equal, then the binary system after the fission event cannot find an escape trajectory. If the mass ratio is very small, then the binary system can find an escape trajectory. For two spheres, the transition mass ratio between these two regimes is $q \sim 0.2$, which matches the cut-off between doubly synchronous binaries and the rest.

The evolution of small binary asteroid systems after a rotational fission event was mostly determined in Jacobson & Scheeres 2011a, but there have been updates. The basic evolutionary flow is shown in Figure 4. Particularly, we highlight alternative formation mechanisms for different morphologies including binaries with the primary synchronous, asteroid pairs that do not follow the mass ratio–primary period relationship in Pravec *et al.* 2010, and triple systems.

2.1. *Low mass ratio track—chaotic binaries and asteroid pairs*

After a low mass ratio fission event, the resulting components enter into a chaotic orbit (Jacobson & Scheeres 2011a). Chaotic because the outcomes can only be described probabilistically. The fission process is likely very dusty (Polishook *et al.* 2014b), although the dust will have nearly no relative velocity once it escapes the system. The first observations of an asteroid undergoing rotational fission are possibly being made today (Jewitt *et al.* 2010, Jewitt *et al.* 2014b, Jewitt *et al.* 2014a), although alternative explanations such as collisions and volatile-driven mass loss exist. Multiple event structures seen in this data are more likely due to repeated pericenter passages of the mutual orbit than repeated YORP-driven accelerations of an asteroid (since the YORP timescale is 100 years at best). During these pericenter passages angular momentum and energy are transferred between three reservoirs: the spin states of both components and the mutual

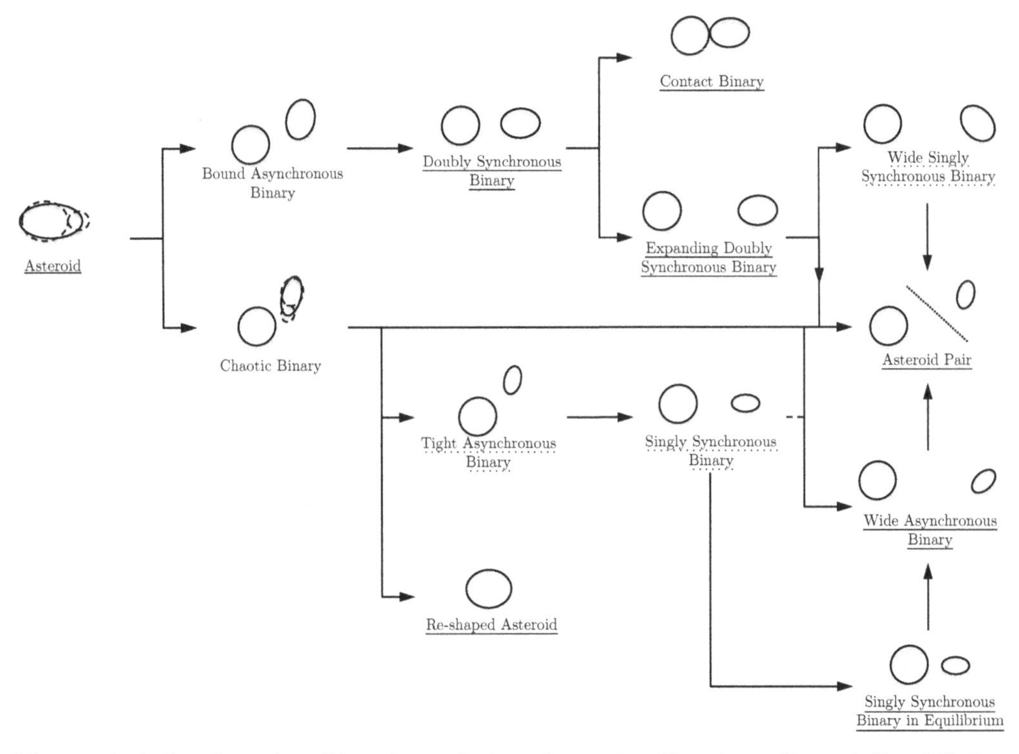

Figure 4. A flowchart describing the evolution of an asteroid system after rotational fission from a single asteroid. All motion along the flowchart is left to right. Each state is described by a cartoon and labeled below. The typical timescale for an asteroid system to occupy that state is indicated by the underline: a solid underline indicates that the state persists until an external impulsive event such as a collision or planetary flyby or until a component undergoes YORP-induced rotational fission, a dashed underline indicates that the state is dynamically evolving according to a slow ($\sim10^4$–10^7 years), regular process such as tides, YORP and/or BYORP effect, and no underline indicates that the state is dynamically evolving on a fast ($\ll10^4$ years), irregular process such as strong spin-orbit coupling between the higher order gravity moments of the asteroid system components.

orbit. These energy exchanges disrupt most systems on 1000 day timescales, and these disrupted systems are observationally consistent with the asteroid pair population—there is a predicted relationship between the mass ratio and the spin rate of the larger pair member (Pravec *et al.* 2010), and they have identical colors and spectrum (Moskovitz 2012, Polishook *et al.* 2014a). Sometimes, they can find stable orbits. Often, the secondary is torqued so much by the spin-orbit coupling during pericenter passage that it undergoes rotational fission itself, i.e. secondary fission. Surface shedding produces the multiple event structures observed by Jewitt *et al.* 2010, Jewitt *et al.* 2014b, Jewitt *et al.* 2014a, but secondary fission produces triple systems. A lot can happen to these triple systems. They may stabilize, a component may collide with the primary, or a component may escape the system. Material leaving the secondary during secondary fission and impacting the primary is likely to collect on the equator creating the equatorial bulge and the ubiquitous diamond shape. Secondary fission is a process that can occur many times.

2.2. *Low mass ratio track—stable binary evolution*

Low mass ratio binary systems stabilize with semi-major axes between 2 and 17 primary radii R_p (Jacobson & Scheeres 2011a); the larger the stable orbit, the more eccentric the orbit is and the largest in numerical simulations were 17 R_p, which is a possible explanation for tight and wide binaries. Theoretically, stable binaries could be formed out to about 34 R_p, but these would be very eccentric (Jacobson *et al.* 2014b). A simple analysis of eccentricity damping due to tides suggests that any mutual orbits larger than 10 R_p are unlikely to damp, and so are still eccentric today. This may explain why some low mass ratio binaries have mild eccentricities.

The fastest tidal process according to theory is synchronization of the secondary (Goldreich & Sari 2009), and this is very consistent with the majority of tight binary systems being singly synchronous. Tight asynchronous binaries are likely younger than the average singly synchronous binary and the YORP torque on their smaller component is likely in the opposite direction of the synchronization tides. The combination of these two effects can explain the presence of a small population of tight asynchronous binary asteroids amongst a mostly singly synchronous population. Wide binary systems are asynchronous because synchronization tides fall off as the separation distance to the sixth power, and so are incredibly weak at these distances. Tides on the primary are very weak due to the low mass ratio compared to the YORP effect. Since the YORP effect likely has a positive bias from the tangential YORP component (Golubov & Krugly 2012), the primary maintains a spin period near the critical disruption period. If the primary accelerates enough, it could undergo rotational fission again and create a ternary system. This evolution has yet to be modeled in detail. However, spin-up in the presence of a tightly orbiting satellite is significantly different than companionless spin-up. Unlike companionless spin-up, the secondary's gravitational potential is constantly moving through the primary. This produces tides, but it also can set-off mini-avalanches and mass shedding events that maintain and perfect the diamond-shape. This process may be similar to those proposed in Harris *et al.* 2009. Companionless asteroids are more likely to evolve catastrophically as they try to relax to the Maclaurin-Jacobi ellipsoid shapes (Holsapple 2010). This would argue that the diamond-shape is not a result of YORP-driven rotational acceleration alone as suggested by Walsh *et al.* 2008, but a combination of YORP acceleration and the at least temporary presence of a companion.

2.3. *Low mass ratio track—singly synchronous binary evolution*

Singly synchronous binaries evolve due to both tides and the BYORP effect. The BYORP effect is a radiative torque on the mutual orbit of any synchronous satellite (Ćuk & Burns 2005, McMahon & Scheeres 2010b). It operates much like the YORP effect, and can substantially change the orbit of small binary systems in only $\sim 10^5$ years (McMahon & Scheeres 2010a). Since the BYORP effect can contract or expand the mutual orbit, there are two possible evolutionary paths. First, if the BYORP effect contracts the mutual orbit, then the singly synchronous system can reach a tidal-BYORP equilibrium (Jacobson & Scheeres 2011b). In this equilibrium, the mutual orbit circularizes and then no longer evolves. If it is, then the tidal-BYORP equilibrium can be used to determine important geophysical parameters. This binary state is stable and lasts until either the primary undergoes rotational fission, a possibility mentioned above, or a collision or planetary flyby occurs. It is possible that a planetary flyby could de-synchronize the secondary and expand the mutual orbit. If the expansion is great enough, then tides cannot synchronize the secondary and the system becomes a wide asynchronous system. Likely though, planetary flybys disrupt the binary forming an asteroid pair (Fang & Margot 2012).

If the BYORP effect expands the mutual orbit, then it is working with the tides. These binary systems may expand all the way to the Hill radius and disrupt forming asteroid pairs (Jacobson & Scheeres 2011b). These asteroid pairs will not follow the mass ratio–spin period relationship discovered in Pravec *et al.* 2010, since the primary can spin back up after the rotational fission event but before the system disruption. Alternatively, an adiabatic invariance between the orbital period and the libration amplitude of the secondary can de-synchronize the secondary, when the orbit is very large. This creates wide asynchronous binaries on large circular orbits (Jacobson *et al.* 2014b). This is in contrast to other formation mechanisms such as planetary flybys, direct formation from rotational fission or the escaping ejecta binary mechanism which all predict moderate to extreme mutual orbit eccentricity. Wide asynchronous binaries do not evolve by either tides (too wide an orbit) or the BYORP effect (asynchronous), so they can only be destroyed by collision or planetary flyby.

2.4. *High mass ratio track*

The high mass ratio track after rotational fission is much simpler. Since the two components have nearly the same mass, the tidal de-spinning timescales are nearly the same (Jacobson & Scheeres 2011a). Once both members are tidally locked, the system is a doubly synchronous binary, which agrees exactly with observations that all doubly synchronous binaries have high mass ratios. The BYORP effect does evolve these binaries, but the BYORP torques on each binary are independent. Likely, the observed binaries represent the population that have opposing BYORP torques, and so the mutual orbit is very slowly evolving. If the BYORP torques drain angular momentum from the system, then the binary may collapse (Ćuk 2007, Taylor & Margot 2014) forming a contact binary, but the BYORP torques can also expand the orbit. If they do so then, it can reach the Hill radius and form an asteroid pair. In this case, the asteroid pairs would have a mass ratio greater than $q \sim 0.2$. If the system evolved directly to the Hill radius then both would be rotating slowly, if one were to de-synchronize due to the adiabatic invariance but the other continues to drive the mutual orbit to the Hill radius from the BYORP effect, then the asteroid pair might have a rapidly rotating member, a slowly rotating member, and a mass ratio greater than 0.2.

References

Bottke, W. F., Nolan, M. C., Greenberg, R., & Kolvoord, R. A., 1994, *Hazards due to comets & asteroids* p. 337

Ćuk, M., 2007, *ApJ* 659, L57

Ćuk, M. & Burns, J. A., 2005, *Icarus* 176, 418

Duddy, S. R., Lowry, S. C., Wolters, S. D., Christou, A., Weissman, P. R., Green, S. F., & Rozitis, B., 2012, *A&A* 539, A36

Durda, D. D., Bottke, W. F., Enke, B. L., Merline, W. J., Asphaug, E., Richardson, D. C., & Leinhardt, Z. M., 2004, *Icarus* 170, 243

Durda, D. D., Enke, B. L., Merline, W. J., Richardson, D. C., Asphaug, E., & Bottke, W. F., 2010, *Abstracts of the Lunar & Planetary Science Conference* **41**, 2558

Fang, J. & Margot, J.-L., 2012, *AJ* 143, 25

Farinella, P., Vokrouhlický, D., & Hartmann, W. K., 1998, *Icarus* 132, 378

Goldreich, P., & Sari, R., 2009, *ApJ* 691, 54

Golubov, O. & Krugly, Y. N., 2012, *ApJ Letters* 752, L11

Harris, A. W., Fahnestock, E. G., & Pravec, P., 2009, *Icarus* 199, 310

Holsapple, K. A., 2010, *Icarus* 205, 430

Jacobson, S. A., Marzari, F., Rossi, A., Scheeres, D. J., & Davis, D. R., 2014a, *MNRAS Letters* L15

Jacobson, S. A. & Scheeres, D. J., 2011a, *Icarus* 214, 161

Jacobson, S. A. & Scheeres, D. J., 2011b, *ApJ Letters* 736, L19

Jacobson, S. A., Scheeres, D. J., & McMahon, J. W., 2014b, *ApJ* 780, 60

Jewitt, D., Agarwal, J., Li, J.-Y., Weaver, H., Mutchler, M., & Larson, S., 2014a, *arXiv.org* p. 1237

Jewitt, D., Ishiguro, M., Weaver, H., Agarwal, J., Mutchler, M., & Larson, S., 2014b, *AJ* 147, 117

Jewitt, D., Weaver, H., Agarwal, J., Mutchler, M., & Drahus, M., 2010, *Nature* 467, 817

Kaasalainen, M., Ďurech, J., Warner, B. D., Krugly, Y. N., & Gaftonyuk, N. M., 2007, *Nature* 446, 420

Marchi, S., Bottke, W. F., Cohen, B. A., Wünnemann, K., Kring, D. A., McSween, H. Y., de Sanctis, M. C., O'Brien, D. P., Schenk, P., Raymond, C. A., & Russell, C. T., 2013, *Nature Geoscience* 6, 303

Margot, J.-L., Nolan, M. C., Benner, L. A. M., Ostro, S. J., Jurgens, R. F., Giorgini, J. D., Slade, M. A., & Campbell, D. B., 2002, *Science* 296, 1445

McMahon, J. W. & Scheeres, D. J., 2010a, *Icarus* 209, 494

McMahon, J. W. & Scheeres, D. J., 2010b, *Cel. Mech. Dyn. Astr.* 106, 261

Moskovitz, N. A., 2012, *Icarus* 221, 63

Polishook, D., Moskovitz, N. A., Binzel, R. P., Demeo, F. E., Vokrouhlický, D., Žižka, J., & Oszkiewicz, D. A., 2014a, *arXiv.org*

Polishook, D., Moskovitz, N. A., DeMeo, F., & Binzel, R. P., 2014b, *arXiv.org* p. 2837

Pravec, P. & Harris, A. W., 2000, *Icarus* 148, 12

Pravec, P. & Harris, A. W., 2007, *Icarus* 190, 250

Pravec, P., Harris, A. W., Scheirich, P., Kusnirák, P., Šarounová, L., Hergenrother, C. W., Mottola, S., Hicks, M. D., Masi, G., Krugly, Y. N., Shevchenko, V. S., Nolan, M. C., Howell, E. S., Kaasalainen, M., Galad, A., Brown, P., Degraff, D. R., Lambert, J. V., Cooney, W. R., & Foglia, S., 2005, *Icarus* 173, 108

Pravec, P., Kusnirák, P., Hornoch, K., Galad, A., Krugly, Y. N., Chiorny, V., Inasaridze, R., Kvaratskhelia, O., Ayvazian, V., Parmonov, O., Pollock, J., Mottola, S., Oey, J., Pray, D., Zizka, J., Vraštil, J., Molotov, I. E., Reichart, D. E., Ivarsen, K. M., Haislip, J. B., & Lacluyze, A. P., 2013, *IAU Circ.* 9268, 1

Pravec, P. & Vokrouhlický, D., 2009, *Icarus* 204, 580

Pravec, P., Vokrouhlický, D., Polishook, D., Scheeres, D. J., Harris, A. W., Galad, A., Vaduvescu, O., Pozo, F., Barr, A., Longa, P., Vachier, F., Colas, F., Pray, D. P., Pollock, J., Reichart, D. E., Ivarsen, K. M., Haislip, J. B., Lacluyze, A. P., Kusnirák, P., Henych, T., Marchis, F., Macomber, B., Jacobson, S. A., Krugly, Y. N., Sergeev, A. V., & Leroy, A., 2010, *Nature* 466, 1085

Rossi, A., Marzari, F. & Scheeres, D. J., 2009, *Icarus* 202, 95

Rubincam, D. P., 2000, *Icarus* 148, 2

Sánchez, D. P. & Scheeres, D. J., 2012, *Icarus* 218, 876

Sánchez, D. P. & Scheeres, D. J., 2014, *Meteoritics & Planetary Science* 49, 788

Scheeres, D. J., 2007a, *Icarus* 189, 370

Scheeres, D. J., 2007b, *Icarus* 188, 430

Taylor, P. A. & Margot, J.-L., 2014, *Icarus* 229, 418

Taylor, P. A., Margot, J.-L., Vokrouhlický, D., Scheeres, D. J., Pravec, P., Lowry, S. C., Fitzsimmons, A., Nolan, M. C., Ostro, S. J., Benner, L. A. M., Giorgini, J. D., & Magri, C., 2007, *Science* 316, 274

Vokrouhlický, D., 2009, *ApJLetters* 706, L37

Vokrouhlický, D. & Nesvorný, D., 2008, *AJ* 136, 280

Vokrouhlický, D. & Nesvorný, D., 2009, *AJ* 137, 111

Walsh, K. J. & Richardson, D. C., 2008, *Icarus* 193, 553

Walsh, K. J., Richardson, D. C., & Michel, P., 2008, *Nature* 454, 188

Warner, B. D., Harris, A. W., & Pravec, P., 2009, *Icarus* 202, 134

Weidenschilling, S. J., 1980, *Icarus* 44, 807

Complex Planetary Systems
Proceedings IAU Symposium No. 310, 2014
Z. Knežević & A. Lemaitre, eds.

doi:10.1017/S1743921314008011

Ranking in-orbit fragmentations and space objects

Alessandro Rossi,[1] Giovanni B. Valsecchi[2,1] and Elisa Maria Alessi[1]

[1]IFAC-CNR
Via Madonna del Piano 10, 50019, Sesto Fiorentino, Italy
email: a.rossi@ifac.cnr.it, em.alessi@ifac.cnr.it

[2]IAPS-INAF
via Fosso del Cavaliere 100, 00133, Roma, Italy
email: giovanni@iaps.inaf.it

Abstract. The future space debris environment will be dominated by the production of fragments coming from massive fragmentations. In order to identify the most relevant parameters influencing the long term evolution of the environment and to assess the criticality of selected space objects in different regions of the circumterrestrial space, a large parametric study was performed. In this framework some indicators were produced to quantify and rank the relevance of selected fragmentations on the long term evolution of the space debris population. Based on the results of the fragmentation studies, a novel analytic index, the Criticality of Spacecraft Index, aimed at ranking the environmental criticality of abandoned objects in LEO, has been devised and tested on a sample population of orbiting objects.

Keywords. celestial mechanics, space debris, fragmentation, space objects ranking

1. Introduction

The simulations of the long term evolution of the space debris population show how the future environment will be dominated by the production of fragments coming from massive fragmentations. A first warning in this respect came from the well known collision between Iridium 33 and Cosmos 2251 in 2009 which injected thousands of fragments larger than 1 cm into long lasting orbits in the most crowded zone of LEO, around 800-900 km of altitude. In this respect, it is of paramount importance to understand the consequences on the environment of a possible future large fragmentation of a given spacecraft in different regions of space.

This kind of study was performed in the framework of the ESA-ESOC Contract "*Assessment Study for Fragmentation Consequence Analysis for LEO and GEO Orbits*", taking the present distribution of intact objects as a proxy to identify the space objects more prone to future catastrophic collisions. Using SDM 4.2, the latest version of the long term LEO to GEO debris environment evolution model developed by our research group at the Italian National Research Council (CNR) in the past decades (Rossi *et al.* 2009), a reference long term evolution scenario, where the space activities are performed in a way similar to the one adopted in the last decade, was simulated for a time span of 200 years. On top of this reference scenario a number of different spacecraft were supposed to fragment in selected epochs. Comparing the long term evolution in the cases with and without the additional fragments generated by the artificially introduced fragmentation, the effect of this particular fragmentation on the environment was evaluated.

Performing this kind of comparisons over many spacecraft with different masses and orbital elements allows to understand the effects of selected *typical* fragmentations on the long term evolution of the debris population, as a function of the main driving

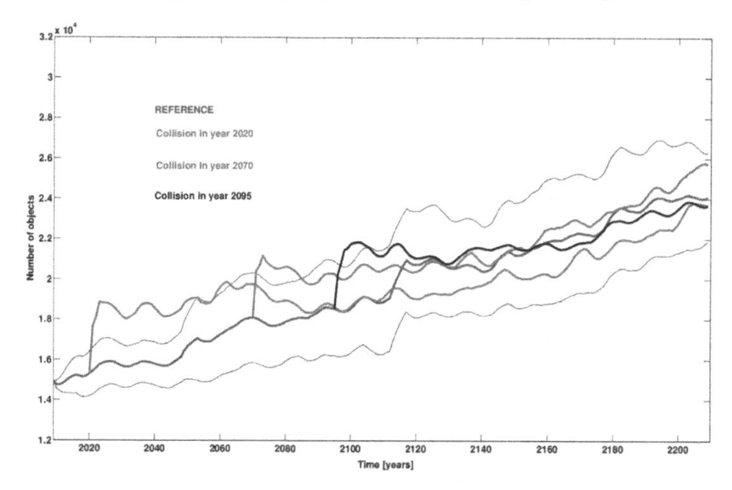

Figure 1. Number of objects larger than 10 cm in the reference scenario (blue line) and in 3 cases where a fragmentation of Envisat happens at three different epochs: 2020 (red line), 2070 (magenta line) and 2095 (black line). The thin blue lines represent the $\pm 1\sigma$ standard deviation of the reference case.

parameters. The final goal is to measure the danger represented by *typical* classes of space objects, in order to be able to rank the abandoned space objects in terms of the possible effects on the environment of the spacecraft and, conversely, in terms of the effect of the environment on the spacecraft itself.

2. Fragmentation ranking

The full results of the above mentioned Contract are under final review and cannot be shown here, also for lack of space. Here we intend to present the analysis methods devised for this study, along with a few sample results.

Figure 1 shows the number of objects larger than 10 cm in LEO produced in the reference case (thick blue line), averaged over 50 Monte Carlo runs. The thin blue lines represent the $\pm 1\,\sigma$ standard deviation of the Monte Carlo simulations. The other three curves represent the number of objects obtained in scenarios where the fragmentation of a spacecraft resembling the Envisat polar platform (mass $\simeq$ 8000 kg) is artificially introduced in the simulation. In particular it was assumed that this additional fragmentation would take place in the year 2020 (red line) or in year 2070 (magenta line) or in year 2095 (black line). The Envisat fragmentation is supposed to take place in a decaying orbit, i.e., the same spacecraft is always fragmented but at different altitudes according to the epoch, from about 760 km of altitude in the year 2020 to about 695 km in 2095.

Contrary to what one might expect, the final number of objects in all the four cases is statistically the same (i.e., all well within the $\pm 1\sigma$ standard deviation bounds). This means that, *in the long run*, even the fragmentation of a very large spacecraft leaves no noticeable signature on the environment. The reason for this outcome is that the reference evolution is highly stochastic and is dominated by a large number of fragmentations (on average one every 5 years). Therefore, the effects of our additional Envisat-like fragmentation get soon "diluted" in the vast number of background fragments and leave almost no trace after 200 years. On the other hand, the situation can be different in the "interim" regime, in the orbital regions in the vicinity of the Envisat fragmentation, during the few decades following the event; these shorter term effects have been studied too and will be described elsewhere.

In order to analyze many different long term evolution scenarios, it is necessary to find an evaluation norm to order the events in terms of their danger for the environment. For this purpose we devised a norm able to quantify and easily visualize these results. Given an underlying "reference" scenario and a "fragmentation" scenario in which the simulation of a particular fragmentation is added, the growth of the population of the "fragmentation" scenario w.r.t. the "reference" one can be quantified by:

$$C_i = \left| \frac{n_{FRAG}(i) - n_{REF}(i)}{\sigma_{REF}} \right| \tag{2.1}$$

where n_{FRAG} is the number of objects in the fragmentation case, n_{REF} is the number of objects in the reference case and σ_{REF} is the standard deviation of the reference Monte Carlo runs.

As an example, Fig. 2 plots the value of C_i in the case of the Envisat fragmentation scenarios shown in Fig. 1. The decreasing relative importance of the fragmentations happening in later years (both due to the larger number of "background" fragments and, mainly, to the lower altitude of the event) is clearly highlighted here.

A ranking of the danger represented by selected fragmentations can be easily expressed with a single number. In fact, the sum of the differences, weighted by the time interval, gives an indication of the criticality:

$$C^* = \sum_{i=1}^{N} \frac{C_i}{N} = \sum_{i=1}^{N} \left| \frac{n_{FRAG}(i) - n_{REF}(i)}{\sigma_{REF}} \right| /N \tag{2.2}$$

where N is the number of years in the simulation. In the ESA study a full ranking of a large number of different fragmentations is being elaborated and will be soon published in a forthcoming paper.

3. The Criticality of Spacecraft Index

Similarly to what is currently done in the Near Earth Object field with the Palermo Scale (Chesley *et al.* 2002), a quantitative measure of the criticality of the artificial objects in Earth orbit would be important for a number of reasons. In particular, it would help

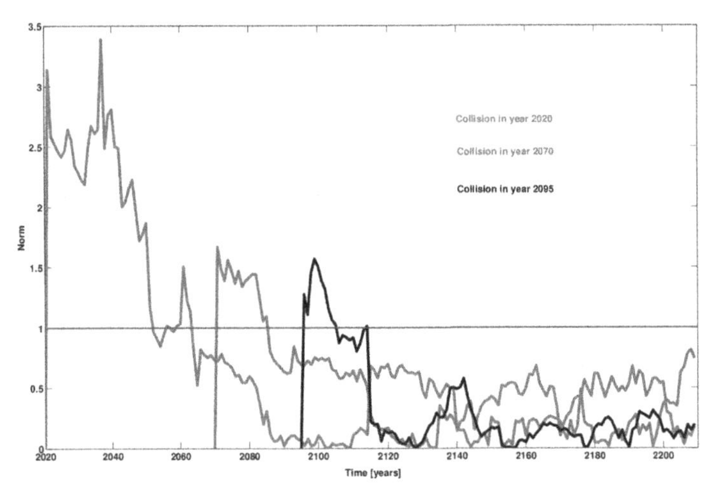

Figure 2. The time evolution of the norm of Eq. 2.1, computed for the three Envisat-like fragmentations, shown in Fig. 1.

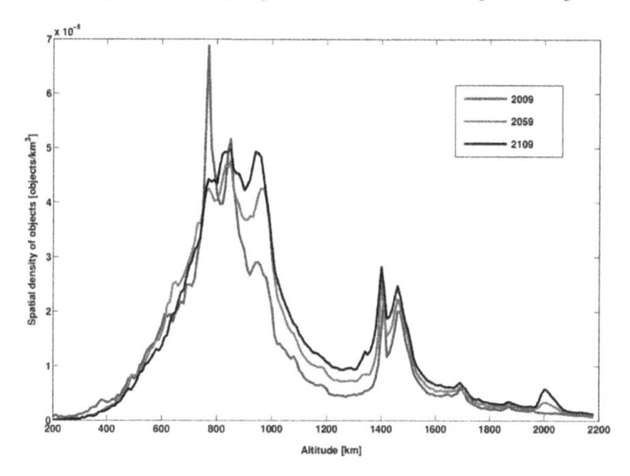

Figure 3. Spatial density of objects as a function of altitude in three different epochs: 2009 (blue line), 2059 (red line) and 2109 (black line).

spacecraft operators to easily assess the current collision risk faced by a given asset in space; moreover it would be a measure to evaluate the potential danger posed to the environment in case an object would be abandoned in space and therefore could be used as a ranking to establish active removal priorities. Finally, as is the case of the Palermo Scale, it could be useful to spread the public awareness of the danger posed by space debris by allowing a wider non-specialist audience to catch, with a single number, the environmental criticality of a given spacecraft.

For this purpose we developed an index, called the Criticality of Spacecraft Index (hereafter, CSI). In the following sections the factors entering in the definition of the CSI are described and discussed.

Environment dependence. The environment is considered in terms of the spatial density of objects as a function of time and altitude. It is well known that the collision probability is higher in regions where a higher concentration of objects is found. For this purpose, a reference simulation of the evolution of the space debris environment, spanning 200 years (considering the population of objects larger than 10 cm from the MASTER 2009 population), was performed with SDM 4.2 (Rossi *et al.* 2009).

For this reference case, a scenario where the space activities are performed in a way similar to the one adopted in the last decade is simulated. In particular, the traffic launch repeats an 8-year cycle representing the current launch pace, an 8-year lifetime is assumed for future spacecraft, no new explosions are considered and no avoidance maneuvers are performed. A post mission disposal scenario according to the 25-year rule is adopted, with a 60 % compliance to this rule (i.e., only 60% of the spacecraft are actually de-orbited at end-of-life). This reference scenario was simulated with 50 Monte Carlo runs and averages of the evolutions were computed. In particular, the resulting spatial density of objects as a function of altitude was recorded every year and stored. As an example, Fig. 3 shows the spatial density of objects larger than 10 cm as a function of altitude, for three different epochs.

The way in which the spatial density is taken into account in the CSI is as follows: given an epoch (or interval of time) and the orbital altitude, h, of the object under consideration, the spatial density, D, is taken from the stored values and normalized to the value of the maximal spatial density in the initial year 2009, that is the one at the altitude of 770 km, D_0. Therefore, the multiplicative contribution to the CSI accounting

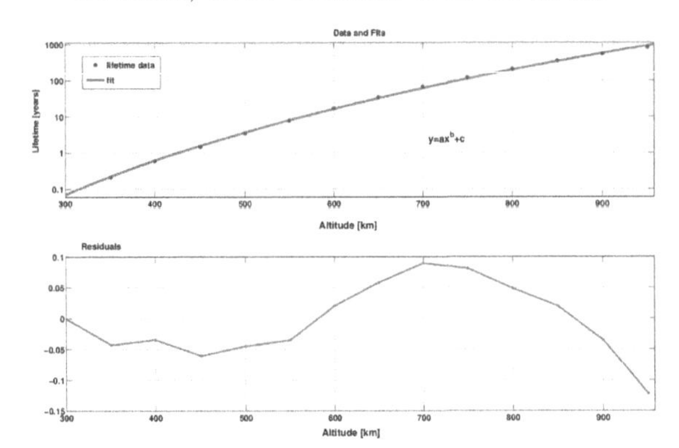

Figure 4. Orbital lifetime of a sample object with $A/M = 0.012\ \mathrm{m^2 kg^{-1}}$ as a function of altitude. The upper panel shows a power law fit to the lifetime values and the bottom panel shows the residuals of the fit (see text for details).

for the environment density is given by:

$$\frac{D(h)}{D_0}.$$

Lifetime dependence. The danger represented by an object left in space and the probability that it will be destroyed by a collision is a function of the time that this object will spend in space. Moreover the long term consequences of a fragmentation are much more severe for events happening at high altitudes where the cleaning effects of the atmosphere are not effective. Therefore the residual lifetime of an object is an important parameter to include in the index computation.

The lifetime of the objects, as a function of the orbital altitude h, is estimated from an average lifetime given by the curve shown in Fig. 4, computed assuming an area over mass ratio, $A/M = 0.012\ \mathrm{m^2kg^{-1}}$, which reflects the average value observed for intact objects. An average solar flux between 110 and 130 units is considered. The lifetime curve was computed as a power law fit of the form:

$$\log(life) = ah^b + c \tag{3.1}$$

where a, b and c are the coefficients of the fit. Therefore, given an object with mean altitude h (note that, for LEO objects having low eccentricity, the semimajor axis can be used as a good approximation of h) the index term accounting for the lifetime is given by:

$$\frac{life(h)}{life(h_0)}$$

where $life(h)$ is computed with Eq. 3.1 and the normalizing value is computed, as a default, for $h_0 = 1000$ km.

Mass dependence. Along with the altitude of the event, the other most influential parameter in determining the environment consequences of a given fragmentation is the mass, M. This fact is taken into account in the index by including the term:

$$\frac{M}{M_0}$$

where the normalizing factor is arbitrarily taken as $M_0 = 10000$ kg. An alternative explored is to use the same exponent found in the NASA breakup model by expressing the mass term as:

$$\left(\frac{M}{M_0}\right)^{0.75}$$

It has been checked that the adoption of the 0.75 exponent does not change significantly the results. Therefore, for the sake of simplicity, in the following the value of the exponent used is 1.

Inclination dependence. It is well known that the collision risk is maximum for high inclination orbits that can cross all the other orbits in their range of altitude and that can lead to very high mutual inclinations (and therefore high impact velocities) due to the precessing orbital planes. For this reason an inclination (i) dependence is included in the CSI, in the form:

$$\frac{1 + k\,\Gamma(i)}{1 + k}$$

where:

$$\Gamma = \frac{(2 - \cos(i)) - 1}{2}$$

and $k = 0.6$ since the typical flux of debris on an almost equatorial orbit is about 60% of the flux on a polar orbit.

Index definition. Combining the terms described above, the final definition of the CSI reads as:

$$\Xi = \frac{M(h)}{M_0}\frac{D(h)}{D_0}\frac{life(h)}{life(h_0)}\frac{1 + k\,\Gamma(i)}{1 + k}, \tag{3.2}$$

where we denote with Ξ the value of the CSI. The definition was kept as simple as possible in order to allow its easy application and understanding by the largest possible community; the larger the value of the CSI, the more dangerous to the environment is *an abandoned object.*

Note that, due to the normalization, for all the space objects in our population $\Xi < 1$, although in theory it is not bound by 1. In order to consider possible time variations in the environment, the CSI could also be computed taking into account the average density of objects over an interval of time (e.g., 10 years) instead of the single value in the year of reference.

In order to get an idea of the values of the CSI that are to be expected, we list in Table 3 the 15 objects having the largest Ξ in our population (MASTER 2009). Given the illustrative purpose of this discussion, in the Table are not listed “names” of objects, but rather physical and orbital characteristics. This allows to identify “families” of objects particularly dangerous for the environment and prone to potential active debris removal missions.

As it can be noticed, all the objects have large mass, well above one metric ton. However, it is also worth noting that the ranking is not just dominated by the mass, given that the semimajor axis (i.e., the mean altitude) plays a significant role, and that all the objects in the Table have high equatorial inclinations.

In Fig. 5 the objects having the highest 100 values of the CSI appear as circles with diameters proportional to their mass, and the panels show their distributions in terms of the mean altitude and inclination. In these planes the most critical objects can naturally be grouped into “families” according to their missions and launching countries.

It is worth noting how the orbital distributions shown in Figs. 5 compare nicely with Fig. 13 of Liou (2011), where the orbital distribution of the existing LEO R/Bs and S/Cs having highest mass and collision probability products (computed with 100 Monte Carlo runs of LEGEND) is shown. This confirms that the CSI can be considered as a reliable

Table 1. List of the 15 objects having the largest values of the index in our MASTER 2009 population. Objects in boldface are upper stages, the others are satellites.

	a [km]	ecc	inc [deg]	Mass [kg]	Ξ
1	7372.2	0.002	99.25	9000.0	**0.313**
2	7365.7	0.003	64.98	4500.0	0.163
3	7343.1	0.003	64.99	4955.0	0.161
4	7342.1	0.004	65.04	4955.0	0.160
5	7355.2	0.006	64.49	4500.0	0.154
6	7346.5	0.007	65.28	4500.0	0.151
7	7342.9	0.006	64.95	4500.0	0.146
8	7349.3	0.005	64.81	4500.0	0.145
9	7332.1	0.005	64.98	4955.0	0.143
10	7222.0	0.001	71.00	9000.0	**0.139**
11	7221.6	0.000	70.98	9000.0	**0.139**
12	7336.6	0.004	64.70	4500.0	0.135
13	7227.3	0.002	70.88	8226.0	**0.135**
14	7335.5	0.006	64.86	4500.0	0.134
15	7333.3	0.009	65.08	4500.0	0.131

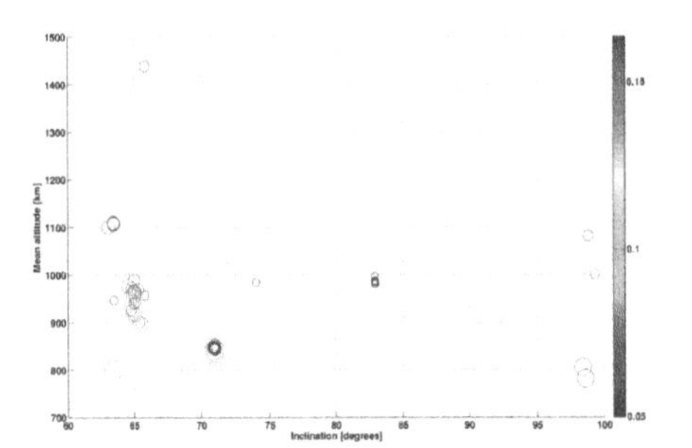

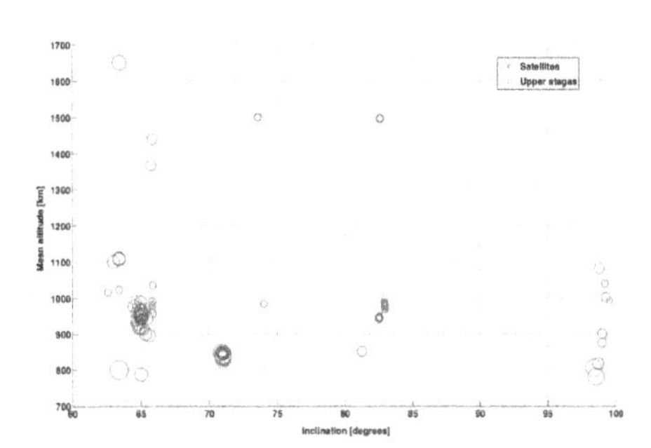

Figure 5. Distribution of the first 100 objects in the index ranking as a function of inclination and of the mean altitude. In the left panel, the color of the circles gives Ξ, according to the color coding on the right. In the right panel, the blue circles represent satellites while the red ones represent upper stages. The size of the circles is proportional to the mass of the object.

indicator of the actual risk faced and posed by objects in LEO and as such is a good analytic, fast and easy-to-compute proxy for active removal strategies planning.

The histograms in Fig. 6 show the distribution in altitude and inclination, respectively, of all the fragmentations recorded in the 50 Monte Carlo runs performed with SDM to derive the spatial density distributions. Therefore these distributions highlight the most dangerous zones in LEO in terms of the catastrophic collision risk. Within these histograms the red dots show the orbital parameters of the top 500 objects in the CSI ranking. It can be noticed how these objects clearly populate the bins where the highest number of collisions occur, once again showing that the CSI is a good proxy for the collision risk faced by a spacecraft in LEO.

We can conclude that the Criticality of Spacecraft Index presented in this paper is a simple, fast, easy-to-compute analytic tool able to rank the abandoned space objects in terms of the danger they can pose to the environment (or, conversely, in terms of the risk they face from the environment), taking into account their orbital and physical characteristics. Moreover, it is able to catch the main known features of the in-orbit collision risk and can be viewed as a good proxy to prioritize active debris removal planning.

Acknowledgements

The study described in the paper was performed in the framework of the contracts: *SPARC-Space Threats and Critical Infrastructures: Risks and Countermeasures*

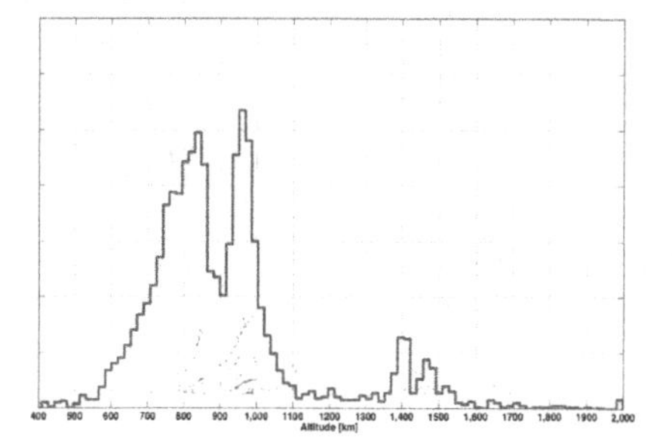

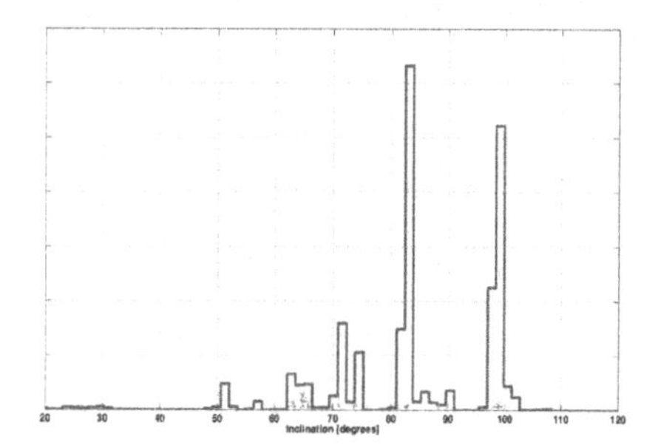

Figure 6. Altitude (left panel) and inclination (right panel) relative distribution of all the fragmentations recorded in the SDM runs of the reference case used to derive the objects spatial density. The red dots show the altitude and inclination distribution of the first 500 objects in the CSI ranking.

(HOME/2011/CIPS/AG/4000002119) and *Assessment Study for Fragmentation Consequence Analysis for LEO and GEO Orbits*, ESA/ESOC No. 4000106534/12/F/MOS. The authors wish to thank L. Anselmo and C. Pardini for useful discussions regarding the index computation.

References

Chesley, S. R., Chodas, P. W., Milani, A., Valsecchi, G. B. & Yeomans, D. K. 2002, *Icarus*, 159, 423

Liou, J.-C. 2011, *Advances in Space Research*, 47, 1865

Rossi, A., Anselmo, L., Pardini, C., Jehn, R., & Valsecchi, G. B. 2009, *Proceedings of the Fifth European Conference on Space Debris, ESA SP-672*, (ESA Communication Production Office, Noordwijk, The Netherlands)

Complex Planetary Systems
Proceedings IAU Symposium No. 310, 2014
Z. Knežević & A. Lemaitre, eds.

doi:10.1017/S1743921314008023

Dynamical evolution of near-Sun objects

Vacheslav V. Emel'yanenko[1] **and Mikhail A. Shelyakov**[1]

[1]Institute of Astronomy, Russian Academy of Sciences,
48 Pyatnitskaya str., 119017, Moscow, Russia
email: vvemel@inasan.ru, mshelyakov@inasan.ru

Abstract. The dynamical evolution of short-period objects having perihelia at small heliocentric distances is discussed. We have investigated the motion of multiple-apparition members of the Marsden and Kracht sungrazing groups. The orbital evolution of these objects on timescales < 10 Kyr is mainly determined by the Kozai-Lidov secular perturbations. These objects are dynamically connected with high-inclination near-Earth objects. On the other hand, we have found several observed near-Earth objects that evolve in the same way, reaching small perihelion distances on short timescales in the past .

Keywords. comets: sungrazing, dynamical evolution, near-Earth objects

---

1. Introduction

It is well known that near-Earth objects evolve frequently to orbits with small perihelion distances (Farinella *et al.* 1994, Gladman *et al.* 2000, Foschini *et al.* 2000, Marchi *et al.* 2009). It is estimated that up to $\sim 70\%$ of near-Earth objects collide with the Sun during their orbital evolution (Marchi *et al.* 2009). The solar tide, thermal stresses and interaction with the solar atmosphere are expected to be severe for objects passing near the Sun.

Modification of the surface composition at past sungrazing conditions may be recorded in spectral reflectance properties of some existing near-Earth asteroids. But only a few very rare cases (e.g., comet P96/Machholz, in ~ 1 Kyr (Bailey *et al.* 1992); 2004 LG, in $\sim$ 3.5 Kyr (Vokrouhlicky & Nesvorny 2012)) have definite predictions about solar encounters of real objects in the past. All objects detected as asteroids had perihelion distances $q > 0.07$ AU at the moment of discovery. However, many comets have been observed near the Sun, and some of them were registered in a few apparitions. These multiple-apparition objects are usually called sunskirting comets (their perihelion distances equal ~ 0.05 AU).

In this paper, we discuss the dynamical connection between these periodic near-Sun comets and typical near-Earth objects. We show that sunskirting comets evolve gradually to orbits with larger perihelion distances that are inherent to observed near-Earth objects. On the other hand, we have found several real near-Earth objects that were near the Sun in the recent past.

2. Dynamical evolution of sunskirting comets

We determined orbits of eight multiple-apparition sunskirting comets with the best observational material (2 members of the Kracht group, 1 member of the Kracht 2 group and 5 members of the Marsden group) (Emel'yanenko & Shelyakov 2014). Two of them have observations in three apparitions, but no significant nongravitational effects were found in our calculations. To study the dynamical evolution of these objects we considered

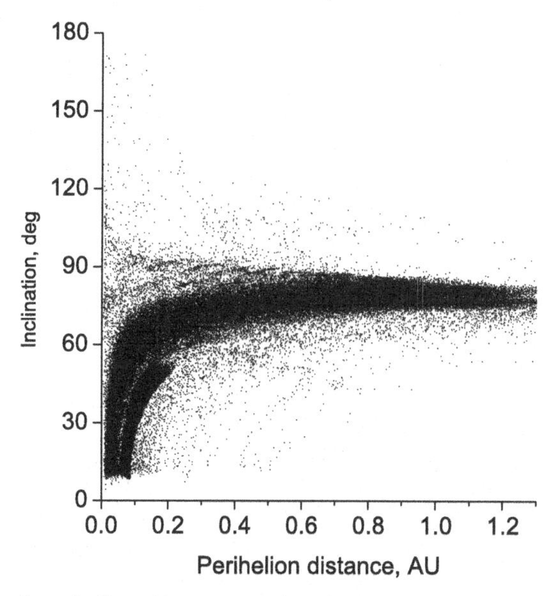

Figure 1. Changes of perihelion distances and inclinations for sunskirting objects. Data for orbits from the confidence region are plotted every 100 years.

a set of 200 initial orbits from the confidence region for each object, using the covariance matrix. These orbits were integrated forward taking account of perturbations from all planets. The dynamical evolution of test particles was calculated using the symplectic integrator (Emel'yanenko 2007). Particles were removed from integrations when they collide with planets or $q < 0.005$ AU or the semimajor axis $a > 50$ AU.

Fig. 1 shows the changes of perihelion distances and inclinations for the studied objects for 10000 years. The planetary perturbations lead to drastic changes of orbits on a relatively short interval of time. The black region in Fig. 1 corresponds to those orbits which are the most probable for the former sunskirting objects in the future.

The systematic increase in perihelion distances and inclinations is caused by the Lidov-Kozai secular perturbations. Fig. 2 shows the evolution of the semimajor axis, eccentricity, inclination and argument of perihelion for the sungrazing object C/2004 R4 = C/2007 Y4, representing a typical example of the Lidov-Kozai circles for changes of orbital elements.

3. Recent approaches of observed asteroids to the Sun

The results on the dynamical evolution of sunskirting comets (Fig. 1) show that the present high-inclination near-Earth objects may have experienced close approaches to the Sun in the recent past. To analyze this question, we integrated back orbits of several observed high-inclination near-Earth objects. We applied the same numerical methods as above for sunskirting comets. Some remarkable examples of asteroids with recent passages near the Sun are shown in Fig. 3, 4. 2003 EH1 had $q \sim 0.09$ AU 1.4 Kyr ago. 2012 FZ23 had $q \sim 0.06$ AU 2.8 Kyr ago. We have found also other observed near-Earth objects that had definitely solar encounters in the recent past (2010 KY27, $q \sim 0.08$ AU in 0.8 Kyr; 2001 AU43, $q \sim 0.12$ AU in 7.9 Kyr; 2008 GV3, $q \sim 0.12$ AU in 9.0 Kyr).

It is natural to assume that the solar tide, thermal stresses and interaction with the solar atmosphere could lead to disruption and surface modification of bodies near the Sun. This may be important in determining the population of near-Earth objects. Unfortunately, there are no direct observational indications how to distinguish objects

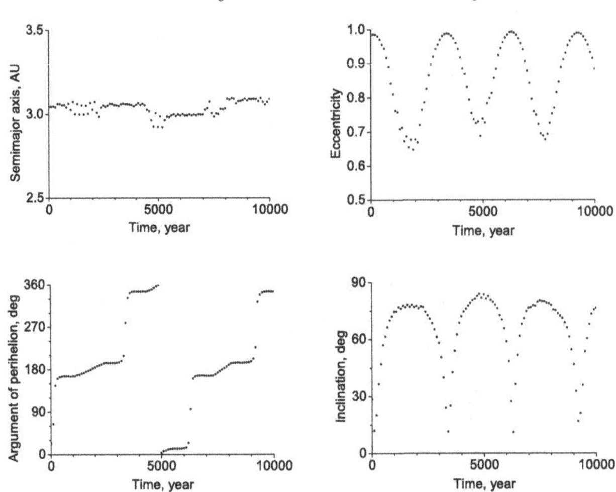

Figure 2. Changes of the semimajor axis, eccentricity, inclination and argument of perihelion for the sungrazing object C/2004 R4 = C/2007 Y4. Data are plotted every 100 years.

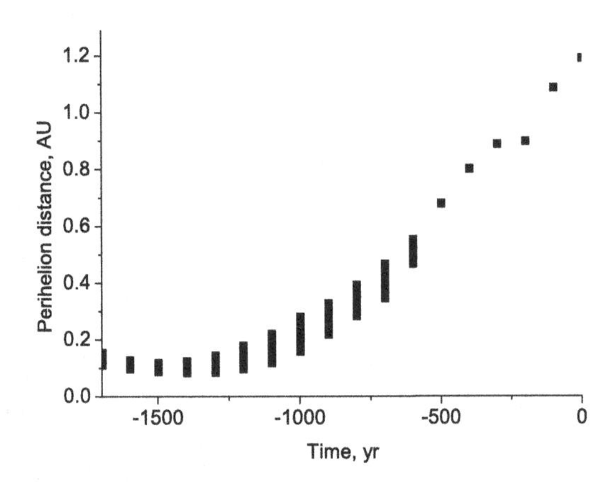

Figure 3. Changes of q for 2003 EH1. Data for orbits from the confidence region are plotted every 100 year.

that have visited the near-Sun region. In particular, Jewitt (2013) concluded that his observations provided no evidence to suggest that physical properties of small-perihelion objects were measurably influenced by the heat of the Sun.

We hope that our results on new observed near-Earth objects, moving in orbits with small perihelion distances in the recent past, will stimulate additional investigations in this direction. Special observations of these objects are very desirable.

4. Conclusions

1. The orbital evolution of the Kracht and Marsden group members on timescales < 10 Kyr is mainly determined by the Kozai-Lidov secular perturbations. These objects are dynamically connected with high-inclination near-Earth objects.

2. We have found several observed near-Earth objects that evolve in the same way, reaching small perihelion distances on short timescales in the past.

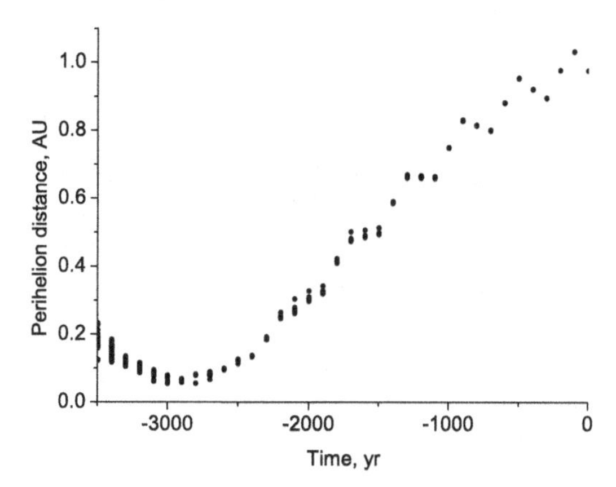

Figure 4. Changes of q for 2012 FZ23. Data for orbits from the confidence region are plotted every 100 years.

References

Bailey, M. E., Chambers, J. E., & Hahn, G. 1992, *A&A*, 257, 315

Emel'yanenko 2007, *Celestial Mechanics & Dynamical Astronomy*, 98, 191

Emel'yanenko, V. V. & Shelyakov, M. A. 2014, *Icarus*, submitted

Farinella, P., Froeschle, Ch., Froeschle, Cl., Gonczi, R., Hahn, G., Morbidelli, A., & Valsecchi, G. B. 1994, *Nature*, 371, 314

Foschini, L., Farinella, P., Froeschle, Ch., Gonczi, R., Jopek, T. J., & Michel, P. 2000, *A&A*, 353, 797

Gladman, B., Michel, P., & Froeschle, Ch. 2000, *Icarus*, 146, 176

Jewitt 2013, *AJ*, 145, 133

Marchi, S., Delbo, M., Morbidelli, A., Paolicchi, P., & Lazzarin, M. 2009, *MNRAS*, 400, 147

Vokrouchlicky, D. & Nesvorny, D. 2012, *A&A*, 541, A109

Complex Planetary Systems
Proceedings IAU Symposium No. 310, 2014
Z. Knežević & A. Lemaitre, eds.

doi:10.1017/S1743921314008035

Automated Classification of Asteroids into Families at Work

Zoran Knežević[1], Andrea Milani[2], Alberto Cellino[3], Bojan Novaković[4], Federica Spoto[2], and Paolo Paolicchi[5]

[1]Astronomical Observatory, Volgina 7, 11060 Belgrade, Serbia
email: zoran@aob.rs

[2]Dept. of Mathematics, University of Pisa, Largo Pontecorvo 5, 56127 Pisa, Italy
email: milani@dm.unipi.it

[3]INAF-Osservatorio astrofisico di Torino, 10025 Pino Torinese, Italy
email: cellino@oato.inaf.it

[4]Dept. of Astronomy, University of Belgrade, Studentski trg 16, 11000 Belgrade, Serbia
email: bojan@matf.bg.ac.rs

[5]Dept. of Physics, University of Pisa, Largo Pontecorvo 3, 56127 Pisa, Italy
email: paolicchi@df.unipi.it

Abstract. We have recently proposed a new approach to the asteroid family classification by combining the classical HCM method with an automated procedure to add newly discovered members to existing families. This approach is specifically intended to cope with ever increasing asteroid data sets, and consists of several steps to segment the problem and handle the very large amount of data in an efficient and accurate manner. We briefly present all these steps and show the results from three subsequent updates making use of only the automated step of attributing the newly numbered asteroids to the known families. We describe the changes of the individual families membership, as well as the evolution of the classification due to the newly added intersections between the families, resolved candidate family mergers, and emergence of the new candidates for the mergers. We thus demonstrate how by the new approach the asteroid family classification becomes stable in general terms (converging towards a permanent list of confirmed families), and in the same time evolving in details (to account for the newly discovered asteroids) at each update.

Keywords. asteroids, asteroid families, automated classification

---

1. Introduction

The classification of asteroids into families becomes increasingly challenging as the asteroid datasets continue to grow at an unprecedented rate. The problem is not only with the need to handle huge amounts of data, but also with different and rapidly changing outcomes of classification attempts which tend to create confusion with the users.

In order to cope with these problems we have recently proposed a new approach to the asteroid family classification by combining the classical HCM method with an automated procedure to add new members to existing families (Milani *et al.* 2014). The basic idea is to setup a classification which can be revised only once in a while, but being automatically updated every time the dataset is significantly increased. In practice, our approach consists of several steps to segment the problem and handle the very large amount of data in the most efficient and accurate manner. We use the proper elements first, thus defining dynamical families, then use information from absolute magnitudes, albedos and colors as either confirmation or rejection.

In the following we briefly present all these steps. We then show some results of the application of our procedure in three subsequent updates making use of only the automated step of attributing the newly numbered asteroids to the known families.

2. Method

Our procedure for family identification is based on the classical Hierarchical Clustering Method (HCM), used in most previous families searches since the pioneering work by Zappalà *et al.* (1990) and later improved in a number of papers (Zappalà *et al.* 1994, Zappalà *et al.* 1995, Milani *et al.* 2010, Novaković *et al.* 2011, Masiero *et al.* (2011), Carruba *et al.* (2013)).

We begin our classification with segmenting the problem, that is by dividing the catalog of asteroid osculating elements into parts that can be considered separately. We first divide the asteroid belt into zones corresponding to different intervals of heliocentric distance, delimited by the Kirkwood gaps wide enough to exclude family classification across the boundaries. Next, we split the most populous zones in the central part of the asteroid belt by the value of proper sin I, between a moderate inclination region $\sin I < 0.3$ and a high inclination region $\sin I > 0.3$. Finally, we split the sample in the low-inclination regions of the same central zones by the absolute magnitude, so that in the first step of our procedure we consider only objects having absolute magnitudes H brighter than H_{comp}, with H_{comp} roughly corresponding to the local completeness limit (see Table 2 in Milani *et al.* 2014). Thus, after this first step, in the central zones of the belt we get the "core" families that consist of only the brightest/largest members (red points in Fig. 1 of Milani *et al.* 2014). They represent the inner skeletons of larger families, whose other members are to be identified in the following steps of the procedure. Note, that in the zones with less objects (Hungarias, zones beyond 2:1 mean motion resonance with Jupiter, and high-inclination zones) we identified families by the direct application of the HCM procedure, without the multistep approach.

The second step of the procedure in the low-I portions of the populous central zones is the classification of faint asteroids not used in the first step, that is attaching them to the previously established family cores. We allowed only single links for this attachment, to avoid chaining which would result in merging most families together. Consequently, in step 2 we attribute to the core families the asteroids having a distance from at least one member of the same core family not larger than the critical (threshold) distance. The result is that the families are extended in the absolute magnitude/size dimension, but not much in proper elements space, especially not in proper semimajor axis (green points in Fig. 1 of Milani *et al.* 2014).

As an input to the third step we use the intermediate background asteroids, defined as the set of all the objects not attributed to any family in steps 1 and 2. Families identified at this step are formed by the population of asteroids left after removing from the proper elements data set the already identified family members. We can distinguish two possible outcomes of this step: families can either be fully independent new families having no relation with the families identified previously, or they may be found to overlap step 1+2 families and form satellite families of smaller objects surrounding family cores (yellow points in Fig. 1 of Milani *et al.* 2014).

With the same algorithm of step 2, in step 4 we repeat a single-link attribution to all the families in the extended list of families formed by adding the step 3 families to the list of core families of step 1. Note that with this procedure a small number of asteroids with double classification is unavoidable; if an asteroid is found to be attributed to more than one family, it belongs to an intersection. The multiple intersections between particular families could be due to the occurence of families at the boundaries between high and low inclination regions in central zones where there is no natural gap (due e.g. to a secular resonance) between these regions, like in the case of family of (729) Watsonia. This is an artifact of our decomposition in zones and needs to be corrected by merging the intersecting families.

More importantly, family intersections occur also when a new family appears as an extension of a family already identified at steps 1 and 2, with intersections near the mutual boundary. Again, a remedy for such a situation is merging of "satellite" families, where, in general, for the merging of two families we require multiple intersections. Visual inspection of the three planar projections of the intersecting families in terms of the proper elements is used to assess the ambiguous cases.

The merging of families constitutes step 5 of our procedure. As an example, let us quote results of the first attempt to family classification by means of our procedure, as given in Milani *et al.* (2014): out of 77 families generated in step 3, 34 have been considered to be satellite families (even 2 core families of step 1 have been found to be satellite of other core families and thus merged); the other 43 families have been left as independent families, consisting mostly of smaller asteroids. There were of course dubious cases, with too few intersections to perform merger. In principle, as the list of asteroids attached to established families grows, the intersection can increase. In some cases the new intersections will support merge previously not implemented, some will certainly open new problems. In any case to add a new merger is a delicate decision which at the moment remains the only step of the procedure we are unable to automatize.

The final step (step 6) of our procedure of asteroid family classification is motivated by the rapid growth of the proper elements database, which results in any family classification becoming quickly outdated. Thus we devised an automatic update of the current family classification, which consists in repeating the attribution of asteroids to the existing families every time the catalog of synthetic proper elements is updated. What we repeat is actually step 4, thus the lists of core families members (found in step 1), of members of smaller families (from step 3), and also the list of already implemented mergers (from step 5) are kept unchanged.

Let us emphasize here that the purpose of this final step is to maintain the general validity of the classification for many years, without the need for repeating the entire procedure. With time, the new data will require to repeat also step 5, to reconsider the list of small families, confirm some of them as statistically significant, discard others as statistical flukes, to decide on pending mergers if intersections increase enough or otherwise new data give enough reason for such a decision, and so on. In brief, we must monitor as the classification is updated and perform non-automated changes whenever we believe there is enough evidence to justify them.

In the next section we shall show how our classification upgrade works, by reviewing the results of the non-automated step 5 application after 3 iterations of automated step 6, during which the catalog of proper elements increased by 48,117 objects or $\sim 14.3\%$.

3. Classification upgrade

As a result of the automatic attribution step in the previous upgrades, the number of family members increased by 10 355 members to 97 440. In the same time, the number of intersections among the 128 families increased from 29 to 48. Thus we had to analyse these intersections, to decide if in some cases the number of intersections among two families has increased enough to suggest a merge.

The case in which a merge was suggested most convincingly was that of the families of (1040) Klumpkea and of (3667) Anne-Marie. In the initial classification there were

10 intersections, now they have grown to 19, which means that out of the total of 30 members of 3667, the majority of members of the smaller family has been attached also to the larger one. Since the inspection of projections in the three proper elements planes did not contradict the proposed merger, we merged the two families and removed 3667 from the list. Note that there is also a new intersection between families 1040 and 29185, but this is not enough to perform a merger at this update.

Another case we found significant enough for a merger involves the families of (375) Ursula and (2967) Vladisvyat, now with 3 intersections. In the original classification there were no intersections, but this had already been pointed out as a future satellite family case by the overlapping box method (see Milani *et al.* 2014, [Section 4.3.2]). The third case is that of the small satellite family 6138 with the family of (135) Hertha: the intersections have grown from 2 to 4, which in this case was enough to decide on merger.

As a result of the 3 mergers described above, the number of families has decreased to 125 and the number of intersections to 22. The remaining ones are mostly due to possible satellite families of the large families 15, 221, 135, 10, and 2076.

An interesting case is an intersection of the family of (5) Astraea with the small family 4945, which was already suggested by the overlapping box method. This latter family appears to form, together with some other small families, a structure that extends along the secular resonance $g - 2\,g_6 + g_5$ in which most of the big family is locked, as predicted by Milani & Knežević (1994) [Figure 9].

An increase of the number of members in a proposed small family can be considered as confirmation of its statistical significance. This growth occured for almost all the families in our classification, with only 5 exceptions. The worst case in this sense is family 3460, which remained unchanged at its original 52 members: this family could be a satellite of family of (24) Themis at higher e and strongly affected by the 2/1 mean motion resonance with Jupiter. Four tiny families (with < 30 members) have also failed to grow: 20494, 1101, 6355, 10654; however, they all belong to the high inclination region in which the total number of asteroids is growing slowly, thus the significance of the lack of increase is dubious. Therefore, we are not removing them from the list yet.

In conclusion, we would like to stress that through the above described procedures and example results we demonstrate the automated classification at work and show how, by converging to a permanent list of confirmed families, the asteroid family classification becomes stable in general terms with the new approach, while evolving in the same time in details (to account for the newly discovered asteroids) at each update.

Acknowledgement

ZK and BN gratefully acknowledge support from the Ministry of Education, Science and Technological Development of Serbia through the grant OI176011.

References

Carruba, V., Domingos, R. C., Nesvorný, D., Roig, F., Huaman, M. E., & Souami, D. 2013, *MNRAS*, 433, 2075

Masiero, J. R., Mainzer, A. K., Grav, T., *et al.* 2011, *Astrophys. J.* 741, 68

Milani, A. & Knežević 1994, *Icarus*, 107, 219

Milani, A., Knežević, Z., Novaković, B., & Cellino, A. 2010, *Icarus*, 207, 769

Milani, A., Cellino, A., Knežević, Z. *et al.* 2014, *Icarus*, 239, 46

Novaković, B., Cellino, A., & Knežević, Z. 2011, *Icarus*, 216, 69

Zappalà, V., Cellino, A., Farinella, P., & Knežević, Z. 1990, *Astron. J.*, 100, 2030

Zappalà, V., Cellino, A., Farinella, P., & Milani, A. 1994, *Astron. J.*, 107, 772

Zappalà, V., Bendjoya, P. H., Cellino, A., Farinella, P., & Froeschlé, C. 1995, *Icarus*, 116, 291

Complex Planetary Systems
Proceedings IAU Symposium No. 310, 2014
Z. Knežević & A. Lemaitre, eds.

doi:10.1017/S1743921314008047

Hill Stability in the Full 3-Body Problem

D. J. Scheeres

Abstract. Hill stability cannot be easily established in the classical 3-body problem with point masses, as sufficient energy for escape of one of the bodies can always be extracted from the gravitational potential energy. For the finite density, so-called Full 3-body problem the lower limits on the gravitational potential energy ensure that Hill stability can exist. For the equal mass Full 3-body problem this can be easily established, with the result that for any equal mass, finite density 3-body problem in or near a contact equilibrium, none of the components of the system can escape in the ensuing motion.

Keywords. celestial mechanics, methods: analytical, 3-body problem, Hill stability

1. Introduction

In Scheeres 2012 the Full N-body problem is introduced, distinguished from the traditional N-body problem in that each body has a finite density, and hence two bodies cannot come arbitrarily close to each other. The finite-sized bodies are assumed to be rigid, and only exhibit contact forces between each other of friction and coefficient of restitution. This means that relative equilibria must also include resting configurations, with all relative motion between components zeroed. This both results in new possible relative equilibrium configurations of an N-body system, with components resting on each other, and also places lower limits on the gravitational potential energy. This model has been developed to describe the relative mechanics and dynamics of self-gravitating rubble pile asteroids, where the relative forces between components are weak enough for the rigidity of the components to not be compromised. In Scheeres 2012 it is shown how this simple change provides a drastic modification of the stable states of the 2 and 3 body problems. For the 3-body problem under the restriction that the bodies are equal mass spheres these relative equilibria are shown in Fig. 1, and in addition to the classical Euler and Lagrange solutions also include the so-called Resting Euler, Resting Lagrange, Transitional, Transverse and Aligned. The energetic stability of these solutions are indicated on the figure, with light being stable and dark being unstable. In Fig. 2 we show the energy / angular momentum chart of these equilibria, which traces out the total (normalized) energy as a function of angular momentum for each relative equilibrium. Wherever a line stops or the ends of two lines meet each other indicates a bifurcation point where the equilibrium may not longer exist or may transition to a different type. We note that the Lagrange and Transverse configuration lines are not shown although they are indicated at the top of the chart, and that these bifurcate into existence once the Lagrange Resting equilibrium loses its stability. Similarly the Euler Resting configuration line is not shown, but comes into existence when the Euler Resting equilibrium loses its stability. The current paper explores one aspect of this problem, namely whether Hill stability can be proven to exist for such a Full 3-body problem.

2. Background

The main results are established by using the minimum energy function defined in Scheeres 2012, which can be derived from Sundman's Inequality:

$$\mathcal{E} = \frac{1}{2}\frac{H^2}{I_H} + \mathcal{U} \leqslant E \tag{2.1}$$

where E and H are the total energy and angular momentum of the system, including translational and rotational motion, I_H is the total moment of inertia of the system about the fixed total angular momentum vector of the system, and $\mathcal{U}$ is the gravitational potential energy of the system. The function $\mathcal{E}$ is intimately related to Smale's Amended Potential (Smale 1970) and can be used to discover all of the previously mentioned relative equilibria as outlined in Scheeres 2012.

If we assume that the three bodies are of equal mass and density the system can be normalized such that

$$I_H = 0.3 + \frac{1}{3}\left(r_{12}^2 + r_{23}^2 + r_{31}^2\right) \geqslant 1.3 \tag{2.2}$$

$$\mathcal{U} = -\left[\frac{1}{r_{12}} + \frac{1}{r_{23}} + \frac{1}{r_{31}}\right] \geqslant -3 \tag{2.3}$$

where r_{ij} denotes the distance between bodies i and j. For a finite density system, the distance between the bodies has a lower limit such that $r_{ij} \geqslant 1$. This implies that the potential energy and moment of inertia have lower limits, as indicated above. For a point mass 3-body problem the lower limit on relative distance is $r_{ij} \geqslant 0$, meaning that there is no lower limit on the potential energy for that case.

Finally, we define the stability concept of interest for this paper.

Definition: *A system is Hill Stable if* $r_{ij} < C < \infty$ *for all time, both future and past.*

3. Hill Stability of the equal mass Full 3-Body Problem

Hill stability tells us whether the components of a gravitating system can escape with respect to each other. In the two body problem, both full and point mass, this is simply established by considering the total energy of the system: If it is positive the system is not Hill stable in general, while if it is negative it is Hill stable. For multi-particle systems this simple result no longer holds. In general, for the point mass N-body problem it can be shown that if a system has a positive energy, then the system *is not* Hill stable, and that at least one body must escape (Pollard 1976). Thus, a Necessary Condition for Hill Stability in the N-body problem is that $E < 0$.

Thus, while Hill Stability can exist in the point mass 3-body problem, it cannot be rigorously shown based on simple inequalities, such as is possible in the restricted 3-body problem. At best, as discussed in Marchal and Saari 1975, constraints can be given for when exchanges cannot occur. This difficulty of establishing a sufficient condition for Hill Stability in the point mass 3-body problem can be understood by considering the minimum energy function $\mathcal{E} \leqslant E$. If one of the bodies escapes from the other two, say body 1, then $r_{12}, r_{31} \to \infty$ while $r_{23} < \infty$ (note, since $E < 0$, r_{23} cannot become arbitrarily large without violating Sundman's Inequality). Then $\mathcal{E} \to -1/r_{23} \leqslant E$, leading to $r_{23} \leqslant -1/E$. In the point mass problem the only restriction is that $r_{23} \geqslant 0$, thus this inequality is always satisfied, providing no definite restrictions.

For the Full 3-body problem, with our constraint $r_{ij} \geqslant 1$, we note that the potential energy between each body, $\mathcal{U}_{ij} = -1/r_{ij}$ has a lower limit, $\mathcal{U}_{ij} \geqslant -1$. With this observation, we can present the following Theorem and Proof.

Theorem: *Consider an equal mass, finite density 3-body problem that has been normalized as in Eqns. 2.2 and 2.3. If $E < -1$ then it is Hill Stable.*

To prove, we first establish the result by contradiction and then show that systems that satisfy the constraint exist.

Proof. Given $E < -1$, assume that one of the bodies, say body 1, undergoes escape. Then $r_{12}, r_{31} \rightarrow \infty$, and $\mathcal{E} \rightarrow -1/r_{23}$. However, the Sundman inequality $\mathcal{E} \leqslant E < -1$ must still hold, leading to the inequality $r_{23} < 1$, a contradiction since in the normalized Full Body Problem we must have $r_{23} \geqslant 1$.

To show that systems with $E < -1$ can exist in the equal mass Full 3-body problem, consider Fig. 2, modified from Scheeres 2012. From this figure it is evident that all relative equilibrium configurations that have a resting component have an energy less than -1, for both stable and unstable configurations. Only orbital relative equilibria can have energies above this limit. Finally, note that the energy of the stable Aligned Configuration has an energy value < -1, for all finite distance between the components. □

Note that if a system is Hill stable, one cannot conclude that $E < -1$, making this only a sufficient condition. In fact, the particular solutions of the Lagrange and Euler relative equilibria are both Hill Stable (even though they are unstable to small variations in general), and both can exist at energies above -1.

The theorem leads directly to the following corollary.

Corollary: *All motions starting close enough to relative equilibrium in the Full 3-body problem that have resting elements are Hill Stable.*

Proof. The complete set of resting relative equilibria is presented in Fig. 2, and established through proof in Scheeres 2012. All of these equilibria have a total energy strictly less than -1. Thus, any deviation from a relative equilibrium can be made small enough such that the energy remains less than -1, and thus by the previous theorem are Hill stable. □

4. Discussion

The results presented in the above Theorem and Corollary have implications. Any equal mass 3-body system that undergoes rotational fission due to exogenous effects that increase its angular momentum over time will be Hill stable, and the components will not be able to escape from each other. This mirrors a similar result in Scheeres 2009 for the Full 2-body problem, where an equal mass resting configuration spun to fission is dynamically unstable, yet is Hill stable and cannot mutually escape. The implication (somewhat limited due to the strong model assumptions), is that an equal mass 'contact triple' asteroid will remain a contact triple, mirroring the more general results for two component bodies described in Jacobson & Scheeres 2011.

In analogy with the Full 2-body problem, these Hill stability results are not expected to exist in the more general problem where the bodies do not have equal mass. Thus, similar to the Full 2-body problem described in Scheeres 2009 there should exist limits on the relative masses of a Full 3-body problem for the relative equilibrium configurations to be Hill stable. Future research will probe this.

Equilibria in the Spherical, Full 3 Body Problem with Equal Size

Configuration	Name	Energetic Stability	Conditions
	Lagrange Resting	Stable	
	Euler Resting	Stable	For high enough H
	Aligned	Stable	Outer Solution
	Lagrange	Unstable	
	Euler	Unstable	
	Euler Resting	Unstable	For low H
	Aligned	Unstable	Inner Solution
	Transverse	Unstable	
	Transitional	Unstable	

Figure 1. Relative equilibria in the equal mass, finite density 3-body problem.

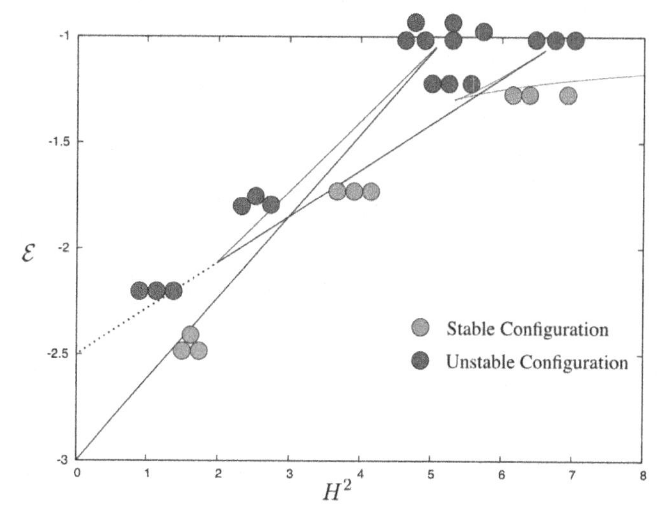

Figure 2. Chart showing the energy and angular momentum for all classes of relative equilibria in the equal mass Full 3-Body Problem. Cusps occur when the lines meet each other and indicate a bifurcation either creating or destroying relative equilibria.

References

S. A. Jacobson and D. J. Scheeres. *Icarus*, 214:161–178, July 2011.

C. Marchal and D. G. Saari. *Celestial mechanics*, 12(2):115–129, 1975.

H. Pollard. The Carus Mathematical Monographs, Providence: Mathematical Association of America, 1976.

D. J. Scheeres. *Celestial Mechanics and Dynamical Astronomy*, 104(1):103–128, 2009.

D. J. Scheeres. *Celestial Mechanics and Dynamical Astronomy*, 113(3):291–320, 2012.

S. Smale. *Inventiones mathematicae*, 10(4):305–331, 1970.

Complex Planetary Systems
Proceedings IAU Symposium No. 310, 2014
Z. Knežević & A. Lemaitre, eds.

doi:10.1017/S1743921314008059

Fragmentation of colliding planetesimals with water content

Thomas I. Maindl[1], Rudolf Dvorak[1], Christoph Schäfer[2] and Roland Speith[3]

[1]Department of Astrophysics, University of Vienna, Türkenschanzstraße 17, A-1180 Vienna, Austria
email: thomas.maindl@univie.ac.at, rudolf.dvorak@univie.ac.at

[2]Institut für Astronomie und Astrophysik, Eberhard Karls Universität Tübingen, Auf der Morgenstelle 10, 72076 Tübingen, Germany
email: ch.schaefer@uni-tuebingen.de

[3]Physikalisches Institut, Eberhard Karls Universität Tübingen, Auf der Morgenstelle 14, 72076 Tübingen, Germany
email: speith@pit.physik.uni-tuebingen.de

Abstract. We investigate the outcome of collisions of Ceres-sized planetesimals composed of a rocky core and a shell of water ice. These collisions are not only relevant for explaining the formation of planetary embryos in early planetary systems, but also provide insight into the formation of asteroid families and possible water transport via colliding small bodies. Earlier studies show characteristic collision velocities exceeding the bodies' mutual escape velocity which—along with the distribution of the impact angles—cover the collision outcome regimes 'partial accretion', 'erosion', and 'hit-and-run' leading to different expected fragmentation scenarios. Existing collision simulations use bodies composed of strengthless material; we study the distribution of fragments and their water contents considering the full elasto-plastic continuum mechanics equations also including brittle failure and fragmentation.

Keywords. minor planets, asteroids; planets and satellites: formation; solar system: general

---

1. Introduction

Most simulations of giant impacts use a strengthless material model (e.g., Canup et al. 2013) based on the fact that beyond a certain size (400 m in radius, Melosh & Ryan 1997) self-gravity dominates the material's tensile strength. In Maindl et al. (2014) we study colliding objects of a size close to this limit and compare strengthless material simulation ("hydro model") to a model including material strength leading to elasto-plastic effects and the possibility of brittle failure ("solid model"). These investigations are done with our own smoothed particle hydrodynamics (SPH) code as introduced in Maindl et al. (2013, 2014) implementing the Grady & Kipp (1980) fragmentation model (see also Benz & Asphaug 1994) and ensuring first-order consistency (tensorial correction, described in Schäfer et al. 2007). Our scenarios include two objects, the target consisting of a basalt core (70 mass-%) and a shell of water ice (30 mass-%) and the projectile consisting of solid basalt. Projectile and target have a mass of M_{Ceres} each. In total we simulate 42 scenarios (each in both the hydro and solid models) defined by collision velocities v_0 between 0.95 and 5.88 two-body escape velocities v_{esc} (which were found representative for such impacts in Maindl & Dvorak 2014) and impact angles α between 0° (head-on) and a flyby. We find that while the collision outcome in terms of merging/erosion/hit-and-run for the solid model is similar to the hydro case we observe a significantly higher degree of fragmentation and more water loss in the solid case.

v_0	α		α		α		α		α		α	
$[v_{esc}]$	[°]	N_{frag}	[°]	N_{frag}	[°]	N_{frag}	[°]	N_{frag}	[°]	N_{frag}	[°]	N_{frag}
0.95	0	1	14	1	23	1	25	1	31	1	62	7
1.32	0	1	11	1	21	1	40	1	48	2	-	2
1.36	0	1	11	1	20	1	40	2	48	2	-	2
2.12	0	1	12	1	25	5	50	2	62	6	-	2
3.04	0	39	12	52	25	22	53	2	67	2	-	2
3.97	0	41	13	67	27	35	55	2	72	2	-	2
5.88	0	45	13	60	28	61	58	2	72	2	-	2

Table 1. Number of significant fragments after 2000 min simulation time. The scenarios are characterized by the collision velocity v_0 given in units of the two-body escape velocity upon impact v_{esc} and the impact angle α. The v_0 (averaged) and α values are taken from Maindl et al. (2014); no impact angle can be given for flyby scenarios (see text).

Here we will focus on analyzing "significant fragments" surviving the collision in the solid case regarding their number and water content after the impact as we are interested in water transport mechanisms in the early solar system (Dvorak et al. 2012).

2. Results

SPH-based simulations resolve continuous bodies into discrete SPH particles carrying the physical properties of interest and contributing to the physical properties of their respective volume elements. Because of the high degree of fragmentation after impacts in the solid model and in order to get a statistically significant number of these discrete particles we limit our study to fragments with masses m_{frag} corresponding to at least 20 basalt particles. With our scenarios' single basalt SPH particle mass of 1.2×10^{17} kg this translates to $m_{frag} \geqslant 2.4 \times 10^{18}$ kg which limits the resolution of our results. Based on the 42 different collision configurations simulated in Maindl et al. (2014) we get the numbers N_{frag} of such "significant fragments" as listed in Table 1. Each α-N_{frag} column pair corresponds to one initial impact parameter. Some of the latter lead to a flyby and hence cannot be assigned an impact angle and show two surviving fragments which are exactly the original bodies. As projectile and target were placed five mean diameters apart at the start of the simulation the impact parameters and angles α changed during the approach depending on mutual gravitational interaction resulting in the values given in the α columns.

2.1. *Fragment properties*

In Fig. 1 we relate the numerical results to an analytic model for the outcome of strengthless planet collisions as presented by Leinhardt & Stewart (2012). In the erosion regime we find many surviving fragments, which is consistent with the analytic prediction. Also, most of the one-fragment and two-fragment outcomes are in the appropriate regions. There is indication however, that the onset of hit-and-run events happens at a higher impact angle than predicted in the analytic strengthless model: while at an impact angle of 40° a velocity of $v_0 = 1.32\, v_{esc}$ leads to a merge, a slightly higher $v_0 = 1.36\, v_{esc}$ produces a hit-and-run outcome (cf. the "1–2" bubble in Fig. 1). At $\alpha = 25°$ there is a $N_{frag} = 5$ outcome that suggests the border between merging and erosion to actually be at lower velocities for colliding smaller bodies. A closer look at that particular scenario reveals two major surviving fragments amounting for about 80 % of the system mass and several smaller ones. At $\alpha = 62°$, the $N_{frag} = 6$ outcome corresponds to a hit-and-run where the two main survivors amount for not quite the total mass of the system (around 90 mass-%) and several minor fragments. The $N_{frag} = 7$ case however marks the border

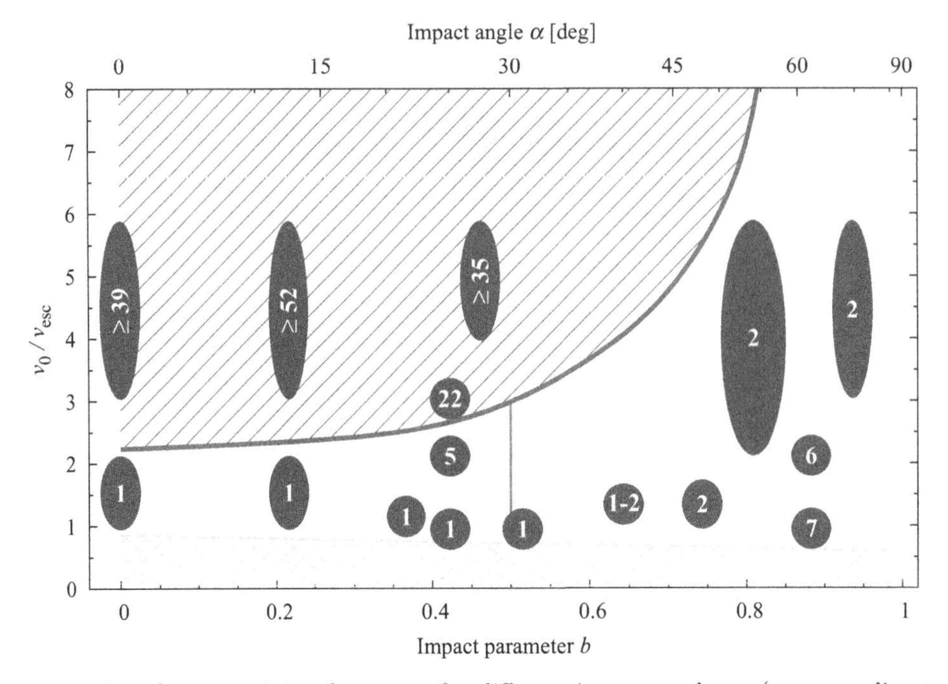

Figure 1. Significant surviving fragments for different impact angles α (corresponding to a dimensionless impact parameter $b = \sin\alpha$) and collision velocities v_0 in units of the two-body escape velocity v_{esc}. The different areas correspond to erosion (red, shaded), partial accretion (white, $\alpha \lesssim 30°$), hit-and-run (white, $\alpha \gtrsim 30°$), and perfect merging (blue, hashed) given by the analytic model for strengthless planet collisions in Leinhardt & Stewart (2012), Fig. 11A. See text for discussion.

to the merging area: among the surviving significant fragments there is only one major survivor ($\geqslant 90$ mass-%). For more details see the systematic discussion of the collision outcomes using a three-biggest-fragment approach in Maindl et al. (2014).

2.2. *Water content*

In order to get an estimate on the loss of volatiles such as water ice on the surface of colliding bodies we track the overall water content of significant fragments after the collision. As initially all the water is present as surface ice on the target amounting to 30 mass-% the system's total water fraction is 0.15. Figure 2 shows how this fraction develops between collision scenarios and reveals that in general more water is retained for "less violent" impacts—for $\alpha \lesssim 20°$ and $v_0 \lesssim 1.3\,v_{\mathrm{esc}}$ almost all water ice stays on the survivor, for strongly inclined hit-and-run collisions most of the water stays as well. Generally, an increasing amount of water ice gets lost for smaller collision angles and higher velocities with a stronger dependency on velocity (impact energy) than on the angle.

Notable features in Fig. 2 are the downward spikes occurring in the $v_0 \leqslant 1.36\,v_{\mathrm{esc}}$ curves. These are due to a single surviving body that spins very quickly spraying debris into space and therefore losing large portions of its surface water ice. Again, we refer to fragment counts and individual scenario descriptions in Maindl et al. (2014) for more detailed discussions.

3. Conclusions and further research

We further evaluated results from an earlier study comparing the "hydro" and "solid" models for simulating small to mid scale collisions of planetesimals at moderate energies

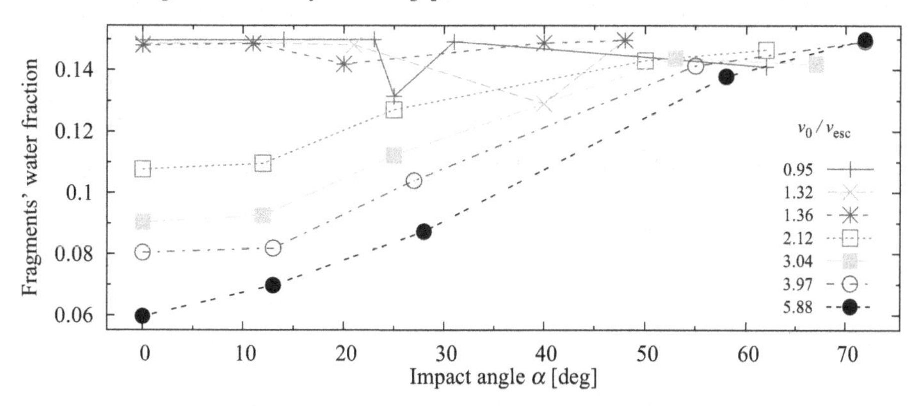

Figure 2. Cumulated water content of significant fragments after the collision (mass fraction). The total water in the system is 15 mass-%. The different curves correspond to different collision velocities v_0; $v_{\rm esc}$ is the two-body escape velocity. See text for discussion.

(Maindl et al. 2014). Rather than focusing on hydro-solid differences analyzing total fragment counts and three-biggest-fragment properties we were looking at the number of surviving "significant fragments" and their volatile content. At a qualitative level the collision outcomes agree with analytic models for (giant) collisions of strengthless planets (as given e.g., in Leinhardt & Stewart 2012) with some minor shifts of boundaries between the merging, erosion, and hit-and-run regimes. As expected there is more water ice surviving the collision for smaller velocities and more inclined impacts. Future studies will focus on (a) fragment dynamics determining their ability to escape the system's Hill sphere and (b) the fate of subsurface water ice inclusions as opposed to an icy shell.

Acknowledgements

This research is produced as part of the FWF Austrian Science Fund project S 11603-N16. In part the calculations for this work were performed on the hpc-bw-cluster—we gratefully thank the bwGRiD project† for the computational resources.

References

Benz, W. & Asphaug, E. 1994, *Icarus*, 107, 98

Canup, R. M., Barr, A. C., & Crawford, D. A. 2013, *Icarus*, 222, 200

Dvorak, R., Eggl, S., Süli, Á., *et al.* 2012, in American Institute of Physics Conference Series, Vol. 1468, American Institute of Physics Conference Series, ed. M. Robnik & V. G. Romanovski, 137–147

Grady, D. E. & Kipp, M. E. 1980,*International Journal of Rock Mechanics and Mining Sciences & Geomechanics Abstracts*, 17, 147

Leinhardt, Z. M. & Stewart, S. T. 2012, ApJ, 745, 79

Maindl, T. I. & Dvorak, R. 2014, in IAU Symposium, Vol. 299, IAU Symposium, ed. M. Booth, B. C. Matthews, & J. R. Graham, 370–373

Maindl, T. I., Dvorak, R., Speith, R., & Schäfer, C. 2014, ArXiv e-print arXiv:1401.0045

Maindl, T. I., Schäfer, C., Speith, R., *et al.* 2013, *Astronomische Nachrichten*, 334, 996

Melosh, H. J. & Ryan, E. V. 1997, *Icarus*, 129, 562

Schäfer, C., Speith, R., & Kley, W. 2007, *A&A*, 470, 733

† bwGRiD (http://www.bw-grid.de), member of the German D-Grid initiative, funded by the Ministry for Education and Research (Bundesministerium fuer Bildung und Forschung) and the Ministry for Science, Research and Arts Baden-Wuerttemberg (Ministerium fuer Wissenschaft, Forschung und Kunst Baden-Wuerttemberg).

Complex Planetary Systems
Proceedings IAU Symposium No. 310, 2014
Z. Knežević & A. Lemaitre, eds.

doi:10.1017/S1743921314008060

Trajectory and physical properties of near-Earth asteroid 2009 BD

D. Farnocchia[1], M. Mommert[2], J. L. Hora[3], S. R. Chesley[1], D. Vokrouhlický[4], D. E. Trilling[2], M. Mueller[5], A. W. Harris[6], H. A. Smith[3], and G. G. Fazio[3]

[1]Jet Propulsion Laboratory, California Institute of Technology
Pasadena, CA 91109, USA
email: Davide.Farnocchia@jpl.nasa.gov

[2]Department of Physics and Astronomy, Northern Arizona University
Flagstaff, AZ 86011, USA

[3]Harvard-Smithsonian Center for Astrophysics
Cambridge, MA 02138, USA

[4]Institute of Astronomy, Charles University
CZ-18000, Prague 8, Czech Republic

[5]SRON Netherlands Institute for Space Research
9700-AV Groningen, The Netherlands

[6]DLR Institute of Planetary Research
D-12489 Berlin, Germany

Abstract. We analyze the trajectory of near-Earth asteroid 2009 BD, which is a candidate target of the NASA Asteroid Redirect Mission. The small size of 2009 BD and its Earth-like orbit pose challenges to understanding the dynamical properties of 2009 BD. In particular, nongravitational perturbations, such as solar radiation pressure and the Yarkovsky effect, are essential to match observational data and provide reliable predictions. By using Spitzer Space Telescope IRAC observations and our model for the thermophysical properties and the nongravitational forces acting on 2009 BD we obtain probabilistic derivations of the physical properties of this object. We find two physically possible solutions. The first solution shows 2009 BD as a 2.9 ± 0.3 m diameter rocky body with an extremely high albedo that is covered with regolith-like material, causing it to exhibit a low thermal inertia. The second solution suggests 2009 BD to be a 4 ± 1 m diameter asteroid with albedo 0.45 ± 0.35 that consists of a collection of individual bare rock slabs. We are unable to rule out either solution based on physical reasoning. 2009 BD is the smallest asteroid for which physical properties have been constrained, providing unique information on the physical properties of objects in the size range smaller than 10 m.

Keywords. Radiation mechanisms: general, celestial mechanics, methods: analytical, techniques: image processing, astrometry, ephemerides, minor planets, asteroids, infrared: solar system

---

1. Introduction

Near-Earth asteroid 2009 BD was discovered on 2009 January 16, at a geocentric distance of 10 million kilometers (Buzzi *et al.* 2009), and it is on an Earth-like orbit.

The orbital geometry makes 2009 BD one of the asteroids with the lowest relative velocity with respect to Earth and it is thus considered a potential target for the Asteroid Redirect Mission (ARM, Mazanek *et al.* 2103). The current design of ARM requires that the target asteroid has a diameter between 7 m and 10 m, and a mass of about 500 t.

Little is known about the physical properties of 2009 BD. The absolute magnitude $H = 28.43 \pm 0.12$ (Micheli *et al.* 2012) suggests that 2009 BD is a small object, with a diameter around 10 m or less. However, the lack of albedo information prevents a more accurate estimate. Tholen *et al.* (2013) found that 2009 BD has a rotation period somewhat larger than 3 hours.

Due to its small size, indirect information on 2009 BD physical properties comes from the action of nongravitational perturbations, namely the Yarkovsky effect (Vokrouhlický *et al.* 2000) and solar radiation pressure (Vokrouhlický & Milani 2000). In this paper we combine the constraints coming from nongravitational perturbations and observations from the Spitzer Space Telescope (Werner *et al.* 2004) to estimate the physical properties of 2009 BD. For more details see Mommert *et al.* (2014).

2. The orbit of 2009 BD

The osculating orbit of 2009 BD at epoch 2010 January 4 is very close to that of the Earth. The semimajor axis is 1.01 au, the eccentricity 4%, and the inclination 0.4°. Because of its orbital configuration 2009 BD stayed in the Earth's neighborhood from 2009 to 2011 thus allowing astronomers to collect astrometric observations. Then, the Earth encounter of June 2011 changed the semimajor axis to 1.06 au and 2009 BD started drifting away from Earth to get closer again only around 2022.

Though the observed arc is quite short, nongravitational perturbations are needed in order to fit the observational data (Micheli *et al.* 2012; Farnocchia *et al.* 2013) as the small size of 2009 BD amplifies the magnitude of nongravitational perturbations, which are inversely proportional to the diameter D.

Similarly to Marsden *et al.* (1973), we modeled nongravitational perturbations as

$$\mathbf{a}_{NG} = \left(A_1\hat{\mathbf{r}} + A_2\hat{\mathbf{t}}\right)(1\ \mathrm{au}/r)^2 \tag{2.1}$$

where $\hat{\mathbf{r}}$ and $\hat{\mathbf{t}}$ are the radial and transverse directions, respectively, and r is the heliocentric distance. A_2 is related to the transverse component of the Yarkovsky effect while A_1 accounts for solar radiation pressure and the radial component of the Yarkovsky effect.

We included A_1 and A_2 in the list of parameters estimated from the orbital fit and obtained $A_1 = (57 \pm 8) \times 10^{-12}$ au/d^2 and $A_2 = (-113 \pm 8) \times 10^{-14}$ au/d^2. The values of A_1 and A_2 along with their uncertainties affect the ephemeris predictions of 2009 BD and provide constraints on its physical properties. In particular, since A_2 is proportional to $\cos\gamma$, where γ is the obliquity (Farnocchia *et al.* 2013), the negative value of A_2 implies that 2009 BD is a retrograde rotator.

3. Spitzer observations

To obtain additional information on its physical properties, we observed 2009 BD starting on 2013 October 13 for a total of 25 hours with IRAC (Fazio *et al.* 2004) on the Spitzer Space Telescope. We selected the observation window according to Spitzer observability, the predicted flux density of the asteroid, and we chose the IRAC channel 2 to maximize our chances of detecting 2009 BD (Mommert *et al.* 2014).

To make sure that 2009 BD was in the field of view, we computed plane-of-sky position along with the corresponding 3σ uncertainty for different settings of the dynamical model: 1) gravitational-only solution, 2) accounting for solar radiation pressure, and 3) including both solar radiation pressure and the Yarkovsky effect. Moreover, we tested different statistical treatments of the astrometry by selecting different outlier rejection thresholds

(Carpino *et al.* 2003). All these different predictions fell within 20" in right ascension and 2" in declination, well inside the Spitzer field of view, i.e., 312" × 312".

Though we did not detect 2009 BD in the Spitzer field of view, we derived a 0.78 μJy 3σ upper bound to the flux density. The resulting constraints on the physical properties of 2009 BD are discussed below.

4. Nongravitational perturbations and thermophysical modeling

The lack of a clear detection of 2009 BD in our observations precludes a direct determination of its physical properties. To indirectly constrain the physical properties of 2009 BD, we take a probabilistic approach that combines a thermophysical model with a model of the nongravitational effects on the asteroid's orbit.

For solar radiation pressure we adopt the model by Vokrouhlický & Milani (2000) whereas for the Yarkovsky effect we use the model approach described by Vokrouhlický *et al.* (2000). The asteroid is assumed to be spherical and the heat transfer is solved analytically using the linearized heat transfer equation. Using the dependence of solar radiation pressure and the Yarkovsky effect on the physical properties of 2009 BD, the model derives bulk density ρ and thermal inertia Γ as a function of γ, and D as a result of the fit to the available astrometric data.

The thermophysical model approximates the surface temperature distribution to determine the thermal-infrared emission from the surface as a function of the physical properties of 2009 BD, e.g., rotation state, thermal inertia, and surface roughness. We assume a spherical shape of 2009 BD, i.e., the derived diameter is the one of a sphere with the same cross-sectional area as the real shape of 2009 BD. The model numerically solves the heat transfer equation and computes the IRAC channel 2 in-band flux density. The contribution from reflected solar light is added to the calculated flux density as described by Mueller *et al.* (2011), assuming an infrared/optical reflectance ratio of 1.4.

In both the orbital and the thermophysical models we adopt an absolute magnitude $H = 28.43 \pm 0.12$ (Micheli *et al.* 2012), a photometric slope parameter $G = 0.18 \pm 0.13$ (derived as the average from all G measurements of asteroids in the JPL Small-Body Database), and a rotation period $P = 2^{(2\pm0.5)}$ hr.

5. Results

The mutual dependence among physical properties used by the orbital and the thermophysical model require an iterative solution of the problem. We first sampled the obliquity γ from 90° to 180°. We tested the possible diameters using the thermophysical model, based on an upper-limit flux density measurement (0.78 Jy), and obtained that the diameter of 2009 BD is smaller than 8 m.

For each $D < 8$ m and $90° < \gamma < 180°$ we find two solutions for (Γ, ρ) from the orbital fit to the astrometry. The first solution displays a low Γ of about 10 J m^{-2} s$^{-0.5}$ K^{-1} with a higher bulk density, whereas the second solution stands out with $\Gamma \sim 1000$ J m^{-2} s$^{-0.5}$ K^{-1} and a lower bulk density. Based on the orbital fit χ^2, we can also further constrain the obliquity (see Table 1) and confidently rule out that 2009 BD is smaller than 2.6 m.

We use our intermediate results to derive diameter distributions for both solutions according to the thermal inertia constraints and the Spitzer upper-limit flux density measurement. We generate a sample of synthetic objects with pairs (D, Γ) and we sample the parameters (H, G, P, γ). For the surface roughness we randomly pick one of four different roughness models (see Mueller 2007). We model each synthetic sample object and derive its IRAC in-band flux density combined with contributions from reflected

Table 1. Physical properties of 2009 BD.

	Low thermal inertia	**High thermal inertia**
Diameter [m]	2.9 ± 0.3	4 ± 1
Albedo	$0.85^{+0.20}_{-0.10}$	$0.45^{+0.35}_{-0.15}$
Obliquity [°]	170^{+10}_{-15}	180^{+0}_{-5}
Density [g cm $^{-3}$]	2.9 ± 0.5	$1.7^{+0.7}_{-0.4}$
Mass [t]	36^{+10}_{-8}	55^{+30}_{-25}
Thermal inertia [J m^{-2} s$^{-0.5}$ K^{-1}]	30^{+20}_{-10}	2000 ± 1000

solar light, which we then compare with the upper-limit flux density as derived from our observations. The final solutions for the diameter are shown in Table 1.

Based on the solution-specific diameter ranges, we finally constrain the other physical properties of 2009 BD using the nongravitational perturbation model and the orbital fit to the astrometry. Table 1 shows the derived physical properties of 2009 BD for both the low and high thermal inertia solutions. The first solution shows 2009 BD as a massive rock body covered with regolith-like material, causing it to exhibit a low thermal inertia. The second solution suggests 2009 BD to be a rubble-pile asteroid that consists of individual bare rock slabs. We are unable to rule out either solution at this stage.

Acknowledgements

The authors thank Tom Soifer, Director of the Spitzer Space Telescope, for the time allocation to observe 2009 BD. The work of D. Farnocchia and S. R. Chesley was conducted at the Jet Propulsion Laboratory, California Institute of Technology under a contract with NASA. D. Vokrouhlický was partially supported by the Grant Agency of the Czech Republic (grant P209-13-01308S). J. L. Hora and H. A. Smith acknowledge partial support from Jet Propulsion Laboratory RSA No. 1367413. This work is based on observations made with the Spitzer Space Telescope, which is operated by the Jet Propulsion Laboratory, California Institute of Technology under a contract with NASA.

References

Buzzi, L., Hormuth, F., Bittesini, L., *et al.* 2009, *MPEC*, 2009-B14

Carpino, M., Milani, A., & Chesley, S. R. 2003, *Icarus*, 166, 248

Farnocchia, D., Chesley, S. R., Vokrouhlický, D., *et al.* 2013, *Icarus*, 224, 1

Fazio, G. G., Hora, J. L., Allen, L. E., *et al.* 2004, *ApJS*, 154, 10

Marsden, B. G., Sekanina, Z., & Yeomans, D. K. 1973, *AJ*, 78, 211

Mazanek, D. D., Brophy, J. R., & Merrill, R. G. 2013, *Planetary Defense Conference*, IAA-PDC13-04-14

Micheli, M., Tholen, D. J., & Elliott, G. T. 2012, *New Astron.*, 17, 446

Mommert, M., Hora, J. L., Farnocchia, D., *et al.* 2014, *ApJ*, 786, 148

Mueller, M. 2007, *arXiv*, 1208.3993

Mueller, M., Delbo', M., Hora, J. L., *et al.* 2011, *AJ*, 141, 109

Tholen, D. J., Micheli, M., Bauer, J., & Mainzer, A. 2013, *AAS/Division for Planetary Sciences Meeting Abstracts*, 45, #101.08

Vokrouhlický, D. & Milani, A. 2000, *A&A*, 362, 746

Vokrouhlický, D., Milani, A., & Chesley, S. R. 2000, *Icarus*, 148, 118

Werner, M. W., Roellig, T. L., Low, F. J., *et al.* 2004, *ApJS*, 154, 1

Complex Planetary Systems
Proceedings IAU Symposium No. 310, 2014
Z. Knežević & A. Lemaitre, eds.

doi:10.1017/S1743921314008072

Orbit computation of the TELECOM-2D satellite with a Genetic Algorithm

Florent Deleflie[1], David Coulot[2,1], Alain Vienne[1], Romain Decosta[1], Pascal Richard[3] and Mohammed Amjad Lasri[2]

[1]Institut de Mécanique Céleste et de Calcul des Ephémérides / GRGS, Univ. Lille1, UPMC
77 Av. Denfert Rochereau, 75014 Paris, France
email: florent.deleflie@imcce.fr

[2]IGN LAREG, Université Paris Diderot, Sorbonne Paris Cité,
5 rue Thomas Mann, 75205 Paris Cedex 13, France

[3]Centre National d'Etudes Spatiales,
18 Avenue Edouard Belin, 31400 Toulouse, France

Abstract. In order to test a preliminary orbit determination method, we fit an orbit of the geostationary satellite TELECOM-2D, as if we did not know any a priori information on its trajectory. The method is based on a genetic algorithm coupled to an analytical propagator of the trajectory, that is used over a couple of days, and that uses a whole set of altazimutal data that are acquired by the tracking network made up of the two TAROT telescopes. The adjusted orbit is then compared to a numerical reference. The method is described, and the results are analyzed, as a step towards an operational method of preliminary orbit determination for uncatalogued objects.

Keywords. Orbit computation, Genetic algorithm, Analytical Propagation

1. Introduction

We aim at fitting an orbit to tracking data, provided as a time series of angular coordinates on the sky, when no *a priori* information on the trajectory is available at all. In that case, classical methods such as least-squares can not be used any more, since the function to be minimized can not be linearized in the neighborhood of the *a priori* values of the parameters. Moreover, the usual methods of preliminary orbit determination may suffer from many drawbacks which can make them be unappropriate: the well-known Gauss, Laplace, Escobal... approaches are not valid for all dynamical configurations in case of singularities due to orbital planes alignments; additionally, they are often based on motion theories accounting only for the Keplerian motion, and can hence not be applied over time scales longer than a couple of hours, since in that case a propagator has to account for the main perturbations, at least for the secular ones. A general comparison between these methods could be the subject of a forthcoming paper.

On the contrary, even if other kinds of difficulties have to be managed, methods based on genetic algorithms are supposed to be valid for all dynamical configurations, since the algorithm itself is independent from the orbit propagator used to compute the cost function. With an efficient dynamical modeling, they can be used over different periods of time, from a couple of minutes (for Too-Short Arcs, TSA) up to a couple of days or weeks.

The starting point is the system of the equations of motion, that can be written in an usual way:

$$\frac{\mathrm{d}^2 \boldsymbol{r}}{\mathrm{d}t^2} = \boldsymbol{F}(\boldsymbol{r}, \dot{\boldsymbol{r}}, t, \sigma) \qquad \text{with} \qquad \boldsymbol{r}(t_0) = \boldsymbol{r}_0 \qquad \dot{\boldsymbol{r}}(t_0) = \dot{\boldsymbol{r}}_0$$

and where the initial positions and velocities to be estimated at an epoch t_0 are denoted $\boldsymbol{r}(t_0)$ and $\dot{\boldsymbol{r}}(t_0)$. The right-hand side describes the force model through the vector $\boldsymbol{F}$, that is characterized with a set of parameters σ. Genetic algorithms allow a way to find satisfying initial conditions $\boldsymbol{r}(t_0)$ and $\dot{\boldsymbol{r}}(t_0)$, without testing all the possibilities in a space of dimension 6, once the frame is roughly defined.

Following (Deleflie *et al.*, (2013)), we provide the finalized results that we obtain after some refinements of the method.

2. Orbital modeling

To keep a reasonable computation time, since many iterations are tested, we use an analytical approach to get orbital element time series. Since the method is supposed to be valid in all dynamical configuration (whatever the values of the eccentricity and the inclination, in particular), the model is written in a set $\bar{\boldsymbol{E}}$ of equinoctial elements (Deleflie & Decosta (2013)), namely: a, $\xi = \Omega + \omega + M$, $e\cos(\Omega + \omega)$, $e\sin(\Omega + \omega)$, $\sin i/2 \cos\Omega$, $\sin i/2 \sin\Omega$, where a, e, i, Ω, ω, M stand for the classical Keplerian elements. The whole analytical modeling is governed by the set of mean initial conditions, whereas it is the corresponding osculating initial conditions that are adjusted by the genetic algorithm. The relation between mean (denoted $\bar{\boldsymbol{E}}(t_0)$) and osculating initial elements (denoted $\boldsymbol{E}(t_0)$) is merely obtained by setting the time t to the initial epoch t_0 in the equation defining the shape of the analytical solution, that is: $\boldsymbol{E}(t) = \bar{\boldsymbol{E}}(t) + \mathcal{L}(\bar{\boldsymbol{E}})\frac{\partial W}{\partial \bar{\boldsymbol{E}}}(\bar{\boldsymbol{E}}(t))$, the matrix 6×6 $\mathcal{L}(\bar{\boldsymbol{E}})$ standing for the Lagrange Planetary Equations and W being a generating function of the short periodic terms, both depending of the mean equinoctial elements at a given epoch. Here, the force model is the central gravity field developed up to degree 10, to strike a balance between the accuracy and the total required CPU time.

3. Multi-Objective Genetic Algorithm (MOGA) used

3.1. *Description*

The criteria to be optimized (maximized or minimized) are defined as functions of the initial conditions, and they are optimized through a large number of iterations that make the process converge to a set of optima. For the computation of the TELECOM-2D orbit, the double-objective of getting a minimum on the two components of the data, independently, has been set up. Though more time consuming than a standard Genetic Algorithm, a multi-objective Genetic Algorithm does not require any weighting choice as it appears when using a solely aggregated objective.

The Multi-Objective Genetic Algorithm (MOGA) used here is ϵ-MOEA (Deb *et al.* (2003). In the course of successive iterations, some vectors of initial conditions are replaced by other ones and the best ones are archived. The evolution through the iterations is governed by mutations (random small changes in vectors of possible initial conditions) and by crossover (mix two vectors of possible initial conditions). At the end of the iteration procedure, a set of solutions is supplied (Coello Coello *et al.* (2007)).

As in many orbit determination algorithms, an evaluation is made up of several steps:

- ϵ-MOEA provides a vector of initial conditions, randomly chosen among a large set of possible (osculating) initial values;

• These initial conditions are used to propagate an analytical orbit over the period when tracking or astrometric data are available;

• The analytical orbit, as time series of orbital elements, is used to compute predicted measurements, that can be compared to the available data sets, at the same epochs of the observations;

• These predicted measurements are compared to the true data;

• The cost functions are computed, with the n observations (Obs) and the corresponding theoretical quantities carried out by the Genetic Algorithm (GA), for the two angles elevation (el) and azimut (az): $\sqrt{\frac{1}{n}\sum_{i=1}^{n}(\mathrm{el}_i^{\mathrm{obs}}-\mathrm{el}_i^{\mathrm{GA}})^2}$, and $\sqrt{\frac{1}{n}\sum_{i=1}^{n}(\mathrm{az}_i^{\mathrm{obs}}-\mathrm{az}_i^{\mathrm{GA}})^2}$.

The process is then iterated until optimal values of initial conditions are found, optimal values being defined as set of values (not necessary unique) providing the minimum distance in the objective space.

3.2. *Parameterization*

The chromosomes represent the initial state of the solution: each one is then made up of a vector with six components, which determines an unique orbit.

We choose here an initial population of 400 chromosomes, with fixed intervals for each of the initial components of the state vector, natively written through the set of equinoctial elements:

• semi–major axis $a \in [40000; 45000]$ km for Telecom-2D

• eccentricity vector assuming $e \in [0; 0.1]$, i.e. the two components of the eccentricity vector $\in [-0.1; 0.1]$

• inclination $i \in [0; 180°[$, i.e. the two components of the eccentricity vector $\in [-1; 1]$

• the angle $\xi = \Omega + \omega + M \in [0; 360°$

Let us note that to reduce computation time, the search for the initial eccentricity has been reduced to an interval with a wideness of 0.1, and the search of the initial semi major axis to intervals large of a few thousands of kilometers. But, the results that are shown hereafter would not have been worse if we have kept all the possibilities ($e \in [0; 1]$, and $a \in [6\,500; 45\,000]$ km for instance) for these two elements as well. But the computation time would have been significantly larger.

The crossover probability has been set up, classically, to $p_C = 0.9$, and the mutation probability to $p_m = 1/6 \simeq 0.16667$. The stop condition is the total number of iterations (here set up to 100 000), corresponding to a total CPU time of the order of 10 hours.

4. Genetic algorithm handled with angular data: computation of the TELECOM-2D satellite orbit

This example is based on the assimilation of altazimutal data obtained after astrometric reductions from images acquired by the two TAROT telescopes, respectively located in France and Chile, on the geostationary satellite Telecom-2D. We use data provided by CNES, which has an agreement to benefit from 15% of the available time each night, for space debris activities. The upper reachable magnitude is of the order of 15 within the GEO region, and the measurement accuracy is of the order of 700 m in GEO. The data set includes nine days of angular data, acquired in Sept. 2012 from the two TAROT-telescopes. The total number of measurements is 235 (97 for la Silla, Chile, and 138 for Calern, France).

The reference orbit was computed with the CNES s/w Romance.

Figure 1 shows how the G. A. coupled with the analytical propagation converges to the numerical solution seen as a reference.

Figure 1. Convergence of the iterations for the TELECOM-2D orbit computation by the G. A. Left: global objective; Right: convergence of the initial semi-major axis

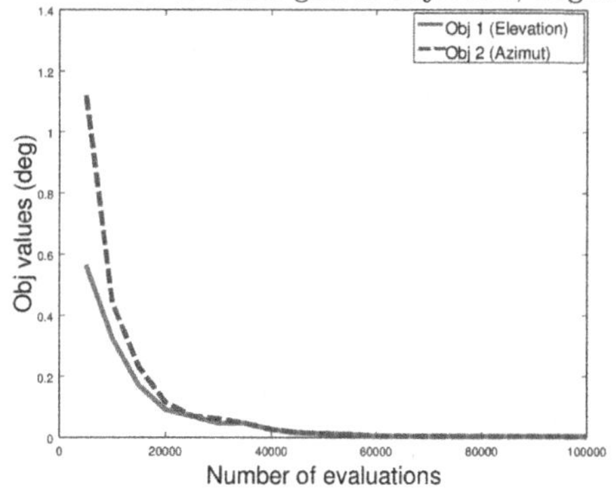

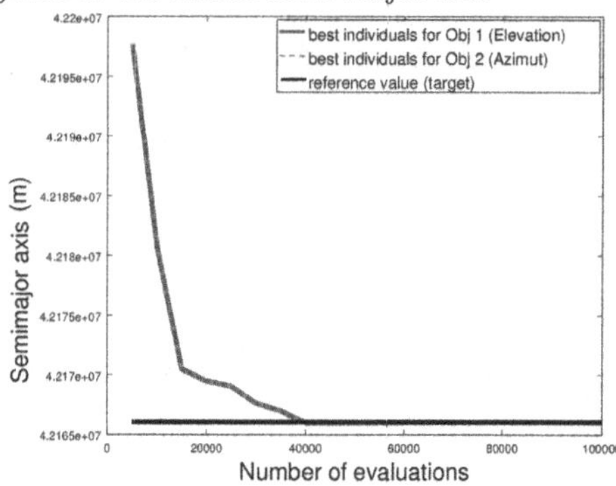

Table 1. Numerical references for the initial state vector, expressed in classical keplerian elements, and best candidates found by the G. A.. Let us remind that, once projected into classical keplerian elements, the values found by the G. A. are not that accurate for the argument of perigee ω and the mean anomaly M, because of a poor geometry defining these two angles. It is only an effect of projection, with no consequence on the computation.

	num. reference	**ref. elements - G. A. best candidate**
s.m.a.	$a = 42\,165\,286$ m	$\Delta a = 773$ m
eccentricity	$e = 0.000\,14$	$\Delta e = 0.000\,147$
inclination (deg)	$i = 5.705°$	$\Delta I = 0.003\,4°$
R. A. A. N. (deg)	$\Omega = 62.066°$	$\Delta\Omega = 0.049°$
arg. perigee (deg)	$\omega = 220.266°$	$\Delta\omega = 15.106°$
mean anomaly (deg)	$M = 29.129°$	$\Delta M = 15.111°$
orbital longitude (deg)	$\omega + M = 249.396°$	$\Delta(\omega + M) = 0.005°$

5. Conclusions

We combine a MOGA and an analytical satellite motion theory to adjust an orbit on tracking data, without any *a priori* knowledge of the initial conditions to be retrieved. The next steps lie (i) in a test of the capabilities of the algorithm in downgraded conditions (data sparse in time, very few number of data), (ii) in a reduction of the required CPU time, by reducing the width of the intervals of research with *a priori* more realistic values.

Acknowledgments

We thank CNES for financial support, in the framework of a GRGS project (Groupe de Recherche de Géodésie Spatiale).

References

Deleflie, F., Coulot, D., Decosta, R., Fernier, A., & Richard, P. , 2013, *Proc. of the 6th European Conference on Space Debris*, ESOC, Darmstadt, Germany, Apr. 2013

Coello Coello, C. A., G. B. Lomont, & D. A. Von Veldhuizen, *Evolutionary Algorithms for Solving Multi-Objective Problems*, Second Edition, Springer

Deb, K., M. Mohan & S. Mishra (2003), 2003, *Kanpur Genetic Algorithms Laboratory Report (KanGAL)* 2003002.

&Deleflie, F., R. Decosta, 2013, *CNES Internal Report*

Complex Planetary Systems
Proceedings IAU Symposium No. 310, 2014
Z. Knežević & A. Lemaitre, eds.

doi:10.1017/S1743921314008084

New methods for space debris collision assessment

Daniel Casanova[1], Chiara Tardioli[2] and Anne Lemaître[1]

[1]University of Namur, naXys - Department of Mathematics
8 Rempart de la Vierge, 5000, Namur, Belgium
email: daniel.casanova@unamur.be
[2]University of Strathclyde, Department of Mechanical & Aerospace Engineering
75 Montrose Street, Glasgow, UK

Abstract. Collisions between two pieces of space debris or between a piece of debris and an operative satellite is a real problem. Furthermore, collisions are responsible for the creation of new space debris systematically. The way to exclude the possibility of a collision consists of analysing the trajectories and looking for a time of coincidence. However, the analysis of all pairs of objects collected in a large orbit catalogue is unfeasible. The proposed method consists of reducing the possible pairs of candidates for a collision into a short list of pairs at real risk of collision. The method is based on a three-filter sequence: the first two filters are based on the geometry of the orbits, while the third one searches for a time of coincidence. This new method is tested resulting into an efficient tool for space debris collision assessment.

Keywords. Celestial Mechanics, Earth, ephemerides

1. Introduction

Since the launch of the first satellite Sputnik I in October 1957, more than 7,125 spacecraft have been launched into space. In addition to this population of satellites, we should include debris generated by the uncontrolled collision between Cosmos 2251 and Iridium 33, or by the intentional explosion of the Fengyun 1C. These events increased the population of space debris of about 40% in just two years (Pardini & Anselmo (2011)). According to NASA, more than 21,000 pieces of space debris larger than 10 cm have been tracked and catalogued.

In this work, the initial conditions of space debris trajectories are determined by two different techniques. The first one uses a new orbit determination method based on the first integrals of the Kepler problem (Gronchi, Dimare & Milani (2010)). The method produces a catalogue of fictitious orbits (Dimare *et al.* (2011)). The second method uses the Two Line Element (TLE) catalogue, which provides information of about 21,000 objects, grouping together active satellites and space debris. In particular, this work considers the TLE of the pieces of space debris generated by the previously mentioned catastrophic events.

Space debris collisions are a real problem. Space debris may cause serious damages to operative satellites and they have potentially devastating repercussions for communications and security. The goal of this work is to present and test an effective and realistic tool for the detection of all possible collisions between couples of objects in a large orbit catalogue. The method has been introduced in Casanova, Tardioli & Lemaitre (2014) and consists of applying a sequence of three filters to significantly reduce the number of pairs candidates for a collision. Indeed, an all-to-all approach is unfeasible, since the computation of all possible pairs and the search for possible collisions between them has computational complexity of order N^2, where N is the number of objects in the catalogue.

In this work, we compute the ephemerides of each piece of space debris using recent models developed by members of the Namur Center for Complex Systems (naXys): Valk, Lemaitre & Anselmo (2008), Lemaitre, Delsate & Valk (2009), Valk, Lemaitre & Deleflie (2009), Valk *et al.* (2009), Delsate *et al.* (2010), Hubaux *et al.* (2012), Casanova & Lemaitre (2014). The ephemerides are stored using direct access files.

The work is organized as follows: Section 2 briefly describes the ephemeris interpolation table. Section 3 focuses on the three-filter sequence method to assess space debris collisions. Afterwards, in Section 4 numerical experiments show the reliability of the filter sequence, and finally, we present the conclusions.

2. Ephemeris interpolation table

The evolution of each space debris object is computed numerically by a symplectic integration scheme (Hubaux *et al.* (2012)), which includes the Earth's gravitational potential, the luni-solar and planetary gravitational perturbations, and the direct solar radiation pressure. We propagate each object with a fixed time step, and we store all the ephemeris information in a direct access file (see Casanova, Tardioli & Lemaitre (2014) for a detailed explanation). This technique allows us to compute the ephemerides of any object at any instant of time by a linear interpolation.

3. Three-filter sequence

The original three-filter sequence was established by Hoots, Crawford & Roehrich (1984) to determine future close approaches between satellites. The method proposed in Casanova, Tardioli & Lemaitre (2014) improves the previous one by using the ephemeris table instead of the osculating orbital elements, but also by applying some novel techniques. In the following we summarize the main features of the new three-filter sequence.

3.1. *Filter I*

Thanks to the ephemeris interpolation table, for each piece of space debris, it is possible to compute its geocentric distance $\vec{r}(t)$ and its time derivative $\dot{\vec{r}}(t)$ at any time t. We are interested in computing the maximum and minimum of this distance, which is equivalent to find the values where the derivative function $\dot{\vec{r}}(t)$ is equal to zero. This problem is solved by using a 'regula-falsi' algorithm similar to the one used in Milani *et al.* (2005). So, given a pair of objects of index s_1, s_2, we compute the maximum and minimum values of their geocentric distances, named $\max(r_{s_j})$ and $\min(r_{s_j})$, for $j = 1, 2$. Then, we set $q := \max\{\min(r_{s_1}), \min(r_{s_2})\}$ and $Q := \min\{\max(r_{s_1}), \max(r_{s_2})\}$. If the condition $|q - Q| > D$, with D a fixed threshold distance, holds, then there is no possibility for an orbit crossing and the selected couple is rejected. Filter I improves the one presented by Hoots since a smaller number of false positives pass to the second filter.

3.2. *Filter II*

This filter is based on the definition of the orbit distance map introduced by Gronchi & Tommei (2007). For each pair of objects it is possible to compute their orbit distance with sign at any instant of time of the interpolation process. If there is a change of sign in this distance function, an orbit crossing occurs. Hoots' second filter ensures that the Minimum Orbit Intersection Distance (MOID) is close to the line of nodes. However, the minimal points can be far away from the mutual nodes, especially for low mutual inclinations (Gronchi (2002)). Contrary, the orbit distance with sign is independent of the mutual nodes and it is proven to be more regular than the MOID in a neighbourhood of most crossing configurations.

3.3. *Filter III*

Filter III, also named time filter, computes the position of the two pieces of space debris looking for a time of coincidence. If there is a time where the distance between the two pieces is smaller than a threshold distance D, the pair is considered at real risk of collision, otherwise the possibility of a collision for the pair under consideration is excluded.

4. Numerical experiments

The effectiveness and reliability of the three-filter sequence has been tested through different populations of space debris. The procedure is the following. First, we compute the orbital evolution of all objects during 1 day, with a time span of 15 min. Then, we create a direct access file with the ephemerides of all the objects at all time nodes (ephemeris interpolation table). Afterwards, we apply the three-filter sequence considering an interpolation time step of 7.5 min and a threshold distance of 2 km.

The first space debris population consists of 864 objects in low-Earth orbit (LEO) region, i.e. with semimajor axis less than or equal to 8,600 km. The initial orbital conditions have been determined by a new orbit determination method based on the first integrals of the Kepler problem (Gronchi, Dimare & Milani (2010)). Table 1 illustrates the number of objects that have been rejected by each filter, concluding that there are no pairs at real risk of collision.

Table 1. 864 pieces of space debris. Ratio $A/m = 0.01$.

N. of objects	Possible pairs	Filter I	Filter II	Filter III	Pairs at risk
864	372, 816	236, 625	116, 506	19, 685	0

The second population is composed of 5,000 pieces of space debris, whose initial conditions have been obtained from the previous set of 864 objects by slightly modifying their inclinations. Table 2 shows the results of the filter sequence, concluding that 2,461 pairs of objects are at real risk of collision.

Table 2. 5,000 fictitious pieces of space debris. Ratio $A/m = 0.01$

N. of objects	Possible pairs	Filter I	Filter II	Filter III	Pairs at risk
5, 000	12, 497, 500	7, 843, 579	3, 984, 627	666, 833	2, 461

An innovative part of this paper is that the three-filter sequence can be applied to a set of initial conditions taken from the Two Line Elements (TLE) catalogue. In particular, we consider the catastrophic events already mentioned in the introduction; the uncontrolled collision between Cosmos 2251 and Iridium 33, and the intentional explosion of the Fengyun 1C. These two events generated in total 4,194 pieces of space debris. We consider that these objects have an area-to-mass ratio equal to 0.01 m^2/Kg. After applying the filter sequence to the population, we conclude that there are no pairs at real risk of collision, as we illustrate in Table 3.

Finally, we consider the same population of 4,194 pieces of space debris, but we randomize the ratio A/m of each object in the range $[0.01, 10]$. The results are shown in

Table 3. 4, 194 pieces of space debris. Ratio $A/m = 0.01$

N. of objects	Possible pairs	Filter I	Filter II	Filter III	Pairs at risk
4,194	8,792,721	4,875,138	3,194,869	722,714	0

Table 4 and they are quite similar to the previous population, concluding that there are no pairs at real risk of collision.

Table 4. 4, 194 pieces of space debris. Random ratio A/m

N. of objects	Possible pairs	Filter I	Filter II	Filter III	Pairs at risk
4,194	8,792,721	4,875,176	3,194,784	722,761	0

5. Conclusions

This work proves the effectiveness and reliability of the three-filter sequence through four different experiments. The proposed sequence is a powerful method to exclude pairs of objects that are not at real risk of collision. At the same time, the pairs at real risk of collision are identified for possible collision avoidance analysis. Moreover, this filter sequence can be applied to any population of space debris whose initial conditions are obtained from the TLE catalogue. Future works will take into account the orbit uncertainty, in order to increase the reliability of the results.

Acknowledgements

This paper presents research results of the Belgian Network DYSCO (Dynamical Systems, Control, and Optimization), funded by the Interuniversity Attraction Poles Programme, initiated by the Belgian State, Science Policy Office. The scientific responsibility rests with its author(s).

References

Casanova, D., Tardioli, C., & Lemaitre, A. 2014, *MNRAS*, 442 (4), 3235–3242

Casanova, D. & Lemaitre, A. 2014, Submitted to *Celest. Mech. Dyn. Astr.*

Delsate, N., Lemaitre, A., Carletti, T., & Robutel, P. 2010, *Celest. Mech. Dyn. Astr.*, 108 (3), 275–300

Dimare, L., Farnocchia, D., Gronchi, G. F., Milani, A., Bernardi, F., & Rossi, A. 2011, *Advanced Maui Optical and Space Surveillance Technologies Conference*, 1, 51

Gronchi, G. F. 2002, *SIAM J. Sci. Comput.*, 24 (1), 61–80

Gronchi, G. & Tommei, G. 2007, *Discret. Contin. Dyn-B.*, 7 (4), 755–778

Gronchi, G. F., Dimare, L., & Milani, A. 2010, *Celest. Mech. Dyn. Astr.*, 107 (3), 299–318

Hoots, F. R. & Crawford, L. L., Roehrich, R. L. 1984, *Celest. Mech.*, 33 (2), 143–158

Hubaux, C., Lemaitre, A., Delsate, N., & Carletti, T. 2012, *Adv. Sp. Res.*, 49 (10), 1472–1486

Lemaitre, A., Delsate, N., & Valk, S. 2009, *Celest. Mech. Dyn. Astr.*, 104 (4), 383–402

Milani, A., Tommei, G., Chesley, S., Sansaturio, M., & Valsecchi, G. 2005, *Icarus*, 173 (2), 362–384

Pardini, C. & Anselmo, L. 2011, *Adv. Sp. Res.*, 48 (3), 557–569

Valk, S., Lemaitre, A., & Anselmo, L. 2008, *Adv. Sp. Res.*, 41 (7), 1077–1090

Valk, S., Lemaitre, A., & Deleflie, F. 2009, *Adv. Sp. Res.*, 43 (7), 1070–1082

Valk, S., Delsate, N., Lemaitre, A., & Carletti, T. 2009, *Adv. Sp. Res.*, 43 (10), 1509–1526

Complex Planetary Systems
Proceedings IAU Symposium No. 310, 2014
Z. Knežević & A. Lemaître, eds.

doi:10.1017/S1743921314008096

Comparison of different methods to compute a preliminary orbit of Space Debris using radar observations

Hélène Ma[1] **and Giovanni F. Gronchi**[2]

[1]Dipartimento di Matematica, Università di Pisa,
Largo B. Pontecorvo 5, Pisa, Italy
email: helenema@mail.dm.unipi.it

[2]Dipartimento di Matematica, Università di Pisa,
Largo B. Pontecorvo 5, Pisa, Italy
email: gronchi@dm.unipi.it

Abstract. We advertise a new method of preliminary orbit determination for space debris using radar observations, which we call *Infang*†. We can perform a linkage of two sets of four observations collected at close times. The context is characterized by the accuracy of the range ρ, whereas the right ascension α and the declination δ are much more inaccurate due to observational errors. This method can correct α, δ, assuming the exact knowledge of the range ρ. Considering no perturbations from the J_2 effect, but including errors in the observations, we can compare the new method, the classical method of Gibbs, and the more recent Keplerian integrals method. The development of *Infang* is still on-going and will be further improved and tested.

Keywords. celestial mechanics, preliminary orbit determination, space debris, radar observations, infinitesimal angles.

---

1. Introduction

In the last years, the new method *Infang* started to be implemented by the University of Pisa and SpaceDys, see Gronchi *et al.* (2015). We would like to introduce it and compare with two existing methods using data collected from radar observations. We consider two sets of four radar observations to perform a linkage and compute a preliminary orbit. For each set, the times of consecutive observations are very close‡. We denote by $\bar{t}_j, j = 1, 2$ the average epoch for each set.

<u>*Radar observations*</u>. Each observation is composed by the topocentric distance of the observed object ρ, the right ascension α and the declination δ. We assume that ρ is accurate¶, so that we can obtain a good interpolation of $\dot{\rho}, \ddot{\rho}$. However, the angles are not precisely determined‖. An orbit can be expressed in spherical coordinates by the vector $(\alpha, \delta, \dot{\alpha}, \dot{\delta}, \rho, \dot{\rho})$: therefore here $\dot{\alpha}, \dot{\delta}$ are the unknowns of the preliminary orbit determination problem. This chapter will describe roughly the features of the methods.

† *Infang* stands for *infinitesimal angles*
‡ $\Delta t = 10s$
¶ i.e. RMS $\approx 10m$
‖ i.e. RMS $\approx 0.2°$

2. Features of the methods

Method of Gibbs. It consists in computing the velocity starting from three position vectors at consecutive times $t_1 < t_2 < t_3$, assuming the observed object follows a Kepler motion. In our case, we can select three observations from the first or the second set.

Method of Keplerian integrals. This method, presented in Taff and Hall (1977) (see also Farnocchia *et al.* (2010)), uses the orbital elements in spherical coordinates gathered in a *radar attributable*

$$\mathcal{A} = (\alpha, \delta, \rho, \dot{\rho}).$$

Instead of $\dot{\alpha}$, $\dot{\delta}$, we consider as unknowns the quantities

$$\xi = \rho\dot{\alpha}\cos\delta, \qquad \zeta = \rho\dot{\delta},$$

which are the components of the topocentric velocity in the plane orthogonal to the direction of the line of sight. Assuming the object moves according to the 2-body dynamics, the energy and the angular momentum are conserved, giving a system of linear equations with four unknowns $(\xi_1, \zeta_1, \xi_2, \zeta_2)$, where the indexes refer to epochs $\bar{t}_1$, $\bar{t}_2$.

The infinitesimal angles method "Infang". The range ρ is precise, but α, δ are not accurate. However, the deviations $\Delta\alpha, \Delta\delta$ of the angles from the true values are assumed to be small: therefore they can be treated as "*infinitesimal angles*". We use the following sets of attributable coordinates:

$$(\rho, \alpha, \delta, \dot{\rho}, \dot{\alpha}, \dot{\delta})_j \text{ at } \bar{t}_j \text{ with } j = 1, 2.$$

This method considers as unknowns $\xi, \zeta, \Delta\alpha, \Delta\delta$ at the two epochs, and uses the equations of motion projected in the direction of the line of sight, the algebraic integrals of Kepler's problem, and Lambert's theorem (see Gronchi *et al.* (2015) for the details).

Future work. The development of the *Infang* method is still a work in progress. We intend to investigate the new method with large scale simulations, possibly adding the J_2 effect.

Acknowledgements

The work is supported by the Marie Curie Initial Training Network Stardust, FP7-PEOPLE-2012-ITN, Grant Agreement 317185.

References

L. G. Taff, D. L. Hall. 1977, The use of angles and angular rates, I-initial orbit determination, *Cel. Mech. &Dyn. Astr.*, 16, 481488

G. F. Gronchi, L. Dimare, D. Bracali Cioci, H. Ma 2015, in preparation

Herrick, Samuel, Astrodynamics, Vol. I, [2, 7, 8] 1971, London, Van Nostrand Reinhold Co.

D. Farnocchia, G. Tommei, A. Milani, A. Rossi 2010, Innovative methods of Correlation and orbit determination for space debris, Cel. Mech. & Dyn. Astr., 107 no 1, pp. 169–185.

Complex Planetary Systems
Proceedings IAU Symposium No. 310, 2014
Z. Knežević & A. Lemaitre, eds.

doi:10.1017/S1743921314008102

Orbit identification for large sets of data: preliminary results

Stefano Marò[1] and Giovanni F. Gronchi[2]

Dipartimento di Matematica, Università di Pisa
Largo Bruno Pontecorvo,5 , 56127 Pisa - Italy
[1]email: maro@mail.dm.unipi.it
[2]email: gronchi@dm.unipi.it

Abstract. We propose a strategy to attack the problems of orbit determination arising from the large number of short arcs. The method uses a solution of the linkage problem depending on the first integrals of the Keplerian motion.

Keywords. celestial mechanics, orbit determination, linkage, attributable

---

1. Introduction

This brief note is devoted to a discussion on one of the problems arising in orbit determination of asteroids. The identification problem consists of finding among a set of detections of celestial bodies, those belonging to the same object. The problem is of crucial importance nowadays, due to the large amount of data coming from the asteroid surveys. The Catalina Sky Survey and the Pan-STARRS project, in particular, search for moving objects in the sky every night collecting many new detections. Due to the improvements of the technology, these telescopes are able to observe many small objects, that in general are visible only during a small span of time. The AstDys website monthly computes the orbits of numbered and multiopposition asteroids whose observations are provided by the Minor Planet Center. Anyway, if the described arc is too short and the observations that are collected are few, then a single apparition orbit determination is not possible. Even if the least squares algorithm succeeded, the uncertainty would be very large. It means that the predicted portion of the sky where the next apparition should take place is larger than the field of view of the telescope and the recovery may fail. In this case we would have a lost asteroid. Hence there exists a database of designations without a good orbit that deserves investigation and we are going to concentrate on it.

Since a single arc is not enough to have a good orbit, the usual procedure is to try to find the couples of arcs that can be joined to produce a better orbit. This problem is known as linkage problem and has been considered by several authors (Granvik *et al.* (2005), Granvik & Muinonen (2008), Milani *et al.* (2004), Milani *et al.* (2005), Taff & Hall (1977)). The problem is complicated as one has to join the information on two different arcs that, in principle, may not belong to the same orbit.

There is not just one accepted method, as every proposal has some strength and weakness. In particular, the case in which the time distance between the two arcs is large is the most challenging. In Gronchi *et al.* (2010) a method based on the first integrals of Kepler's motion is proposed. It was tested on simulated data, giving good results. Subsequently, in Gronchi *et al.* (2011) the method has been improved but it has not been tested for the asteroid case.

In the following we are going to concentrate on the application of the improved method to real data. Generally, the orbit coming from a linkage procedure is not good enough to be considered reliable. Hence some work has to be done in order to test and improve it. In the next section we will recall the Keplerian integrals method and briefly describe the procedure that we are going to use in order to improve the results coming from the linkage. We will apply the procedure to a reduced database of real data and show some partial results.

2. The strategy

An observed arc is a set of observations (α_i, δ_i) of an object at time t_i. Here we indicate with α the right ascension and with δ the declination. If the arc is too short and the observations are few, the least squares algorithm (if it converges) generally gives orbits with very large uncertainty. The only information that one can get through interpolation corresponds to the values of right ascension, declination and their corresponding rates at a mean epoch of observations. The vector formed by these quantities is called attributable and is denoted as

$$\mathcal{A} = (\alpha, \delta, \dot{\alpha}, \dot{\delta}).$$

The linkage problem can be formalized as the problem of finding an orbit compatible with two given attributables $\mathcal{A}_1$ and $\mathcal{A}_2$ at time t_1 and t_2 respectively. This means to find the missing range (indicated with ρ) and range rate (indicated with $\dot{\rho}$) either at time t_1 or t_2. Among the various methods we concentrate on the method based on the Keplerian integrals. The basic idea is the following. If we suppose that $\mathcal{A}_1$ and $\mathcal{A}_2$ correspond to the same object and that the motion between times t_1 and t_2 is Keplerian, then the integrals of the two body problem must be preserved. Expressing position and velocity in spherical coordinates (ρ, α, δ) we can equate respectively the values that energy, angular momentum and Lenz vector take at t_1 and t_2. In this way we get a system of seven equations in the four unknowns $(\rho_1, \dot{\rho}_1, \rho_2, \dot{\rho}_2)$. The system is over-determined, and just four equations are enough. In Gronchi *et al.* (2010) the authors consider the angular momentum and the energy. By squaring, a polynomial system of total degree 48 is obtained. Having polynomial equations allows to compute the solutions in an efficient way. At this point we have two cases. If the system has no solutions, then the assumption that the two attributables belong to the same object is false and we discard the couple. If we have solutions, then they are tested with some compatibility conditions. These conditions are also based on the equations that we are not using to solve the system. If this control is passed we have a preliminary orbit that can act as a starting guess for the differential corrections. Note that the output orbits are endowed with covariance matrices. A test on simulated data has been performed, giving good results. Anyway, the high degree of the system could represent a problem when handling large databases. Therefore, in Gronchi *et al.* (2011), the method was improved choosing a suitable projection of the Lenz vector instead of the energy. By squaring, a polynomial system of total degree 20 was obtained.

We are going to consider a database of 80140 arcs of type 2. According to the definition given in Milani *et al.* (2007) these arcs show a significant curvature and can be split in at least two disjoint arcs, each without a significant curvature. A single apparition orbit determination with this kind of arcs does not give reliable results. Hence we split every arc in order to have tracklets that do not show curvature and from them we derive the corresponding attributable. In the end we are left with a database of 198947 attributables.

A brute force approach would lead to a N^2 computational complexity, where N is the total number of attributables. This implies a very long computational time. With some algorithm of sorting the complexity can be reduced to $N \log N$. Anyway as N is large, the computational time still remains too large. In order to have a reduced database we set some filters as described in Gronchi *et al.* (2010). In particular we set the minimum time span between the attributables at 100 days and the maximum at 365 days. In this way, we are not considering the identifications given by the reassembling of the previously splitted arcs.

A proposed orbit, to be considered reliable, needs to be compared with the observations. Only if we have enough observations and the residuals are sufficiently low, the orbit can be considered reliable. Hence, after the linkage, we try and find new observations that can fit the proposed orbit. Once we have more observations, the differential corrections can be performed and new residuals appear. This procedure is generally called attribution and is described in Milani *et al.* (2001).

From a list of attributables, the linkage procedure gives a new list of orbits and a left-over database of attributables. The attribution procedure will try to match orbits and attributables, improving the first ones. This latter list cannot be considered as a final result. Indeed duplicates and contradictions can be created. A method to deal with this problem has been introduced in Milani *et al.* (2005). A list is considered complete when there is no couple of identifications sharing the same attributable. The way to treat discordant identifications is crucial. We remember that two identifications are called discordant if they share some tracklets and not all the tracklets constituting one identification are contained in the other. On one side we can choose the one that, according to some quality parameters, is the best. On the other hand, we can try to merge two discordant identifications into a longer one. In this latter case we would have a new identification composed by more observations than the previous ones. The merging process could introduce new duplicates and contradictions, hence a new control has to be done, without merging, in order to have a normalized list and conclude the identification procedure.

3. Results and conclusion

The results of the application to the reduced database is a set of 580 identifications that can be summarized in the following table.

number of attributables	2	3	4	5	6	7
number of identifications	228	136	119	42	4	3

The identifications are cataloged according to the number of joined tracklets, e.g. we have 42 identifications with 5 tracklets. Those constituted by 5 or more tracklets can be considered good candidate orbits and should be submitted to a more stringent control on the residuals. The results of this control, together with an application to a larger database of tracklets will be presented in a forthcoming paper.

Acknowledgments

The work is supported by the Marie Curie Initial Training Network Stardust, FP7-PEOPLE-2012-ITN, Grant Agreement 317185.

References

Granvik, K., Muinonen, K., Virtanen, J., Delbò, M., Saba, L., De Sanctis, G., Morbidelli, R., Cellino, A., & Tedesco, E. 2005, In *Dynamics of Populations of Planetary Systems,* eds. Knežević, Z and Milani, A. (Cambridge University Press)

Granvik, K. & Muinonen, K. 2008, *Icarus*, 198, 130

Gronchi, G. F., Dimare, L., & Milani, A. 2010, *Cel. Mech. & Dyn. Astr.*, 107, 299
Gronchi, G. F., Farnocchia, D., & Dimare, L. 2011, *Cel. Mech. & Dyn. Astr.*, 110, 257
Milani, A., Sansaturio, M. E., & Chesley, S. R. 2001, *Icarus*, 151, 150
Milani, A., Gronchi, G. F., de' Michieli Vitturi, M., & Knežević, Z. 2004, *Cel. Mech. & Dyn. Astr.*, 90, 59
Milani, A., Gronchi, G. F., & Knežević, Z. 2007, *Earth, Moon and Planets*, 100, 83
Milani, A., Gronchi, G. F., Knežević, Z., Sansaturio, M. E., & Arratia, O. 2005, *Icarus*, 179, 350
Taff, L. G. & Hall, D. L. 1977, *Cel. Mech. & Dyn. Astr.*, 16, 481

Complex Planetary Systems
Proceedings IAU Symposium No. 310, 2014
Z. Knežević & A. Lemaitre, eds.

doi:10.1017/S1743921314008114

Averaged changes in the orbital elements of meteoroids due to Yarkovsky-Radzievskij force

Galina O. Ryabova[1]

[1]Tomsk State University, 634050 Tomsk, Russian Federation
email: ryabova@niipmm.tsu.ru

Abstract. Yarkovsky-Radzievskij effect exceeds the Poynting-Robertson effect in the perturbing action on particles larger than 100 μm. We obtained formulae for averaged changes in a meteoroid's Keplerian orbital elements and used them to estimate dispersion in the Geminid meteoroid stream. It was found that dispersion in semi-major axis of the model shower increased nearly three times on condition that meteoroids rotation is fast, and the rotation axis is stable.

Keywords. meteoroids, methods: analytical, celestial mechanics

---

To obtain the averaged changes in a meteoroid's orbit we use equations for the derivatives of the orbital elements in the form suggested by Burns (1976), namely $db/dt = F(a, ..., \omega, v, E, m, r, F_r, F_t, F_n)$, where b is one of the standard Keplerian orbital elements a, e, i, Ω, ω; r is the heliocentric distance, m is the meteoroid mass, v is the true anomaly, E is the eccentric anomaly, and F_r, F_t, F_n are the radial, transverse, and normal components of the perturbing force. As in (Burns *et al.* 1979), we average the time variation of the elements over the orbital period under the assumption that a and e are essentially constant over this time interval and that the angular momentum is conserved. The average change in any element b is then:

$$\left\langle \frac{da}{dt} \right\rangle = \frac{1}{PH} \oint \frac{da}{dt} r^2 \, dv, \tag{1}$$

where P is the orbital period and $H = [a\mu(1-e^2)]^{1/2}$ is the orbital angular momentum of the meteoroid per unit mass, μ is the gravitational constant of the sun.

The perturbing Yarkovsky-Radzievskij force for fast rotators, according to Burns *et al.* (1979) is

$$F_Y = k_Y r^{-7/2}, \qquad k_Y = 2.962\, r_m^2 c^{-1} \sigma^{1/4} (1-\alpha)^{1/4} \gamma \omega_m^{-1/2} S_0^{7/4} r_0^{7/2} \cos\xi, \tag{2}$$

where $(1/\gamma) = 300$ J/(m^2·s$^{1/2}$·K) is the thermal inertia, r_m is the radius of the meteoroid, $\alpha = 0.1$ is the albedo, $\omega_m = 10^4$ rad/s is the angular velocity of the meteoroid ($r_m = 1$ mm), S_0 is the solar constant at heliocentric distance $r_0 = 1$ au, ξ is the angle between the rotational axis of the particle and its orbital plane ($\xi = 45°$ for prograde, or $\xi = 135°$ for retrograde rotation), c is the speed of light, and σ is the Stefan-Boltzmann constant.

The values of the parameters in (2) correspond to the model accepted by Olsson-Steel (1987). As in (Olsson-Steel 1987), we assume that $F_r = F_t = F_n = F_Y/\sqrt{3}$ due to the precession of the rotation axis and the frequent changes in the rotation speed. Even now, when computers are much faster, the numerical integration of (1) is still very expensive, so analytic expressions for the averaged changes in the orbital elements were found:

$$< \dot{a} >= k_1 I_{52}, \qquad < \dot{e} >= k_2(I_{32c} + eI_{12} + I_{12c}), \qquad < i >= k_2 I_{12c} \cos\omega,$$
$$< \dot{\Omega} >= k_2 I_{12c} \sin\omega / \sin i, \quad < \dot{\omega} >= -k_2 I_{32c}/e - \cos i < \dot{\Omega} >,$$

Table 1. Coefficients of expansion (3)

Integrals	p_0	p_1	p_2	p_3	p_4	p_5	p_6	p_7	p_8
I_{52}	2	0	$\frac{15}{8}$	0	$-\frac{15}{2\cdot 16^2}$	0	$-\frac{25}{2\cdot 16^3}$	0	$-\frac{1575}{2\cdot 16^5}$
I_{32c}	0	$\frac{3}{2}$	0	$-\frac{3}{4\cdot 16}$	0	$-\frac{15}{2\cdot 16^2}$	0	$-\frac{315}{2\cdot 16^4}$	0
I_{12}	2	0	$-\frac{1}{8}$	0	$-\frac{15}{2\cdot 16^2}$	0	$-\frac{105}{2\cdot 16^3}$	0	$-\frac{15015}{2\cdot 16^5}$
I_{12c}	0	$\frac{1}{2}$	0	$\frac{3}{4\cdot 16}$	0	$\frac{35}{8\cdot 16^2}$	0	$-\frac{1155}{2\cdot 16^4}$	0

where

$$\begin{aligned} k_1 &= k_Y/[\pi\sqrt{3\mu}\, ma^2(1-e^2)^{5/2}], & k_2 &= k_Y/[2\pi\sqrt{3\mu}\, ma^3(1-e^2)^{3/2}], \\ I_{52} &= \oint (1+e\cos v)^{5/2}\, dv, & I_{32c} &= \oint (1+e\cos v)^{3/2} \cos v\, dv, \\ I_{12} &= \oint (1+e\cos v)^{1/2}\, dv, & I_{12c} &= \oint (1+e\cos v)^{1/2} \cos v\, dv. \end{aligned}$$

The integrals I are easily found as series by expanding the integrands using the binomial formula

$$I = \pi \sum_{j=0}^{\infty} p_j e^j. \tag{3}$$

Accuracy to 1% is achieved by keeping terms to eighth order in the eccentricity. The expansion coefficients p_j in (3) are given in Table 1.

Using the formulae we made some estimations for the Geminid meteoroid stream. The model of the stream was like in (Ryabova 2007) and the meteoroid mass was taken as 3×10^{-3} g. Direction of rotation (prograde or retrograde) for each meteoroid was chosen using a pseudorandom number generator. The dispersion of the Geminids is anisotropic, so the total dispersion and the dispersion observed at the Earth differ. The width of the model shower is about $1.5 - 2.5°$ in solar longitude (Ryabova 2007). With YR-addition the width increases by 0.3°, remaining less than the observed width 5° (Fox *et al.* 1983). As to the semi-major axis, YR-effect increases Δa for the model *shower* nearly 3 times. We found that YR-effect, being the mass-dependant, increases the mass separation in the stream about twice.

Acknowledgement

The author acknowledges the financial support from IAU. Some part of funding was provided by Tomsk State University Competitiveness Improvement Program. The author is grateful to the anonymous reviewer.†

References

Burns, J. A. 1979, *Am. J. Phys.*, 44, 944

Burns, J. A., Lamy, P. L., & Soter, S. 1979, *Icarus*, 40, 1

Fox, K., Williams, I. P. & Hughes, D. W. 1983, *MNRAS*, 205, 1155

Olsson-Steel, D. 1987, *MNRAS*, 226, 1

Ryabova, G. O. 1990, *Sol. Syst. Res.*, 24, 169

Ryabova, G. O. 2007, *MNRAS*, 375, 1371

† Part of this research was firstly fulfilled about 25 years ago. It remained unnoticed to meteor astronomers community, because the English translation (Ryabova 1990) of the paper was published in a hard-to-reach journal and, until recently, was absent from SAO/NASA ADS database. The method was carefully revised and some corrections were introduced.

Complex Planetary Systems
Proceedings IAU Symposium No. 310, 2014
Z. Knežević & A. Lemaitre, eds.

doi:10.1017/S1743921314008126

Explosive evolution of small bodies in planetary atmospheres

Subhon Ibadov[1,2] **and Firuz S. Ibodov**[1]
[1]Lomonosov Moscow State University, SAI, Moscow, Russia
email: ibadovsu@yandex.ru
[2]Institute of Astrophysics, TAS, Dushanbe, Tajikistan

Abstract. The entry of small celestial bodies such as cometary nuclei, asteroids and their large fragments, into a planetary atmosphere is accompanied by an "explosion", i.e., sudden rise in brightness and the generation of a "blast" shock wave, like the 2013 Chelyabinsk event. Fully analytic approach to the phenomenon is developed taking into account aerodynamic crushing of the body and transversal expansion of the crushed mass, that leads to impulse generation of hot plasma and a "blast" shock wave in the thin "exploding" layer.

Keywords. planets: atmospheres, comets: general, minor planets, asteroids, aerodynamic crushing, hot plasma

1. Introduction

Explosive evolution of small celestial bodies (cometary nuclei, asteroids, and their fragments, meteoroids) in planetary atmospheres is well known phenomenon. The physical mechanism of this phenomenon has been developed over a long time (see, e.g., Levin 1956; Grigorian *et al.* 2013 and references therein). Explosion of the bodies can be explained by their aerodynamic crushing at entry into the dense atmospheric layers accompanied by lateral expansion of the crushed mass. This idea was presented and justified theoretically by Grigorian (1980). The theory was further developed in connection with the 100th anniversary of the 1908 Tunguska phenomenon (Ibadov *et al.* 2008). Analytico-numerical theories were proposed for the collision of Comet Shoemaker-Levy 9 with Jupiter (Fortov *et al.* 1996 and references therein).

We develop a fully analytical approach to the explosive evolution of small bodies in planetary atmospheres, like the 15 February 2013 Chelyabinsk superbolide explosion.

2. Explosive evolution of small bodies

Celestial bodies enter planetary atmosphere with velocities $V_0 \gtrsim 10-20$ km/s, so that the specific kinetic energy of bodies exceeds the sublimation energy of their material, 10^{10} erg/g, more than 10–100 times. Hence, sudden thermalization of the kinetic energy of bodies will be accompanied by explosive high-temperature phenomena (cf. Ibadov 1987, 1990, 1996, 2012; Ibodov & Ibadov 2014).

The height for start of body's aerodynamic crushing in the atmosphere with density distribution as $\varrho_a(h) = \varrho_0 \exp(-h/H)$, is $h_* = H \ln \varrho_0 V_0^2/\sigma_*$. For the Chelyabinsk-like stony superbolide taking realistic values of the strength $\sigma_* = 10^7$ din/cm^2, entry velocity $V_0 = 20$ km/s, $\varrho_0 = 1.3 \times 10^{-3}$ g/cm^3 and height scale $H = 7$ km we get $h_* = 40$ km.

The height of maximum energy release, manifesting as bolide's "explosion" point, is

$$h_{max} = h_{ex} = h_* - H \ln(1 + C/b); \quad \Delta h_{ex} = 0.7H; \tag{2.1}$$

$$V_{ex} = V_0/e^{1/2}; \quad E_{ex} = \pi \varrho_m R_0^3 V_0^2/3e;$$

$$T_0 = Am_p V_0^2/[12ek(1+z+2x_1/3)]; \quad (2.2)$$

$$C = [3C_x R_0 \sin\alpha/(8H)]^{1/2}; b = 3C_x \varrho_0 H [\exp(-h_*/H)]/(4\varrho_m R_0 \sin\alpha).$$

Here Δh_{ex} is the thickness of the "exploding" layer: within this layer velocity of crushed body decreases from $0.9V_0$ to $0.1V_0$ and generation of a hot plasma takes place; C_x is the drag coefficient; α is the angle between the body entry velocity and the local horizon, ϱ_m and R_0 are the body density and initial radius, A is the mean atomic number, m_p is the proton mass, k is the Boltzmann constant, z is the mean charge multiplicity of plasma ions, x_1 is the mean relative ionization potential (cf. Ibadov *et al.* 2009; Ibodov & Ibadov 2011; Grigorian *et al.* 2013).

The height of flare of the Chelyabinsk superbolide found from observations, h_{ex}(obs.) = 23 km, indicates that the body had asteroidal nature (Marov & Shustov 2013). Using (2.1) and (2.2), we find the theoretical value of the initial radius of the supebolide, R_0(theor.) = 20 m, the energy of explosion, $E_{ex} = 4\times10^{22}$ erg, that corresponds to 400 kt TNT (Grigorian *et al.* 2013; cf. www.nasa.gov/mission_pages/asteroids/news/asteroid20130215.html).

The duration of conversion of the crushed body into plasma is $\tau_{ex} = \Delta h_{ex}/V_{ex}$ = 0.3 sec, while flight time of the superbolide was about 15 sec; the initial temperature of plasma in the exploding layer is $T_0 = 10^4$ K, i.e., more than the temperature of the solar photosphere. Similar processes occurred at the 1908 Tunguska explosion and collision of comet SL 9 with Jupiter on July 16–22, 1994 (Fortov *et al.* 1996; Ibodov *et al.* 2010).

3. Conclusions

The entry of small celestial bodies of both asteroidal and cometary nature into planetary atmosphere leads to impulse generation of a high-temperature plasma as well as intense electromagnetic radiation and "blast" shock wave. These topics are relevant to the complex problem of protection of the Earth from impact hazard as well as habitability of planets.

Acknowledgements

The authors are grateful to IAU for the grant to participate to IAUS 310, to reviewers for constructive comments, to Dr. G. M. Rudnitskij for useful discussions and to E. Suyunova for technical support.

References

Fortov, V. E., Gnedin, Yu.N., Ivanov, M. F. *et al.* 1996, *Sov. Phys. Usp.*, 39, 363

Grigorian, S. S. 1980, *Cosmic Res.*, 17, 724

Grigorian, S. S., Ibodov, F. S., & Ibadov, S. 2013, *Solar Sys. Res.*, 47, 268

Ibadov, S. 1987, *ESA*, SP-278, 655

Ibadov, S. 1990, *Icarus*, 86, 283

Ibadov, S. 1996, *Physical Processes in Comets and Related Objects*, Moscow, CI

Ibadov, S. 2012, *Planetary Nebulae, Proc. IAUS 283, CUP, 392*

Ibadov, S., Ibodov, F. S., & Grigorian, S. S. 2008, *Int. Conf. "100 Years Tunguska Phenomenon: Past, Present, Future", Moscow, RAS*

Ibadov, S., Ibodov, F. S., & Grigorian, S. S. 2009, *Universal Heliophysical Processes. Proc. IAUS 257*, CUP, 341

Ibodov, F. S.,Grigorian, S. S. & Ibadov, S. 2010 *Icy Bodies of the Solar System, Proc. IAUS 263*, CUP, 269

Ibodov, F. S. & Ibadov, S. 2011, *Advances in Plasma Astrophysics. Proc. IAUS 274*, CUP, 92

Ibodov, F. S. & Ibadov, S. 2014, *Nature of Prominences. Proc. IAUS 300*, CUP, 509

Levin, B. Yu 1956, *Physical Theory of Meteors*, Moscow, USSR Acad. Sci.

Marov, M. Ya & Shustov, B. M. 2013, *Geochemistry International*, 51, 587

Complex Planetary Systems
Proceedings IAU Symposium No. 310, 2014
Z. Knežević & A. Lemaitre, eds.

doi:10.1017/S1743921314008138

Close encounters of Near Earth Objects with large asteroids

Anatoliy Ivantsov[1,2], Siegfried Eggl[2], Daniel Hestroffer[2], William Thuillot[2] and Pini Gurfil[1]

[1]Faculty of Aerospace Engineering, Technion - Israel Institute of Technology, Technion City, IL-3200003, Haifa, Israel
email: ivantsov@tx.technion.ac.il

[2]Institut de Mécanique Céleste et de Calcul des Ephémérides, Observatoire de Paris, 77 avenue Denfert Rochereau, F-75014, Paris, France

Abstract. Close encounters of Near Earth Objects (NEOs) with large asteroids are a possible source of systematic errors in trajectory propagations and asteroid mitigation. It is, thus, necessary to identify those large asteroids that have to be considered as perturbers in NEO orbit modeling. Using the Standard Dynamical Model we searched for encounters between the 1649 numbered Near Earth Asteroids (NEAs) and 2191 large asteroids having sizes greater than 20 km. Investigating the 21st century A. D. we have found 791 close encounters with 195 different large asteroids that lead to a substantial scattering of NEOs.

Keywords. asteroids, Near-Earth objects, ephemerides

1. Introduction

Close encounters of NEAs with the nearby planets and the Moon are known to be the source of strong gravitational perturbations on the trajectories of the former. In contrast, the influence of massive asteroids has not received a lot of attention. Yet, close encounters with massive asteroids can be the sources of systematic errors in trajectory propagation, asteroid mitigation and sample return missions, see e.g. Eggl *et al.* (2013), Chesley *et al.* (2014) and references therein. The purpose of the current research is to identify those massive asteroids that should be considered in the trajectory propagation of known NEAs.

2. Dynamical Model and Massive Asteroids

The Standard Dynamical Model for orbit propagation of asteroids introduced by Giorgini *et al.* (2008) includes all relativistic gravitational forces caused by the Sun, the planets, the Moon, as well as Ceres, Pallas and Vesta. However, the present research differs in one important aspect from the standard model.

Since NEAs can have close encounters not only with the Earth, but also with the Moon, Venus and Mars, the additional forces due to asphericity of these bodies should be taken into consideration in some cases. The PHA (99942) Apophis, for instance, will have encounters closer than 0.1 au with other bodies than the Earth within the next 15 years. In fact, it comes close to Venus on April 24, 2016 (distance of 0.0782 au) and the Moon on April 14, 2029 (distance of 0.00064 au). For this reason we argue that it is necessary to include the influence of the J_2 gravitational harmonic for the Earth, the Moon, Venus, and Mars whenever a NEA appeared within the gravitational sphere of influence of one of the above mentioned bodies. The Sun's J_2 gravitational harmonic was considered as well.

Table 1. Sample from the list of closest encounters of NEOs with large asteroids.[1]

Perturbing asteroid	NEA	JD	Calendar date	Minimum distance, au	Relative speed, au/day	Hill's radius	Deflection, arcsec	Visual deflect., arcsec	Solar elong., deg.
1	8013	2454388	2007-10-14	0.02178	0.004119	12.6	0.225	0.185	147
52	184266	2458552	2019-03-09	0.00292	0.003119	4.1	0.127	0.057	13
1	3551	2461874	2028-04-12	0.02554	0.005619	15.7	0.103	0.090	150

Notes:
[1]The complete list encompasses 791 encounters between the numbered NEOs and 195 massive perturbing asteroids and is available `http://www.imcce.fr/~ivantsov/nea21cy.txt`

Cross-referencing asteroid diameters (D) listed in the NEOWISE project Mainzer *et al.* (2011) with JPL's Small-Body Database lead to a preselection of 2191 massive asteroids with D>20 km that were used as potential perturbers in our simulations. The bulk-density adopted for the calculation of gravitational parameters for the perturbing asteroids was 3 g/cm^3.

All asteroid trajectories were propagated using an Adams-Moulton integrator of variable step and order, while the positions and velocities of all the major bodies were interpolated from the recent JPL DE430 ephemerides. Initial conditions for the asteroids were taken from the JPL HORIZONS system. The equations of motion were integrated for a century starting from J2000.

3. Close Encounter Detection

We searched for those close encounters between the preselected asteroids greater than 20 km in size and all known 1649 NEAs that would result in a two-body deflection angle greater than 0.1 arcsec. Table 1 contains a small sample of the encounter parameters and observational conditions for specific close encounters between NEAs and perturbing asteroids. The visual deflection angle represents the deflection as visible from the Earth, thus allowing to estimate whether the deflection event will be or has been observable.

4. Conclusions

An accurate orbit prediction of NEOs requires a more detailed gravitational interactions than suggested in the Standard Dynamical Model. In the case of close encounters between NEOs and the nearby planets as well as the Moon planetary asphericities should be considered. Some encounters with larger asteroids are also non-negligible. Including only the three largest bodies in the main belt (Ceres, Pallas, Vesta) may lead to inaccurate orbit predictions. The gravitational effect of 195 main belt asteroids should be taken into account for the current NEO population. Rectifying the dynamical model is especially important for the orbital prediction of Potentially Hazardous Asteroids, such as (99942) Apophis. In a next step we will investigate the robustness of our results with respect to non-gravitational forces.

Acknowledgments: S. E. was supported by the European Union Seventh Framework Program under grant agreement no. 282703 and the IAU Symposium no. 310 travel grant.

References

Eggl, S., Ivantsov, A., Hestroffer, D., Perna, D., Bancelin, D., & Thuillot, W. 2013, *Proc. Journées SF2A-2013*, 237

Chesley, S. R., Farnocchia, D., Nolan, M. C., & 13 other authors. 2014, *Icarus*, 235, 5

Giorgini, J. D., Benner, L. A. M., Ostro, S. J., Nolan, M. C., & Busch, M. W. 2008, *Icarus*, 193, 1

Mainzer, A., Bauer, J., Grav, T., & 32 other authors. 2011, *ApJ*, 731, 53

Complex Planetary Systems
Proceedings IAU Symposium No. 310, 2014
Z. Knežević & A. Lemaitre, eds.

doi:10.1017/S174392131400814X

Determination of an Intermediate Perturbed Orbit using Multiple Observations

Vladimir A. Shefer

Tomsk State University, 634050 Tomsk, Russia
email: shefer@niipmm.tsu.ru

Abstract. Two methods are briefly stated for finding the preliminary orbit of a small celestial body from its three or more pairs of angular measurements and the corresponding time instants. The methods are based on using the approach that we previously developed for constructing the intermediate orbit from a minimum number of observations. This intermediate orbit allows for most of the perturbations in the motion of the body under study. The methods proposed use the Herget's algorithmic scheme that makes it possible to involve additional observations as well. Using the determination of orbits of some asteroids as examples, we compare the results obtained by applying the Herget multiple-observation algorithm and the proposed methods. The comparison shows that the proposed methods are an efficient means for studying perturbed motion. They are especially advantageous if applied to high-precision observational data covering short orbital arcs.

Keywords. celestial mechanics, methods: analytical; methods: statistical; methods: data analysis; techniques: high angular resolution; ephemerides, asteroids

---

With the advent of modern high-precision optoelectronic position measurements, which are one to three orders of magnitude more accurate than classical astrometric measurements, the problem of preliminary determination of perturbed orbit is particularly relevant. The methods proposed in (Shefer 2003, 2008) successfully resolve this problem from three and four positions of a small body on the celestial sphere. However, posing the problem of the determination of a reliable preliminary orbit from three or four positions only, we may face a number of challenges (Herget 1965; Marsden 1991). Because of this, it is desirable to have such an algorithm for constructing a preliminary perturbed orbit that would be able to use all the n ($n \geqslant 3$) available observations.

In this work, the approach developed by us (Shefer 2013) in respect to the problem of finding the intermediate perturbed orbit from all the available astrometric positions of the object, using Herget's idea (Herget 1965) of the introduction of additional observations, was applied. We used two methods for solving this problem: a two-parameter method and a six-parameter method (Shefer 2013). The methodical errors of orbit computation by the proposed methods are two orders of magnitude smaller than the corresponding errors of the commonly used approach based on the construction of an unperturbed Keplerian orbit. We studied the effectiveness of the methods in comparison with Herget's n-position method. Here, by Herget's method we understand a version of our two-parameter method, which ignore the action of perturbations. For short, the computational programs implementing the Herget's method, the two-parameter and six-parameter methods will be referred to as *H*, *S1*, and *S2*, respectively.

The objects whose orbits were used to compare the programs were the asteroids 1566 Icarus and 2010 TO48. The orbit of Icarus has an unusually large eccentricity for an asteroid, and the orbit of 2010 TO48 has a small inclination. In the motion of the asteroids, we took into account the perturbations from the eight major planets, Pluto, and the Moon based on the DE405/LE405 ephemerides.

We took the first seven observations of Icarus that were made at Palomar Mountain Observatory (California, USA) in 1949, from June 27 to July 24.

The asteroid 2010 TO48 was observed in one opposition only from September 4 to October 12, 2010. We took all the twelve observations of the asteroid that were acquired at the Steward Observatory (Kitt Peak-Spacewatch, Arizona, USA) and cover this time interval.

The basic observations in the *H* and *S1* programs were the first and last observations. The two reference position vectors used in the run of the *S2* program were constructed at times $t_a = t_1^{obs} - 0.01$ days and $t_b = t_n^{obs} + 0.01$ days.

Beyond the real observations, we used the model observations of the asteroids. For dates of the real observations, we constructed 10000 sets of positions on the celestial sphere whose deviations from the corresponding precise positions account for $0''.1$, $0''.01$, and $0''.001$.

For the selected real and model observations, the asteroid orbits were constructed by each of the three compared methods. We then performed the representation of all the observations in the set with full consideration of the perturbations. Among the 10000 representations obtained by each method for each of the errors, we selected the worst representation. Let σ be the mean-square error of the worst representation of a single observation.

In all examples under consideration, the methods proposed are more effective than Herget's method. In the case of Icarus, the *S1* and *S2* programs achieve maximum effectiveness at observation accuracies exceeding $0''.1$. The values of σ are less by a factor of about 10 than the corresponding values calculated using *H*. In the case of the asteroid 2010 TO48, *S1* and *S2* demonstrate high effectiveness for all the selected observation accuracies. Here the values of σ are less by a factor of about 10–40 than the corresponding values obtained using *H*. If the observational errors are less than $0''.01$, the proposed methods give almost identical results in terms of accuracy. When the observational errors are greater than $0''.001$, preference should be given to the *S2* program. The examples considered show that the use of the proposed methods can not only improve the representation of observations covering the reference time interval, but also provide more accurate forecast ephemeris, compared to the traditional methods. The higher the accuracy of the observations and the shorter the arc defined by these observations, the more accurate is the approximation of real motion with orbits constructed by the proposed methods. This is an extremely important advantage of the proposed methods over the algorithms of the traditional approach.

Thus, the numerical examples considered lead us to conclude that our methods are a highly effective research instrument allowing the determination of reliable parameters of perturbed motion already at the stage of preliminary orbit calculations.

Acknowledgement

The work was supported by the Ministry of Education and Science of the Russian Federation, project no. 2014/223(1567).

References

Herget, P. 1965, *AJ*, 82, 19
Marsden, B. G. 1991, *AJ*, 102, 1539
Shefer, V. A. 2003, *Solar Syst. Res.*, 37, 326
Shefer, V. A. 2008, *Solar Syst. Res.*, 42, 405
Shefer, V. A. 2013, *Solar Syst. Res.*, 47, 40

Complex Planetary Systems
Proceedings IAU Symposium No. 310, 2014
Z. Knežević & A. Lemaitre, eds.

doi:10.1017/S1743921314008151

Dynamical evolution of objects on highly elliptical orbits near high-order resonance zones

Eduard D. Kuznetsov and Stanislav O. Kudryavtsev

Kourovskaya Astronomical Observatory, Ural Federal University,
Lenin ave., 51, Ekaterinburg, 620000 Russia
email: eduard.kuznetsov@urfu.ru

Abstract. Both analytical and numerical results are used to study high-order resonance regions in the vicinity of Molnya-type orbits. Based on data of numerical simulations, long-term orbital evolution are studied for objects in highly elliptical orbits depending on their area-to-mass ratio. The Poynting–Robertson effect causes a secular decrease in the semi-major axis of a spherically symmetrical satellite. Under the Poynting–Robertson effect, objects pass through the regions of high-order resonances. The Poynting–Robertson effect and secular perturbations of the semi-major axis lead to the formation of weak stochastic trajectories.

Keywords. celestial mechanics, highly elliptical orbit, high-order resonance, satellite, stochastic trajectory

---

1. Introduction

Region of highly elliptical orbits (HEO) has a very complex dynamics. Both active and passive objects are moved on HEO. There is a problem of protecting active satellites from space debris. It requires high-accuracy propagation of HEO objects motion. These objects have a long-term evolution of eccentricities and inclinations due to the Lidov–Kozai resonance (Lidov (1962) and Kozai (1962)). There are secular perturbations of semi-major axes due to the atmospheric drag. The Poynting–Robertson effect also leads to secular perturbations of semi-major axes for objects with area-to-mass ratio (AMR) more than 1 m^2/kg (Kuznetsov *et al.* (2012)). The dynamical evolution of high AMR objects in the vicinity of Molniya-type orbits was studied by Sun *et al.* (2013). In this paper, a vicinity of Molniya orbit is considered. There are several high-order resonance zones in this region. A stochastic trajectory formation due to objects passage through high-order resonance zones was considered.

2. High-order resonances

Frequencies of perturbations caused by the effect of sectoral and tesseral harmonics of the Earth's gravitational potential are linear combinations of a satellite's mean motion n_M, angular velocities of pericenter motion n_g and node motion n_Ω of it's orbit, and angular velocity of the Earth ω.

Following Allan (1967a) and (1967b), we form the frequencies

$$\nu_1 = p(n_M + n_\Omega + n_g) - q\omega, \quad \nu_2 = p(n_M + n_g) + q(n_\Omega - \omega), \quad \nu_3 = pn_M + q(n_g + n_\Omega - \omega),$$

where p, q are integers.

The condition $\nu_1 \approx 0$ corresponds to $p{:}q$ resonance between the satellite's mean motion n_M and the Earth's angular velocity ω. This condition represents n-resonance. The condition $\nu_2 \approx 0$ corresponds to i-resonance under which a position of the ascending node of an orbit repeats periodically in a rotating coordinate system. The condition $\nu_3 \approx 0$ corresponds to e-resonance at which a position of the line of apsides is considered.

We estimated the values of semi-major axis corresponding to the n-, i- and e-resonances from the conditions $\nu_1 = 0$, $\nu_2 = 0$, and $\nu_3 = 0$ in the vicinity of the Molniya orbit. Initial conditions corresponded to HEO with the eccentricity 0.65 and the critical inclination 63.4°. Semi-major axis values varied from 26000 km to 27100 km. There were 17 high-order resonance relations $p{:}q$ between the mean motion of angular orbital elements and the Earth's angular velocity: $16 \leqslant |p| \leqslant 25$, $33 \leqslant |q| \leqslant 49$. The orders of resonance are $49 \leqslant |p| + |q| \leqslant 74$.

3. Stochastic trajectory formation

To conduct of the long-term orbital evolution research, we used "A Numerical Model of Artificial Satellite Earth Motion" created by our colleagues from the Tomsk State University Research Institute of Applied Mathematics and Mechanics (Bordovitsyna *et al.* (2007)). The model of disturbing forces accounted for the nonsphericity of gravitational field of the Earth (model EGM96, harmonics up to the 27th order and degree inclusive), the attraction of the Moon and the Sun, the tides in the Earth's body, the direct radiation pressure, taking into account the shadow of the Earth (the reflection coefficient of the satellite surface $k = 1.44$), the Poynting–Robertson effect, and the atmospheric drag. The model utilized Everhart's method of the 19th order.

The initial conditions corresponded to HEO with the eccentricity $e_0 = 0.65$ and critical inclination $i_0 = 63.4°$. Initial semi-major axes a_0 values were consistent with resonant conditions arisen from the analytical approximation. AMRs tried were equal to 0.02, 0.2, and 2 m^2/kg. The time span for simulation was 24 years.

The Poynting–Robertson effect causes a secular decrease in the semi-major axis of a spherically symmetrical satellite. Secular decrease in the semi-major axis, for a satellite with AMR = 2 m^2/kg near Molniya orbit, equals approximately 0.5 km/year. The effect weakens slightly, in resonance regions. Under the Poynting–Robertson effect, objects pass through high-order resonances regions. The Poynting–Robertson effect and secular perturbations of semi-major axis lead to the formation of weak stochastic trajectories.

Acknowledgement

Authors would like to thank Mrs. Anna Glazyrina for reading and correcting the paper.

This work was supported by the Russian Foundation for Basic Researches (grant 13-02-00026a).

References

Allan, R. R. 1967, *Planet. Space Sci.*, 15, 53

Allan, R. R. 1967, *Planet. Space Sci.*, 15, 1829

Bordovitsyna, T. V., Baturin, A. P., Avdyushev, V. A., & Kulikova, P. V. 2007, *Izv. Vyssh. Uchebn. Zaved., Fiz.*, 50, 60

Kozai, Y. 1962, *AJ*, 67, 591

Kuznetsov, E. D., Zakharova, P. E., Glamazda, D. V., Shagabutdinov, A. I., & Kudryavtsev, S. O. 2012, *Solar System Research*, 46, 442

Lidov, M. L. 1962, *Planet. Space Sci.*, 9, 719

Sun, R.-y., Zhao, C.-y., Zhang, M.-j., & Hou, Y.-G. 2013, *Adv. Sp. Res.*, 51, 2136

Complex Planetary Systems
Proceedings IAU Symposium No. 310, 2014
Z. Knežević & A. Lemaitre, eds.

doi:10.1017/S1743921314008163

On relative equilibria of mutually gravitating massive point and triangular rigid body

Vasily I. Nikonov

Lomonosov Moscow State University, Department of Mechanics and Mathematics
email: nikon_v@list.ru

Abstract. Planar motions of a triangular body and a massive point under the action of mutual Newtonian attraction are studied. For the first formulation the triangle is assumed to be composed of three massive points. For the second formulation it is constructed with three homogeneous rods. Some partial solutions are observed within the analysis of the geometry of mass distribution.

The investigation is motivated by the problem of motion of spacecrafts near asteroid-like celestial objects possessing irregular mass distribution. Comparison of dynamical effects for two types of mass distribution is another goal of the research.

Problems appearing because of irregularities in mass distributions have been known for a long time. Certain approaches to the description of motions under attraction as well as qualitative particulars of dynamics are discussed by Demin (1963), Burov & Karapetyan (1995), Buchin & Burov & Troger (2008), Burov & Guerman & Sulikashvili (2010), Kholshevnikov & Kuznetsov (2011), Beletsky & Rodnikov (2011) and Scheeres (2012).

Keywords. celestial mechanics, generalized two-body problem, relative equilibria, stability, Poincaré bifurcation diagram, barycentric coordinates

---

Consider a motion of $\Delta P_1 P_2 P_3$ and the point mass P in the fixed plane under the action of mutual attraction. The triangular object $\Delta P_1 P_2 P_3$ is assumed to be composed of masses m_1, m_2 and m_3 located at its vertices ("point triangle", PT), or by homogeneous rods P_1P_2, P_2P_3 and P_1P_3 of masses m_3, m_1 and m_2 respectively ("wire triangle", WT). Let P have mass m.

Denote $M = m_1 + m_2 + m_3$, $|P_1P_2| = \ell_3 > 0$ (1,2,3), where (1,2,3) is the cyclic permutation of indices. Let C and S be the centers of mass of the triangle and the whole system respectively, and C_f be the point where the gravitational forces generated by the the triangle vanish. Then, the following statement is true.

Assertion. If $C = C_f$, then for any value of m there exist steady motions of the triangle with an arbitrary angular velocity, such that $C = C_f = P$.

For PT the condition $C = C_f$ requires C to be center of the circumscribed circle, i.e.

$$m_1 = \frac{\ell_1^2(-\ell_1^2 + \ell_2^2 + \ell_3^2)}{16\mathbb{S}^2} \quad (1,2,3), \quad \mathbb{S} = \sqrt{p(p-\ell_1)(p-\ell_2)(p-\ell_3)}, \quad p = \frac{\ell_1 + \ell_2 + \ell_3}{2}$$

For WT this condition is true if and only if $m_1 = m_2 = m_3$.

In case of PT, positivity of masses implies that all angles of the triangle are acute.

Using the Routh method (see Routh (1877)), one can investigate in general steady motions of the systems and their degrees of instability. If (x_1, x_2, x_3) are barycentric coordinates (BC) of the point P with respect to the triangle $P_1P_2P_3$, then the Routh

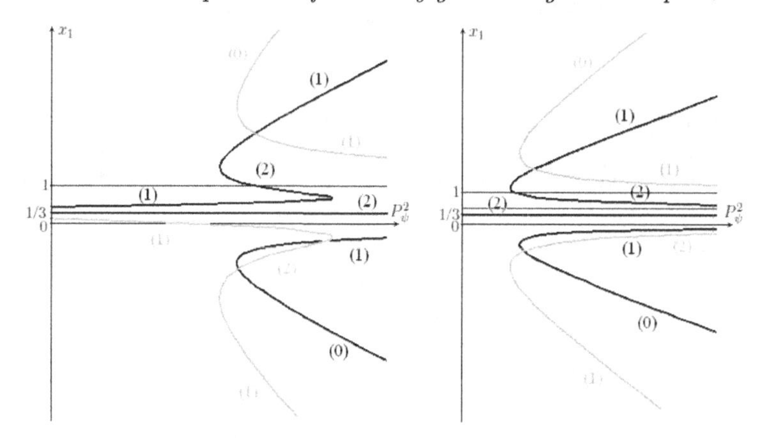

Figure 1. Poincare diagram: point and wire triangles

function leads to $W_\lambda = U + \lambda(x_1 + x_2 + x_3 - 1)$

$$U = \begin{cases} \dfrac{1}{2}\dfrac{P_\psi^2}{J} - mG\displaystyle\sum_{i=1}^{3}\frac{m_i}{r_i} & for\ PT, \\ \dfrac{1}{2}\dfrac{P_\psi^2}{J} - mG\displaystyle\sum_{(1,2,3)}\frac{m_1}{\ell_1}\ln\frac{r_2 + r_3 + \ell_1}{r_2 + r_3 - \ell_1} & for\ WT, \end{cases} \qquad r_i = (\overrightarrow{PP_i}, \overrightarrow{PP_i})^{1/2}$$

where P_ψ is the constant of total angular momentum, G is the gravitational constant, and J is the moment of inertia of the whole system with respect to the axis perpendicular to the plane of motion and passing through the point S.

Steady configurations are determined as critical points of the function $W_\lambda(x_1, x_2, x_3, \lambda)$. Analyzing these equations one can easily discover families of steady motions, such that the point P locates at one of the axes of dynamical symmetry of the triangle (see Nikonov (2014) for PT). Appropriate Poincaré bifurcation diagrams are drawn in Fig 1. There the figures near the curves express degrees of instability. In particular, the figure "0" corresponds to the stability of steady motion in the sense of Lyapunov. The Poincaré bifurcation diagram for PT drawn before in Nikonov (2014) is improved here.

Acknowledgement

The author acknowledges support of IAU and of RFBR (12-08-00637, 12-08-00591).

References

Aksenov, E. P., Grebenikov, E. A., & Demin, V. G. 1963, *Soviet Astron.*, 7, 276

Burov, A. A. & Karapetyan, A. V. 1995, *Mechanics of solids*, 30, 14

Burov, A. A., Guerman, A. D., & Sulikashvili, R. S. 2010, *Dynamics of a Tetrahedral Satellite-Gyrostat*, AIP Conference Proceedings, ICNAAM 2010, 1281, 465

Kholshevnikov, K. V. & Kuznetsov, E. D. 2011, *Celest. Mech. Dyn. Astr.*, 109, 201

Scheeres, D. J. 2012, *Orbital Motion in Strongly Perturbed Environmentss* Berlin: Springer. 390 p.

Beletsky, V. & Rodnikov, A. 2011, *Proc. of 5th PhysCon*, Leon, Spain.

Buchin, V., Burov, A., & Troger, H. 2008, *6th EUROMECH Nonlinear Dynamics Conference. ENDC 2008*, IPME RAS.

Routh, E. J. 1877, *Treatise on the Stability of a Given State of Motion.* L:MacMillan.

Nikonov, V. 2014, *Moscow University Mechanics Bulletin*, 69, 44

Complex Planetary Systems
Proceedings IAU Symposium No. 310, 2014
Z. Knežević & A. Lemaitre, eds.

doi:10.1017/S1743921314008175

The family of Quasi-satellite periodic orbits in the circular co-planar RTBP

Alexandre Pousse[1,2], Philippe Robutel[1] and Alain Vienne[1]

[1]IMCCE - Observatoire de Paris, Université Lille 1, Université Paris 06
77 avenue Denfert Rochereau, 75014 Paris, France
[2]email: apousse@imcce.fr

Abstract. In the circular case of the coplanar Restricted Three-body Problem, we studied how the family of quasi-satellite (QS) periodic orbits allows to define an associated libration center. Using the averaged problem, we highlighted a validity limit of this one: for QS orbits with low eccentricities, the averaged problem does not correspond to the real problem. We do the same procedure to L_3, L_4 and L_5 emerging periodic orbits families and remarked that for very high eccentricities $\mathcal{F}_{L_4}$ and $\mathcal{F}_{L_5}$ merge with $\mathcal{F}_{L_3}$ which bifurcates to a stable family.

Keywords. celestial mechanics, co-orbital motion, quasi-satellites, periodic orbits

1. Introduction

In the framework of the Restricted Three-body Problem (RTBP), we consider a primary whose mass is equal to one, a secondary in circular motion with a mass ε and a massless third body; the three bodies are in coplanar motion and in co-orbital resonance configuration. We actually know three classes of regular co-orbital motions: in rotating frame with the planet, the tadpoles orbits (TP) librate around Lagrangian equilibria L_4 or L_5; the horseshoe orbits (HS) encompass the three equilibrium points L_3, L_4 and L_5; the quasi-satellites orbits (QS) are remote retrograde satellite around the secondary, but outside of its Hill sphere .

Contrarily to TP orbits, the QS orbits do not emerge from a fixed point in rotating frame as these orbits are always eccentric. Thus, we can reformulate QS definition in terms of elliptical heliocentric orbits which librate, in rotating frame, around the planet position. The QS libration center (l.c.) can be materialized in rotating frame by a one-parameter family of periodic orbits. Our goal is to study this family of periodic orbits in the plane.

2. Method

With respect to the primary, we denote (a', λ') the elliptic elements of the secondary and (a, e, ω, λ) those of the third body. To reduce the dimension of the problem to 4, we introduce the resonant angle $\theta = \lambda - \lambda'$ and average the Hamiltonian with respect to λ'. We use the method developed by Nesvorný *et al.* (2002) which is valid for all values of the eccentricity, to compute the averaged Hamiltonian and equations of motions.

In addition to θ and ω, we use the variables $u = (\sqrt{a} - \sqrt{a'})/\sqrt{a'}$ and $\Gamma = \sqrt{a}(1 - \sqrt{1-e^2})$ which is a first integral. The variable ω being ignorable, the problem possesses one degree of freedom – (θ, u) – for a given Γ. Instead of using Γ, it is convenient to introduce e_0, the eccentricity value when $u = 0$.

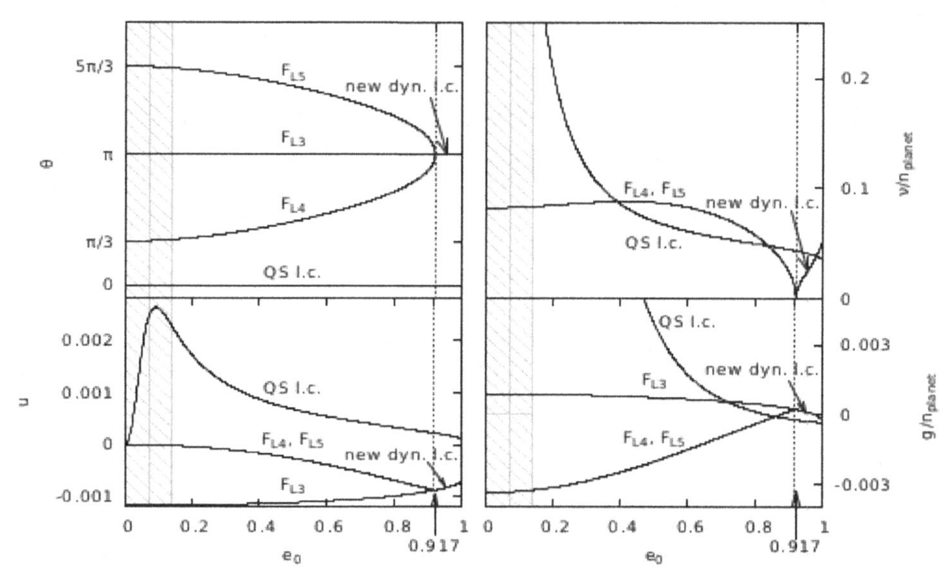

Figure 1. Fixed points and frequencies evolution versus e_0 for $\varepsilon = 0.001$ (Sun-Jupiter system) in the averaged coplanar RTBP in circular case.

In these coordinates, the QS family is represented by a family of fixed points parametrized by e_0. When it is stable, each of these equilibria, processes a eigenfrequency ν that corresponds to the libration frequency around this point in the plan (θ, u), and a second one g in the normal direction associated to (ω, Γ). This last frequency is also the precession rate of the third body's pericenter.

We developed a numerical method to find this fixed point near the origin of the (θ, u) plane in each e_0 and calculated the two frequencies on it. To compare QS with TP and HS, we do the same procedure to $\mathcal{F}_{L_3}$, $\mathcal{F}_{L_4}$ and $\mathcal{F}_{L_5}$ (the family of periodic orbits that originate from L_3, L_4 and L_5). The results obtained are presented in Fig. 1.

3. Discussion

Since QS orbits with low values of eccentricity imply close encounters with the secondary, the averaged problem does not represent the real problem in these conditions (phenomena also observed in Robutel & Pousse (2013) in the planetary Three-body problem). In Fig. 1, we remark that when $e_0 < 0.18$, the frequencies of the QS family are of the same order than the mean motion. This gives us a region where QS motion can not being studied with the averaged problem (hatched domain). We also present a particular orbit for e_0 close to 0.8352 : g crosses zero and highlights a QS frozen ellipse in the fixed frame.

For a very high eccentricity ($e_0 > 0.917$), an unexpected result is that $\mathcal{F}_{L_4}$ and $\mathcal{F}_{L_5}$ merge with $\mathcal{F}_{L_3}$ which bifurcates to a stable periodic orbits family. This implies the disappearance of TP and HS and the appearance of a new type of orbits. These orbits librate around the point diametrically opposed to the secondary, relative to the primary.

References

Nesvorný, D., Thomas, F., Ferraz-Mello, S., & Morbidelli, A. 2002, *Cel. Mech. Dyn. Astr.*, 82, 323–361

Robutel, P. & Pousse, A. 2013, *Cel. Mech. Dyn. Astr.*, 117, 17–40

Complex Planetary Systems
Proceedings IAU Symposium No. 310, 2014
Z. Knežević & A. Lemaitre, eds.

doi:10.1017/S1743921314008187

Excluding interlopers from asteroid families

Viktor Radović and Bojan Novaković

Department of astronomy, Faculty of Mathematics, University of Belgrade,
Studentski trg 16, 11000 Belgrade, Serbia
email: rviktor@matf.bg.ac.rs, bojan@matf.bg.ac.rs

Abstract. To study an asteroid family it is crucial to determine reliably the list of its members, i.e. to reduce the number of interlopers as much as possible. However, as the number of known asteroids increases fast it becomes more and more difficult to obtain robust list of members of an asteroid family. To cope with these challenges we are proposing a new approach that may help to significantly reduce presence of interlopers among the family members.

Keywords. asteroids: general, impact processes, asteroid families

1. Introduction

Asteroid families are first discovered in 1918 by Japanese astronomer Hirayama, and since then a large number of families have been discovered in the main-belt (e.g. Zappala *et al.*, 1990; Nesvorný *et al.*, 2005; Novaković *et al.*, 2011; Brož *et al.*, 2013; Milani *et al.*, 2014). These groups, which are believed to be products of catastrophic collisions between asteroids, are typically identified in the space of proper orbital elements: semi-major axis (a_p), eccentricity (e_p) and sine of inclination ($\sin I_p$). Proper elements have been used because they are nearly constant over time (Knežević and Milani, 2003).

For family identification the Hierarchical Clustering Method (HCM, Zappalà *et al.*, 1990) is often used. HCM identifies an asteroid as part of family if its distance from closest neighbor is smaller than an adopted cut-off distance (v_c). However, there is no strict rule to determine v_c, because it depends on various factors, for example a location of family within the main-belt. Moreover, HCM can not distinguish whether an asteroid is really part of a family or an interloper. These impose important constraints in the ability to reliably identify members of a family. Here we propose a new, HCM based, approach to exclude interlopers from the list of family members.

2. Method

The proposed approach consists of four main steps. In the first step, HCM is applied to the catalog of proper elements using different cut-off values. Starting from $v_c = 5$ m/s, it increases until family merges with background asteroids. For a threshold value we use the one two steps below the distance at which family merges with background. In this way, we get an initial list of the family members.

The second step is to identify interlopers among members of the initial family. As it is well known, to identify interlopers we have to use physical and spectral properties of candidate family members. For this purpose we use the SDSS colors (Ivezić *et al.*, 2001), WISE albedos (Masiero *et al.*, 2011) and other available spectroscopic data. For each data-set we defined criteria to decide if an asteroid is interloper or not. For example, we use average albedo and its standard deviation to distinguish between C and S classes. Three sigma range is used to create confidence interval for each taxonomy class. DeMeo

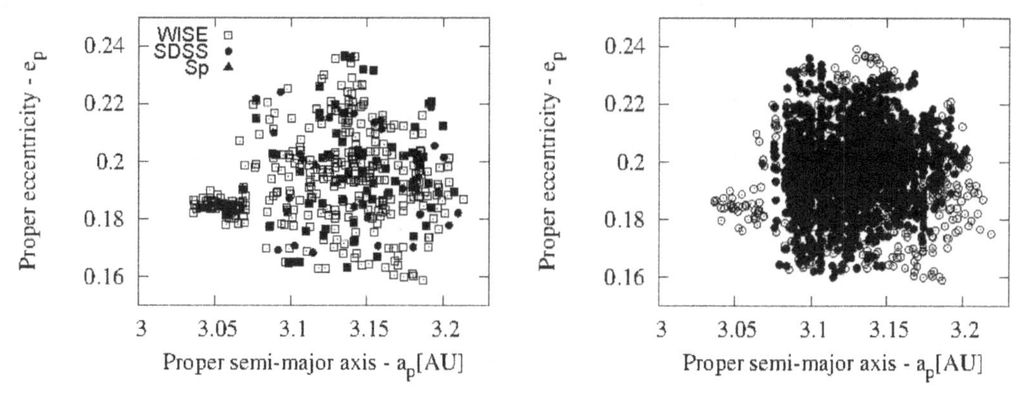

Figure 1. Results for the Klumpkea family. Left image shows interlopers discarded because of different properties. Empty squares are discarded due to its WISE albedo, filled circles due to SDSS colors and filled triangles because of its taxonomy type. Right image shows the Klumpkea family obtained in the first step (empty circles) and in the last (filled circles).

and Curry (2013) found that average albedo of C-type asteroids is 0.06 ± 0.01, thus, the confidence interval is $[0.03, 0.09]$. Than, an asteroid with geometric albedo p_v and standard deviation σ, is not a member of C class if value of $p_v - 3\sigma$ is greater than 0.09.

In third step, asteroids identified as interlopers are excluded from the initial catalog of proper elements. The point of this step is to reduce the chaining effect in the HCM. By removing interlopers from the catalog, we have also removed objects linked to the family through some of these interlopers.

Finally, in fourth step, HCM analysis is performed again using the modified catalog.

3. Results: the case of Klumpkea family

We choose (1040) Klumpkea family as a test case due to its location and spectral type. Klumpkea is situated in the outer part of the main-belt where most of asteroids belong to C-type. However, members of the Klumpkea family are S-type. Hence, a relatively large fraction of the interlopers are expected to be found. Obtained results are shown in Fig. 1. In the first step we identified 2794 asteroids as a family members. Then, in the second step, among these objects we found 447 interlopers. After removing interlopers, using the same v_c as in the first step, HCM identified 2107 family members. Thus, the final family has 687 members less than the initial one.

In conclusion, our approach could significantly reduce number of interlopers in families, and help to get much more reliably the list of family members.

References

Brož M., Morbidelli, A., Bottke, W. F., Rozehnal, J., Vokrouhlicky, D., & Nesvorný D. 2013, *A&A*, 551, A117

DeMeo, F. E. & Carry, B., 2013, *Icarus*, 226, 723

Ivezić, Ž., Tabachnik, S., Rafikov, R., *et al.* 2001, *AJ*, 122, 2749

Knežević Z. & Milani, A. 2003, *A&A*, 403, 1165

Milani, A., Cellino, A., Knežević Z., Novaković B., Spoto, F., & Paolicchi, P. 2014, *Icarus*, 239, 46

Masiero, J. R., Mainzer, A. K., Grav, T., *et al.* 2011, *ApJ*, 741, 68

Nesvorný, D., Jedicke, R., Whiteley, R. J., & Ivezić, Z. 2005, *Icarus*, 173, 132

Novaković B., Cellino, A., & Knežević Z. 2011, *Icarus*, 216, 69

Zappalà, V., Cellino, A., Farinella, P., & Knežević, Z. 1990, *AJ*, 100, 2030

Complex Planetary Systems
Proceedings IAU Symposium No. 310, 2014
Z. Knežević & A. Lemaitre, eds.

doi:10.1017/S1743921314008199

The effect of secular resonances on the long-term orbital evolution of uncontrolled near-Earth objects

T. V. Bordovitsyna, I. V. Tomilova and G. O. Ryabova
Tomsk State University, 634050 Tomsk, Russian Federation
email: tvbord@sibmail.com

Abstract. A set of 29 low order secular resonance relations in the near-Earth satellite dynamics has been obtained and their influence on dynamic evolution of objects has been investigated on time intervals of 100 years. The analysis of the results shows that secular resonances are very common phenomena in the near-Earth orbital space. Sharp resonances are concentrated in the orbital space areas where the semi-major axes $a \geqslant 20000$ km and the inclinations $i \geqslant 50°$. In some areas objects have several secular resonances. The superposition of several secular resonances or orbital and secular resonances are the source of randomness in the motion of objects.

Keywords. celestial mechanics, methods: analytical, methods: numerical

---

To study the effect of secular resonances on uncontrolled near-Earth objects we use a numerical-analytical approach consisting in (Bordovitsyna *et al.* 2012):

– identifying the secular resonances based on an analytical method on condition that the third body orbit is an ellipse with revolving nodal and apsidal lines;

– numerical modeling the long-term orbital evolution using the software package Numerical Model of Motion of Satellite Systems (Bordovitsyna *et al.* 2009) realized in a parallel programming environment at the Tomsk State University computing cluster;

– MEGNO-analysis of dynamic evolution of objects (Cincotta *et al.* 2003).

A complete set of 29 low order resonance relations has been presented in Bordovitsyna *et al.* (2014). This set may be considered as a small modification of the resonance relation list obtained by Cook (1962) and by Hughes (1980, 1981).

The long-term orbital evolution of uncontrolled satellites of systems GLONASS, GPS and BEIDOU IGSO have been simulated numerically using the mentioned software package. It was shown that the superposition of several sharp secular resonances or proximity of the object's mean motion to the tesseral resonance in the presence of a sharp secular resonance are the source of randomness in the motion of objects. This effect appears after the eccentricity increases up to 0.6. See detailes in (Bordovitsyna *et al.* 2014).

To study the distribution of secular lunisolar resonances in the near-Earth orbital space and their effect on the uncontrolled object dynamics, an extensive numerical experiment has been carried out. Numerical values of 29 secular resonance relations have been calculated for the following colections of eccentricity (e), inclination (i) and semi-major axis (a):

$$e = 0.01 \text{ and } 0.1 - 0.9 \text{ with step } 0.1; \quad i = 0° - 90° \text{ with step } 5°;$$
$$a = 8000 \text{ km} - 55000 \text{ km with step 1 km.}$$

The influence of secular resonances on the dynamic evolution of 200 model near-Earth space objects has been investigated on the time interval of 100 years. Objects which go through (1) one or several secular resonances, (2) tesseral and secular resonances simultaneously and (3) have no resonances have been considered. The analysis of the obtained results allows the following preliminary conclusions (Bordovitsyna & Tomilova 2014 a, b).

– Secular resonances are very common phenomena in the near-Earth orbital space, while sharp resonances are concentrated in the orbital space areas, where $a \geqslant 20000$ km and $i \geqslant 50°$.

– Secular resonances where the value of resonance relation $\leqslant 10^{-8}$ rad/s affect the orbital evolution of near-Earth objects distinctly.

– Secular resonances with the mean motion of the Sun are more significant than the similar resonances with the mean motion of the Moon, and this has good agreement with Breiter (1999).

– Secular lunisolar resonances produce growth of e (for any i). This conclusion has good agreement with other authors results (e.g., Chao & Gick 2004; Rossi 2008). Even a single sharp resonance can transform a circular orbit into a high-eccentric one.

– The superposition of several sharp secular resonances or proximity of the object's mean motion to the tesseral resonance in the presence of a sharp secular resonance are the source of randomness in the motion of objects.

– Objects from the region of the orbital space with semi-major axes 40000 km and higher and inclinations more than 55° are subject to the simultaneous influence of a large number of secular resonances; the motion of such objects is irregular and shows longperiod oscillations of the eccentricity and the inclination with large amplitudes; the chaotization of the motion may be considerable and the velocity of the chaotization increases with growth of the initial eccentricity.

– The evolution of the near-polar orbits with inclinations 80° and 90° is very complicated but degrees of the orbital destruction can be various (Bordovitsyna & Tomilova 2014 b). For example, the picture of long-time evolution has a catastrophic character for near-circular orbits with a = 55 000 km and i = 90°. The lifetime of a model object, which goes through 12 secular resonances, is only 40 years, because the orbit eccentricity tends to unit, the change of inclination leads the object away on orbits with the reverse motion, and the semi-major axis rapidly decreases. The chaotization of the motion takes place with rapidly growing MEGNO parameter.

Acknowledgements

We acknowledge the financial support from the Organizers for one of the authors (RGO), which made it possible presentation of this work at the IAUS 310. We thank the anonimous reviewer for careful and useful review of this paper.

References

Bordovitsyna, T. V. *et al.* 2009, *Izv. Vyssh. Uchebn. Zaved., Fiz.*, 52, 10/2, 5 (in Russian)

Bordovitsyna, T. V., Tomilova, I. V., & Chuvashov, I. N. 2012, *Solar System Research.*, 46, 5, 329

Bordovitsyna, T. V., Tomilova, I. V., & Chuvashov, I. N. 2014, *Solar System Research.*, 48, 4, 259

Bordovitsyna, T. V. & Tomilova, I. V. 2014a, *Russian Physical Journal.*, 57, 516

Bordovitsyna, T. V. & Tomilova, I. V. 2014b, *Russian Physical Journal.*, 57, 819

Breiter, S. 1999, *Cel. Mech. and Dyn. Astr.*, 74, 253

Chao, C. & Gick, R. 2004, *Adv. Space Res.*, 34, 1221

Cincotta, P. M., Girdano, C. M., & Simo, C. 2003, *Physica*, 182, 151

Cook, G. E. 1962, *Geophys. Journ. of RAS.*, 6, 3, 271

Hughes, S. 1980, *Proc. R. Soc. Lond. A.*, 372, 243

Hughes, S. 1981, *Proc. R. Soc. Lond. A.*, 375, 379

Rossi, A. 2008, *Celest. Mech. Dyn. Astr.*, 100, 267

Complex Planetary Systems
Proceedings IAU Symposium No. 310, 2014
Z. Knežević & A. Lemaitre, eds.

doi:10.1017/S1743921314008205

Long-term evolution of asteroids in the 2:1 Mean Motion Resonance

Despoina K. Skoulidou, Kleomenis Tsiganis and Harry Varvoglis

Section of Astrophysics, Astronomy & Mechanics, Department of Physics,
Aristotle University of Thessaloniki, 54 124 Thessaloniki, Greece
email: **dskoulid@physics.auth.gr, tsiganis@auth.gr, varvogli@physics.auth.gr**

Abstract. The problem of the origin of asteroids residing in the Jovian first-order mean motion resonances is still open. Is the observed population survivors of a much larger population formed in the resonance in primordial times? Here, we study the evolution of 182 long-lived asteroids in the 2:1 Mean Motion Resonance, identified in Brož & Vokrouhlický (2008). We numerically integrate their trajectories in two different dynamical models of the solar system: (a) accounting for the gravitational effects of the four giant planets (i.e. 4-pl) and (b) adding the terrestrial planets from Venus to Mars (i.e. 7-pl). We also include an approximate treatment of the Yarkovksy effect (as in Tsiganis *et al.* 2003), assuming appropriate values for the asteroid diameters.

Keywords. asteroids, methods: numerical, resonances, Yarkovsky effect

---

1. Introduction

The 2:1 Mean Motion Resonance with Jupiter is one of the famous Kirkwood gaps. In the framework of the 3-body problem, Lemaître & Hernard (1990) found only a small chaotic region in low-eccentricities and Morbidelli & Moons (1993), using an improved 3-body problem, found the borders of secular resonances embedded in the 2:1 MMR. Moons *et al.* (1998) and Nesvorný & Ferraz-Mello (1997) mapped the structure of the 2:1 MMR in detail, using a semi-analytical method and extensive N-body simulations. Long-term evolution of the observed population in the resonance, in the framework of the giant planets model, has been studied by Brož & Vokrouhlický (2008). In this paper, we study the 'stability' of the 2:1 long-lived population in different dynamical models.

2. Long-term evolution in the 2:1 MMR

We performed long-term orbital integrations of the 2:1 'stable' population, for a time equivalent to 3 Gy using the SWIFT package (Levison & Duncan 1994) and various time-steps, taking initial conditions from ASTDYS †. We computed asteroid diameters, assuming the same albedo ($p_V = 0.06$) for all bodies and taking absolute magnitude values (H) from Brož and Vokrouhlický (2008).

Fig. 1 shows a plot of the fraction $N(>t)/N_i$ of asteroids, that remain after time t, vs. t, for all the considered models. We assume a distribution law $N(>t)/N_i - C = e^{-bt}$‡. We find that $C = C_{max}$ is only 19% of the initial population for the 4-pl model, 14% for the 7-pl model, and $\approx 0\%$ for either model that includes the Yarkovsky effect. The e-folding time, $\frac{1}{b}$, decreases from 1.70 ± 0.02 Gy (simple 4-pl model) to 0.89 ± 0.01 Gy (7-pl model with Yarkovsky).

† http://hamilton.dm.unipi.it/astdys/

‡ When correlation coefficient, r^2, reaches its maximum then $C = C_{max}$ represents the percentage of the bodies that is expected to remain in the resonance as t goes to infinity.

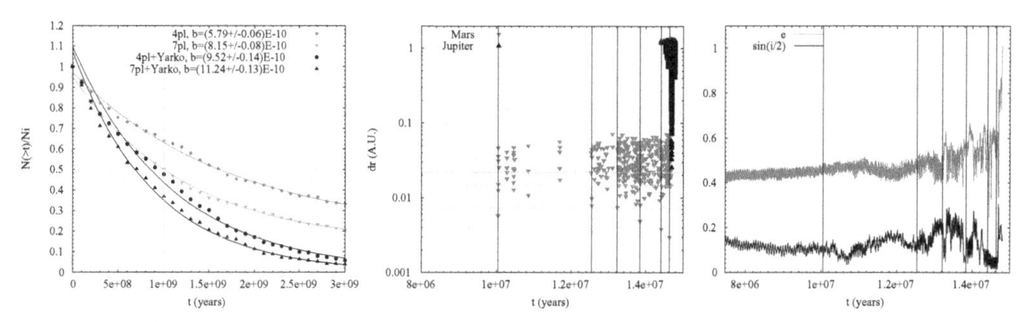

Figure 1. (left) $N\left(>t\right)/N_i$ vs. t for all the considered models and the best-fitting curve for C_{max}, (middle) close encounters with planets of asteroid 2005AO24 in the 7-pl model, (right) e vs. t and $\sin\left(i/2\right)$ vs. t of asteroid 2005AO24 at the same time interval.

3. Close encounters and escape

Asteroids residing in the quasi-stable core of the 2:1 resonance are, in principle, phase-protected by encounters with Jupiter, if their orbits are not strongly modified by the secular resonances (Morbidelli & Moons 1993). We find that many bodies (Fig. 1) have successive encounters with Mars that help in breaking up the phase-protection and extracting bodies from the resonance. The possibility of a 2:1 asteroid becoming Mars-crosser was first mentioned by Wisdom (1987).

4. Conclusions

We found that $\sim 67\%$ of the population manage to escape within 3 Gyr in the 4-pl model, $\sim 80\%$ in the 7-pl model and $\sim 95\%$ in either model that includes the Yarkovsky effect, but the evolution is accelerated when the terrestrial planets are also considered. However, we do not find any significant contribution of the terrestrial planets to the evolution of the size distribution. We find that many asteroids have successive encounters with Mars and the other terrestrial planets, which break the phase-protection mechanism of the 2:1 MMR with Jupiter. The problem of whether the 2:1 quasi-stable population is primordial or captured in later stages of solar system evolution is still open. Clearly, a much larger initial population with a much steeper size frequency distribution is needed to explain the observed population.

Acknowledgments

The work of H. V. is co-financed by the European Union (European Social Fund - ESF) and Greek national funds through the Operational Program "Education and Lifelong Learning" of the National Strategic Reference Framework (NSRF) - Research Funding Program: Thales. Investing in knowledge society through the European Social Fund.

References

Brož M. & Vokrouhlický D. 2008, *MNRAS* 390, 715
Lemaître, A. & Hernard, J. 1990, *Icarus* 83, 391
Levison, H. & Duncan, M. 1994, *Icarus* 108, 18
Moons, M., Morbidelli, A. & Migliorini, F. 1998, *Icarus* 135, 458
Morbidelli, A. & Moons, M. 1993, *Icarus* 102, 316
Nesvorný D. & Ferraz-Mello, S. 1997, *Icarus* 130, 247
Tsiganis, K., Varvoglis, H., & Morbidelli, A. 2003, *Icarus* 166, 131
Wisdom, J. 1987, *Icarus* 72, 241

Complex Planetary Systems
Proceedings IAU Symposium No. 310, 2014
Z. Knežević & A. Lemaitre, eds.

doi:10.1017/S1743921314008217

Dynamical properties of Watsonia asteroid family

Georgios Tsirvoulis[1], **Bojan Novaković**[2], **Zoran Knežević**[1] **and Alberto Cellino**[3]

[1]Astronomical Observatory, Volgina 7, 11060 Belgrade 38, Serbia
[2]Department of Astronomy, University of Belgrade, Serbia
[3]INAF-Osservatorio Astrofisico di Torino, Pino Torinese 10025, Italy

Abstract. In recent years, a rare class of asteroids has been discovered by Cellino *et al.* (2006), with its distinguishing characteristic being the anomalous polarimetric properties of its members. Named Barbarians, after (234) Barbara, the prototype of the class, these asteroids show negative polarization at unusually high phase-angles compared to normal asteroids. Motivated by the fact that some of the few discovered Barbarians seemed to be related to the Watsonia asteroid family, Cellino *et al.* (2014) performed a search for more Barbarians among its members. A positive result of this search led to the conclusion that Watsonia is indeed an important repository of Barbarian asteroids. Based on these findings, we decided to analyze this family in detail.

Keywords. asteroids, methods: numerical, asteroid families, transport mechanisms

1. Identification of family members

The first step in our study is to derive a list of Watsonia family members. To this purpose, we first calculate the synthetic proper elements (Knežević & Milani (2003)) of the numbered, multi- and single-opposition asteroids in a wide region around the family. To this set of proper elements we apply the Hierarchical Clustering Method (HCM) (Zappala *et al.* (1990)) to determine the membership of the family. As seen in Fig. 1a, the well defined plateau ($d_c = 92 - 103\,m/s$) reveals that the family is well separated from the background asteroids. The dynamical family is found to have 83 members.

Afterwards, in order to have a better insight of the family properties, we used the so-called "V-shape" method to determine its age (Vokrouhlicky *et al.* (2006), Milani *et al.* (2010)). The low number of family members did not allow for a precise estimation, so we were only able to achieve a rough result of the order of 1 Gyr.

2. Simulation of fictitious family fragments

The next step of our analysis was to study the orbital evolution of a fictitious population of family members. The location of the family in the proper orbital elements space, close to the 5/2 mean motion resonance (MMR) with Jupiter, with low to moderate eccentricities and high inclinations, suggests that over the age of the family, a significant number of fragments must have entered the resonance. This resonance has been shown by Gladman *et al.* (1997) to be efficient at transporting asteroids to the near-Earth region, therefore we wanted to simulate the inflow of Watsonia family fragments into this resonance, as well as in other resonances relevant for the transport.

We created a population of fictitious family fragments with initial orbital elements near (729) Watsonia, and integrated their orbits for 1 Gyr. The dynamical model included five planets (Mars to Neptune), and the asteroids (1) Ceres and (2) Pallas, and accounted

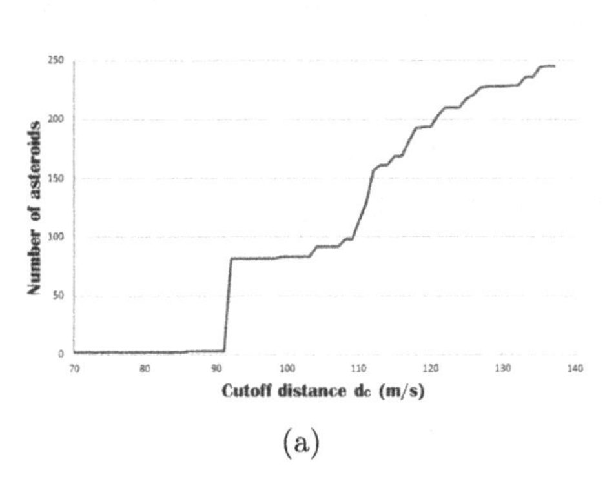

(a)

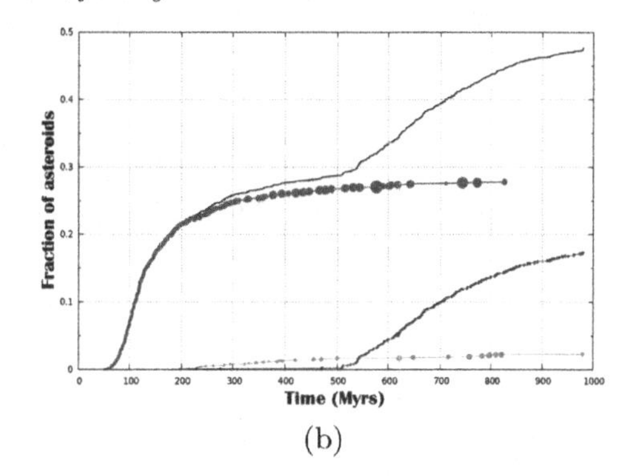

(b)

Figure 1. (a):Number of asteroids associated with (729) Watsonia as a function of cutoff velocity (d_c). The family is defined by the plateau around 100 m/s. (b):Fraction of the initial population asteroids entering the near-Earth region over time. In dark-gray the fraction of asteroids driven through the 3/1 MMR, in medium-gray through the 5/2 MMR and in light-gray through the 8/3 MMR. The black continuous line denotes the total fraction of asteroids becoming NEA.

for the Yarkovsky thermal force (Broz (2006)), set to have maximum drift rate for all objects. The outcome of the simulation is an initial drift of fragments to the 5/ MMR with Jupiter at 2.82 AU, which leads them to terrestrial planet-crossing orbits or out of the solar system. After about 500 Myr, fragments also reach the 3/1 MMR with Jupiter at 2.5 AU, with most of them entering the near-Earth region. A small number of asteroids is also driven to the planet-crossing region via the 8/3 MMR with Jupiter.

As seen in Fig. 1b, the major fraction of fragmens reaching the near-Earth region, is coming through the 5/2 MMR. However, taking into account the age of the family as estimated above, we expect that the 3/1 MMR is responsible for most of the inflow of fragments at the present time. We can thus conclude that the transport of Barbarians to terrestrial planet-crossing orbits is plausible, and will be further studied.

Acknowledgement

The work of G. T. is supported by the European Union [FP7/2007-2013], project: STARDUST-The Asteroid and Space Debris Network". B. N. and Z. K. acknowledge support from the Ministry of Education, Science and Technological Development of the Republic of Serbia through the project OI176011.

References

Broz, M. 2006, *PhD Thesis, Charles University*

Cellino, A., Belskaya, I. N., Bendjoya, Ph., Di Martino, M., Gil-Hutton, R., Muinonen, K., & Tedesco, E. F. 2006 *Icarus*, 180, 565

Cellino, A., Bagnulo, S., Tanga, P., Novaković, B., & Delbo, M. 2014, *MNRAS*, 439, L75

Gladman, B. J., Migliorini, F., Morbidelli, A., Zappala, V., Michel, P., Cellino, A., Froeschle, C., Levison, H. F., Bailey, M., & Duncan, M. 1997, *Science*, 277, 197

Knežević, Z. & Milani, A. 2000, *CeMDA*, 78,17

Milani, A., Knežević, Z., Novaković, B., & Cellino, A. 2010, *Icarus*, 207, 769

Vokrouhlicky, D., Broz, M., Bottke, W. F., Nesvorny, D., & Morbidelli, A. 2006, *Icarus*, 182 118

Zappala, V., Cellino, A., Farinella, P., & Knežević, Z. 1990, *AJ*, 100, 2030

Complex Planetary Systems
Proceedings IAU Symposium No. 310, 2014
Z. Knežević & A. Lemaitre, eds.

doi:10.1017/S1743921314008229

Complex satellite systems: a general model of formation from rings

Aurélien Crida[1] and Sébastien Charnoz[2,3]

[1]Laboratoire Lagrange (UMR 7293)
Université Nice Sophia-antipolis / CNRS / Observatoire de la Côte d'Azur
CS 34229, 06304 NICE cedex 4, FRANCE
email: crida@oca.eu

[2]Laboratoire AIM (UMR 7158) ; Université Paris Diderot / CEA / IRFU-SAp
91191 Gif-sur-Yvette cedex, FRANCE

[3]Institut Universitaire de France; 103 Bd Saint Michel, 75005 Paris, FRANCE.

Summary of the Abstract. *We present our model of formation of regular satellites by the spreading of a cold debris disk beyond the Roche radius. By numerical simulations and analytical calculations, we show that this process explains the peculiar properties of Saturn's small and mid-sized moons, in particular their mass - distance distribution. This process seems to also account for the structures of the satellite systems of Uranus, Neptune, and the Earth, suggesting that they used to have massive rings.*

Abstract. Satellite systems are often seen as mini-planetary systems, and they are as various and complex. The Earth has one single satellite of more than one percent of its own mass, but the giant planets have many, some of them tiny. They constitute fascinating, interacting, evolving, and dynamically complex systems. While irregular satellites, having inclined and eccentric orbits, are supposed to be captured, it was generally admitted that regular satellites, having circular orbits in the equatorial plane of their host planet, formed in the gas and dust circum-planetary disk that surrounded giant planets at the time of their formation, more or less similarly as planets formed in the proto-planetary disk around their host star. Here, we propose an alternative mechanism for the formation of regular satellites: the spreading of a dynamically cold disk of solids beyond the Roche radius of a planet.

The Roche radius is defined as the distance from a planet beyond which self-gravity is stronger than tidal forces, allowing gravitationally bound aggregates to form, and satellites to exist. Saturn's rings are inside the Roche radius, which is why they don't coalesce into one single object. However, like any astrophysical disk in Keplerian rotation, they spread. Once material reaches the Roche limit, it should accrete into moonlets. These moonlets migrate outwards, repelled by tidal interactions with the rings and Saturn. We showed with numerical simulations that the small and mid-sized moons of Saturn formed this way, which explains their young age and composition (Charnoz *et al.* 2010, 2011). Analytical calculations show that the formation of satellites from a ring of solids spreading beyond the Roche limit takes place in 2 phases: (i) the continuous regime, in which only one satellite forms (this regime prevailed for the moon-forming disk around the Earth) ; (ii) the pyramidal regime, in which a series of moons form, migrate outwards, merge... In this regime, the satellites should follow a precise mass-distance relation, which matches that of the Saturnian system, confirming that it formed this way. It also fits the Uranian and Neptunian systems, suggesting that these planets used to have massive rings that gave birth to their regular satellites, and vanished (Crida & Charnoz (2012)).

Refinements of this model, applications to planetary systems, and a recent likely observation of this process will be discussed in conclusion.

Keywords. planets and satellites: formation, planets and satellites: individual (Saturn), planets: rings

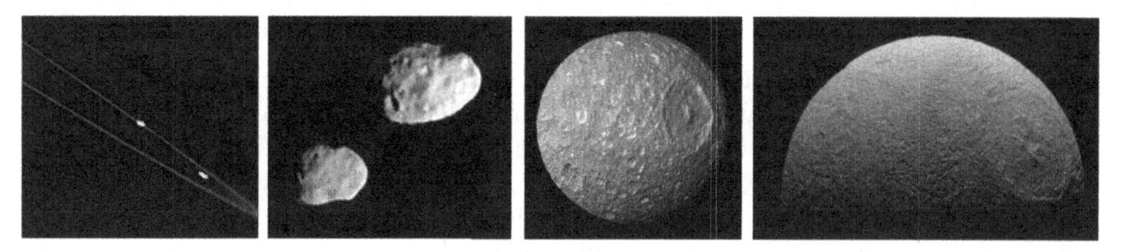

Figure 1. Pictures of Saturn regular satellites, by the Cassini spacecraft. From left to right: Prometheus and Pandora around the F-ring; Janus and Epimetheus; Mimas; Tethys.

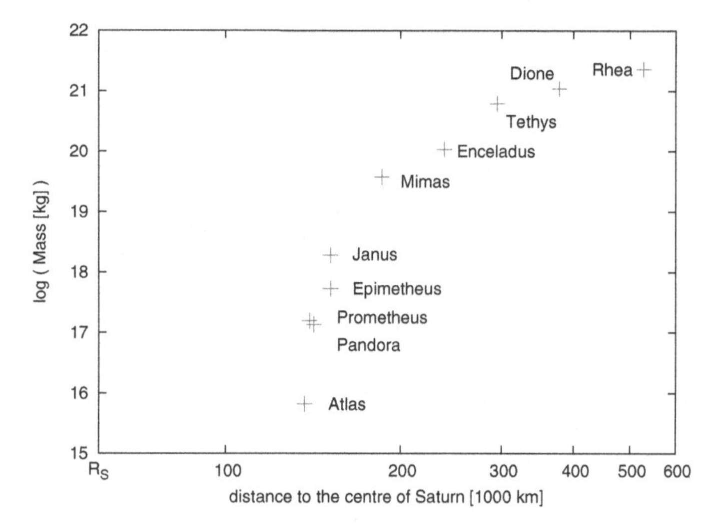

Figure 2. Saturn's regular satellites in a mass – distance diagram (logarithmic scale). A smooth increase of the mass with distance, starting from 0 at 140 000 km from Saturn's centre, appears clearly. The + are just symbols, not errorbars.

1. Introduction

The regular satellites of Saturn are much better known since the Cassini spacecraft brought amazing pictures of these icy worlds (Figure 1). Starting at the outer edge of the rings (140 000 km from Saturn's centre), one finds Prometheus and Pandora on both sides of the F-ring, who are tiny icy bodies, of about 80 km in size and 10^{17} kg. Eleven thousand kilometres further, Janus and Epimetheus share the same orbit, for a total mass of 2.5×10^{18} kg. Then, Mimas stands at 185 000 km from Saturn's centre, and is roughly 400 km in size and 3.7×10^{19} kg. Moving away from Saturn, Enceladus, Tethys, Dione, and Rhea are ranked in size and mass as well, the latter being 390 000 km from the edge of the rings with a mass of 2.3×10^{21} kg. Finally, the main satellite, Titan, has an orbital radius of 1.2×10^{6} km and a mass of 1.3×10^{23} kg. In fact, as shown on Figure 2, there is an intriguing mass – distance relation: a smooth increase of the mass with distance, starting from 0 at 140 000 km from Saturn's centre, *id est* from the Roche radius at the outer edge of the rings. An other noticeable property of this system is that the further from Saturn, the further satellites are from each other.

A similar trend, although less clean, is observed in the regular satellite systems of the other giant planets. Such a remarkable property can't be explained in the frame of satellite formation inside a circum-planetary disc of gas and dust around the young giant planet. In contrast, the accumulation of small bodies just outside the rings suggest that rings and satellites are somehow connected.

2. Evolution of rings

Saturn's rings are made of ice blocks on circular orbit in the equatorial plane of the planet. They constitute a dynamically cold debris disk. Their velocity dispersion is so small that the thickness of the rings is about ten millions times smaller than their radius. In fact, encounters between ice blocks happen all the time, at tiny velocity. The only reason why they don't coalesce into a gravitationally bound aggregate is that Saturn's tidal forces are stronger than the self-gravity of the blocks. But tidal forces weaken with the distance to the planet. The distance at which tides become weaker than self-gravity is called the Roche radius, and given by:

$$r_{\mathrm{Roche}} = \left(\frac{9M_p}{\pi\rho}\right)^{1/3}, \tag{2.1}$$

where M_p is the mass of the planet, and ρ the volume density of the blocks (see for instance Crida & Charnoz (2013)). Around Saturn, for the density of porous ice (600 kg/m^3), $r_{\mathrm{Roche}} = 140\,000$ km. Inside the Roche radius, aggregation is impossible, and the rings have to stay as rings.

In Keplerian rotation around a central body, the angular velocity Ω decreases with the orbital radius r, as given by Kepler's law: $\Omega = \sqrt{GM_p/r^3}$, where G is the gravitational constant and M_p the mass of the central body. Nonetheless, the specific angular momentum $j = r^2\Omega \propto \sqrt{r}$ increases with r. Interactions between the fast inner particles and the slow outer particles slow down the first ones, and accelerate the second ones (this can be modeled as a viscous friction). As a consequence, the inner particles lose angular momentum and move to smaller r, while the outer ones gain angular momentum and move to larger r. The rings spread, while angular momentum flows outwards. This is a well-known result, and has been described for instance by Lynden-Bell & Pringle (1974).

Using a prescription for the viscosity of Saturn's rings given by Daisaka *et al.* (2001), Salmon *et al.* (2010) have computed the evolution of the density profile of the rings with a 1D implicit code. They find that the more massive the rings are, the faster they spread. Two major results derive from their analysis: (i) with their present mass, Saturn's rings can survive for billions of years, thus they can be as old as the Solar System; (ii) they could have been much more massive in the past and lost mass by spreading.

According to Daisaka *et al.* (2001), the viscosity ν of massive rings is proportional to their mass squared. More precisely, the characteristic time for viscous spreading is $t_\nu = r_{\mathrm{Roche}}^2/\nu = \frac{\pi}{92}D^{-2}T_R$, where $D = M_{\mathrm{rings}}/M_p$ and T_R is the orbital period at r_{Roche}. As $dD/dt = -D/t_\nu$, one gets $dD/d\bar{t} \approx -30\,D^3$, with $\bar{t} = t/T_R$. The solution of this differential equation is:

$$D(t) = \frac{1}{\sqrt{60\,\bar{t} + D_0^{-2}}}\,. \tag{2.2}$$

Clearly, the initial condition D_0 is erased as soon as $\bar{t} \gg 1/60D_0^2$, and on long term, $D(t) \approx (60\,\bar{t})^{-1/2}$. Around Saturn, with $t = 4.5$ billon years, this gives $D = 8\times10^{-8}$.

Actually, Canup (2010) suggests that the rings formed from the tidal stripping of a differentiated satellite that migrated inside the Roche radius of Saturn (see also Crida & Charnoz (2010)). In that case, the rings were a few times 10^{22} kg 4.5 billion years ago, while they are only $\sim 4\times10^{19}$ kg now ($D = 7\times10^{-8}$, in agreement with the estimate above).

3. The satellites of Saturn are children of the rings

The rings spread. Material that falls onto Saturn is lost forever, but what about material spreading outwards, beyond the Roche radius ? As soon as this limit is crossed, self-gravity wins over tidal forces. Aggregates form, and new small satellites are produced at r_{Roche}. These satellites have a smaller angular velocity than rings particles ; hence, they are accelerated by them, gain angular momentum, and move further. Lin & Papaloizou (1979) have computed the torque felt by a satellite of mass m orbiting on a circular orbit of radius r with angular velocity Ω outside of rings of surface density Σ :

$$\Gamma = \frac{8}{27}\left(\frac{m}{M_p}\right)^2 \Sigma\, r^4\, \Omega^2\, \Delta^{-3} \tag{3.1}$$

where $\Delta = (r - r_{\mathrm{Roche}})/r_{\mathrm{Roche}}$ is the normalised distance to the outer edge of the rings.

Applying this to Janus, one finds that it is has been moving away from r_{Roche} for $\sim 10^8$ years. In other words, a hundred million years ago, Janus must have been inside the rings. This agrees with the observation that Janus is little cratered, and its surface looks much younger than the Solar System. In fact, Janus, Epimetheus, Prometheus and Pandora share this young aspect, and are all under-dense (~ 600 kg/m^3), and have the same spectrum as the rings. All this suggests very strongly that they have been formed recently, from the spreading of the rings.

Charnoz, Salmon & Crida (2010) performed numerical simulations of the evolution of present day rings, similar to Salmon *et al.* (2010), but with the mass falling beyond r_{Roche} being transformed in small satellites whose evolution is governed by Eq. (3.1). The small satellites are allowed to merge, and the final outcomes of their simulations is in general a system of a handful of satellites of masses in agreement with that of the small moons.

From this spectacular result, it is tempting to generalise to the whole system of regular moons of Saturn. There is a problem though : beyond the 2:1 Mean Motion Resonance with r_{Roche}, located at 222 000 km from the centre of Saturn, the interaction with the rings vanishes. The only process that makes the satellites migrate is the dissipation of tides inside Saturn, with the rate :

$$\frac{dr}{dt} = \frac{3\, k_{2p}\, m\sqrt{G}\, R_p^5}{Q_p\, \sqrt{M_p}\, r^{11/2}} \tag{3.2}$$

where the index p refers to the planet Saturn, k_2 and Q are the Love number and the dissipation factor respectively. With the old value of $Q_{\mathrm{Saturn}} \approx 18\,000$, this gives a too small rate, and Enceladus, Tethys, Dione and Rhea don't migrate significantly over the age of the Solar System ; so they must have been formed at their present location. However, a careful analysis of old photographic plates performed by Lainey *et al.* (2012) reveal that their migration is detectable over a century, and that Q_p must be 10 times smaller. With $Q_p = 1700$, the time it takes for Rhea to migrate from 222 000 km to its present position is about 3.5 billion years. Such a "tidal age" can be calculated for the other mid-sized moons, and they appear to be ranked by increasing tidal age (Charnoz *et al.* (2011)). This opens the possibility of forming Rhea first, then Dione, then Tethys, then Enceladus, and finally Mimas, and have them migrate to their present orbit within the age of the Solar System, without orbit crossing.

Starting simulations with very massive rings as suggested by Canup (2010), using Eq. (3.1) inside 222 000 km and Eq. (3.2) with $Q_p = 1700$ beyond this, Charnoz *et al.* (2011) reproduce successfully the system of the regular satellites of Saturn up to and including Rhea.

Not only is this model in agreement with the mass – distance distribution of the system of the regular satellites of Saturn, but this also solves two problems at the same time. (i) These moons have young crater ages, and anyway should not have survived the Late Heavy Bombardment; forming them not earlier than 3.9 billions years ago is therefore very positive. (ii) These moons have rocky cores, which account for 7% to 70% of the mass of the body; these variations in the bulk composition is hard to explain in a gaseous circum-planetary disk, and the differentiation itself is unexpected for such small objects. In this model, chunks of silicates initially present in the rings coalesce into a handful of rocky cores (because the Roche radius for silicates is only 90 000 km), and are later coated with ice from rings; this naturally forms differentiated bodies, of stochastic bulk composition (Charnoz *et al.* (2011)).

4. A general, analytic model

Crida & Charnoz (2012) describe analytically the spreading of rings beyond the Roche radius, as studied before with numerical simulations. We define

$$\tau = M/FT_R$$

is the dimensionless lifetime of the rings, where M is their mass, F is the flow of mass through the Roche limit, and we remind that T_R is the orbital period at r_{Roche} and $D = M/M_p$. With Daisaka *et al.* (2001)'s prescription for the viscosity, $\tau = r_{\text{Roche}}^2 \,/\, \nu\, T_R = 0.0425 \,/\, D^2$.

4.1. *Continuous regime*

Assume one satellite forms first at r_{Roche} ; its mass is then $m = Ft$ with t the time since its birth. Note $q(t) = m(t)/M_p$ the satellite to planet mass ratio. As it grows, it migrates, following Eq. (3.1). One can then show easily that:

$$q = \left(\frac{\sqrt{3}}{2}\right)^3 \tau^{-1/2}\, \Delta^2 \,. \tag{4.1}$$

We call this the *continuous regime*, in which the satellite's mass increases linearly with time, while its distance to the rings Δ increases as the square root of the time.

This regime is applicable if and only if the Roche limit, where ring material is delivered, falls within the feeding zone of the satellite. This feeding zone is 2 times its Hill radius $r_H = (q/3)^{1/3}\, r$; the condition for the continuous regime to apply reads therefore $\Delta < 2(q/3)^{1/3}$. Input into Eq. (4.1), this gives:

$$\Delta \;<\; \Delta_c = \sqrt{\frac{3}{\tau}} \tag{4.2}$$

$$q \;<\; q_c = \frac{3^{5/2}}{2^3}\,\tau^{-3/2} \,. \tag{4.3}$$

Starting at $\Delta = q = t = 0$, we are in the continuous regime. Hence, as rings begin to spread beyond the Roche radius, one single satellite forms and gathers all the mass.

The smaller τ is, the larger this satellite can become; in the limit of very fast spreading, $q_c \geqslant D$ and only one satellite forms, whose mass is that of the rings. In the case of Saturn's rings though, $q_c \ll D$ and the rings are almost unaffected at the end of the continuous regime.

4.2. *Discrete regime*

When the satellite reaches Δ_c (or q_c), it doesn't accrete material from the rings anymore, but keeps migrating outwards at constant mass. The material flowing through $r_{\rm Roche}$ forms a new satellite at $r_{\rm Roche}$. This new satellite migrates outwards and is accreted by the first one. Another new satellite forms, that is soon accreted by the first one, and so on. In this regime, the first satellite still grows at the same average rate, but step by step, by accreting moonlets formed at the outer edge of the rings. We call this the *discrete regime*. Similarly as before, this holds as long as $\Delta < \Delta_c + 2(q/3)^{1/3}$, which gives:

$$\Delta \quad < \quad \Delta_d = 3.14\,\Delta_c \tag{4.4}$$

$$q \quad < \quad q_d = 9.9 q_c \;. \tag{4.5}$$

Applying this to the Earth's Moon forming disc ($D \approx 0.02$), one gets q_d equal to the mass of the Moon. This explains why one and only one satellite formed around the Earth.

In the case of Saturn's rings, $q_d = 10^{-18}$, so only tiny satellites form.

4.3. *Pyramidal regime*

Satellites of mass q_d are produced by the discrete regime at Δ_d every $t_d = q_d M_p/F$. They then migrate outwards at constant mass. Hence, their migration speed decreases (because $\Gamma \propto \Delta^{-3}$). Consequently, they approach each other, until they enter each other's feeding zone, and merge. This leads to the formation of satellites of mass $2q_d$ every $2t_d$ at a specific location. The products of these mergers migrate outwards, and merge further. And so on, in a hierarchical scheme, that we call the *pyramidal regime.*

Using Eq. (3.1), one finds that a merger occurs (and the mass doubles) every time the distance is multiplied by $2^{5/9}$. Hence, in this regime, $q \propto \Delta^{9/5}$. In addition, the number density of satellites should be proportional to $1/\Delta$, which explains the observed pile-up. Beyond the 2:1 mean motion resonance with $r_{\rm Roche}$ (located at $\Delta = 0.58$), Eq. (3.2) applies, and $q \propto r^{3.9}$.

This mass distance relation fits remarkably well with that of the Saturnian system, shown as red + symbols on figure 3 where the dashed lines follow the above law. From Pandora (bottom left corner) to Titan (top right corner), the masses of all the satellites are within 60% of the theoretical red dashed line. This is the smoking gun proving that these satellites formed from the spreading of Saturn's formerly massive rings. It suggests that maybe even Titan comes from the same process, but its position on the line could be a coincidence, as its tidal age is of the order of ten billion years.

More surprisingly, the regular satellites of Uranus and Neptune follow the same trend. In the case of Neptune, the match is almost perfect except for one outlier, Despina, apparently three times too massive... In the system of Uranus, there is much more scatter, but this is not surprising as this system is chaotic, with frequent collisions and orbit crossing (French & Showalter (2012)). Nonetheless, the background slope is in perfect agreement with the pyramidal regime. As a consequence, we claim that Uranus and Neptune used to have massive rings, from which their regular satellites were born. In contrast with the case of Saturn, these rings almost completely disappeared, but the satellite system kept track of its formation from their spreading beyond their Roche radius. We checked that a ring of 1.5 times the total mass of the regular satellites of these planets would have $D < 2 \times 10^{-4}$, so that q_d would be tiny, and the pyramidal regime should prevail (Crida & Charnoz (2012)).

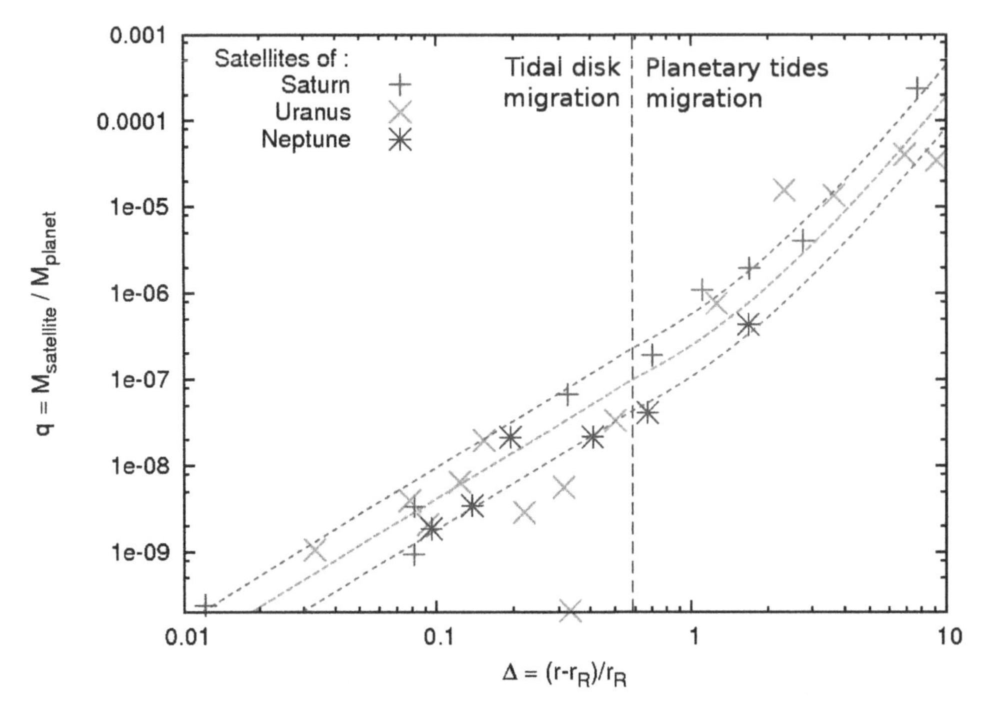

Figure 3. Symbols: Regular satellites in the $q - \Delta$ diagram for the three outer giant planets. Dashed lines: $q \propto \Delta^{9/5}$ for $\Delta < 0.58$, $q \propto r^{3.9}$ for $\Delta > 0.58$.

5. Conclusion and perspectives

We studied the spreading of a dynamically cold disk of debris (rings) beyond its Roche radius, where gravitational aggregation then takes place. At first, a single satellite forms, and migrates outwards while accreting the material that flows beyond the Roche radius (continuous then discrete regimes); then, a series of satellites are produced, whose mass increases with distance following a specific law (pyramidal regime). We have shown that this process

(*a*) explains the physical properties of the icy moons of Saturn,

(*b*) explains the mass – distance distribution of the regular satellites of the giant planets (which can be considered as an observational signature of this process),

(*c*) unifies terrestrial and giant planets in the same paradigm,

(*d*) is the likely origin of most regular satellites in the Solar System, from our Moon to the giant planets.

Although the mathematical model is complete, it relies on a few approximations, and several open questions remain.

• Jupiter's system doesn't fit in this picture. The Galilean satellites probably didn't form this way.

• Our model neglects satellite - satellite interactions, and possible resonance captures. How would this affect the pyramidal regime ? Could the Laplace resonance in the Jovian system be reproduced ?

• If Uranus ans Neptune had massive rings, where did they come from ? And how did they almost disappear ? These questions were, of course, never addressed...

• Could this model explain compact exoplanet systems like Kepler 32 ? One could imagine that a migrating planet gets tidally stripped by its star inside it Roche radius, and that the resulting debris disc spreads, giving birth to a second generation of planets.

These questions open perspective for future research.

To conclude, Murray *et al.* (2014) have observed small bodies at the outer edge of Saturn's A ring. Their inferred size is larger than q_d and their behaviour suggests collisions, evocative of the pyramidal regime. Future observations are needed, and the resonance with Janus will most likely perturb the behaviour of these bodies. Nonetheless, it seems that our theoretical model is possibly at play right now, and the next generation of Saturnian satellites may be built from the rings.

References

Canup, R. M. 2010, *Nature*, 468, 943.

Charnoz, S., Salmon, J., & Crida, A. 2010, *Nature*, 465, 752.

Charnoz, S., Crida, A., Castillo-Rogez, J., Lainey, V., Dones, L., Karatekin, Ö, Tobie, G., Mathis, S., Le Poncin-Lafitte, C., & Salmon, J. 2011, *Icarus*, 216, 535.

Crida, A. & Charnoz, S. 2010, *Nature*, 468, 903.

Crida, A. & Charnoz, S. 2012, *Science*, 338, 1196.

Crida, A. & Charnoz, S. 2013, *SF2A-2013: Proceedings of the Annual meeting of the French Society of Astronomy and Astrophysics*, 57.

Daisaka, H., Tanaka, H., & Ida, S. 2001, *Icarus*, 154, 296.

French, R. S. & Showalter, M. R. 2012, *Icarus*, 220, 911.

Lainey, V., Karatekin, Ö., Desmars, J., Charnoz, S., Arlot, J.-E., Emelyanov, N., Le Poncin-Lafitte, C., Mathis, S., Remus, F., Tobie, G., & Zahn, J.-P. 2012, *ApJ*, 752, 14.

Lin, D. N. C. & Papaloizou, J. 1979, *MNRAS*, 186, 799.

Lynden-Bell, D. & Pringle, J. E. 1974, *MNRAS*, 168, 603.

Murray, C. D., Cooper, N. J., Williams, G. A., Attree, N. O., & Boyer, J. S. 2014, *Icarus*, 236, 165.

Salmon, J., Charnoz, S., Crida, A. & Brahic, A. 2010 *Icarus*, 209, 771.

Complex Planetary Systems
Proceedings IAU Symposium No. 310, 2014
Z. Knežević & A. Lemaitre, eds.

doi:10.1017/S1743921314008230

Spin-orbit coupling and chaotic rotation for eccentric coorbital bodies

Adrien Leleu[1], Philippe Robutel[1] and Alexandre C. M. Correia[2,1]

[1]IMCCE, Observatoire de Paris, CNRS, UPMC Univ. Paris 06, Univ. Lille 1, 77Av.Denfert-Rochereau, 75014 Paris, France
email: aleleu@imcce.fr, robutel@imcce.fr

[2]Departamento de Física, I3N, Universidade de Aveiro, Campus de Santiago, 3810-193 Aveiro, Portugal
email: correia@ua.pt

Abstract. The presence of a co-orbital companion induces the splitting of the well known Keplerian spin-orbit resonances. It leads to chaotic rotation when those resonances overlap.

Keywords. celestial mechanics, coorbitals, rotation, spin-orbit resonance

---

1. Introduction and Notations

Given an asymmetric body on a circular orbit, denoting θ its rotation angle in the plane with respect to the inertial frame, the only possible spin-orbit resonance is the synchronous one $\dot{\theta} = n$, n being the mean motion of the orbit. On an Keplerian eccentric orbit, Wisdom *et al.* (1984) showed that there is a whole family of spin-orbit eccentric resonances, the main ones being $\dot{\theta} = pn/2$ where p is an integer. In 2013, Correia and Robutel showed that in the circular case, the presence of a coorbital companion induced a splitting of the synchronous resonance, forming a family of co-orbital spin-orbit resonances of the form $\dot{\theta} = n \pm k\nu/2$, ν being the libration frequency in the coorbital resonance. Inside this resonance, the difference of the mean anomaly of the two coorbitals, denoted by ζ, librates around a value close to $\pm\pi/3$ (around the L4 or L5 Lagrangian equilibrium - tadpole configuration), around π (encompassing L3, L4 and L5 - horseshoe configuration) or 0 (quasi-satellite) configuration. We generalize the results of Correia and Robutel (2013) from the case of circular co-orbital orbits to eccentric ones.

2. Rotation

The rotation angle θ satisfies the differential equation:

$$\ddot{\theta} + \frac{\sigma^2}{2}\left(\frac{a}{r}\right)^3 \sin 2(\theta - f) = 0, \text{ with } \sigma = n\sqrt{\frac{3(B-A)}{C}}, \tag{2.1}$$

where $A < B < C$ are the internal momenta of the body, (r, f) the polar coordinates of the center of the studied body and a its instantaneous semi-major axis.

Let us consider that the orbit is quasi-periodic. As a consequence, the elliptic elements of the body can be expended in Fourier series whose frequencies are the fundamental frequencies of the planetary system. In other words the time-dependent quantity $\left(\frac{a}{r}\right)^3 e^{i2f}$ that appears in equation (2.1) reads:

$$\left(\frac{a}{r}\right)^3 e^{i2f} = \sum_{j \geqslant 0} \rho_j \; e^{(i\eta_j t + \phi_j)}. \tag{2.2}$$

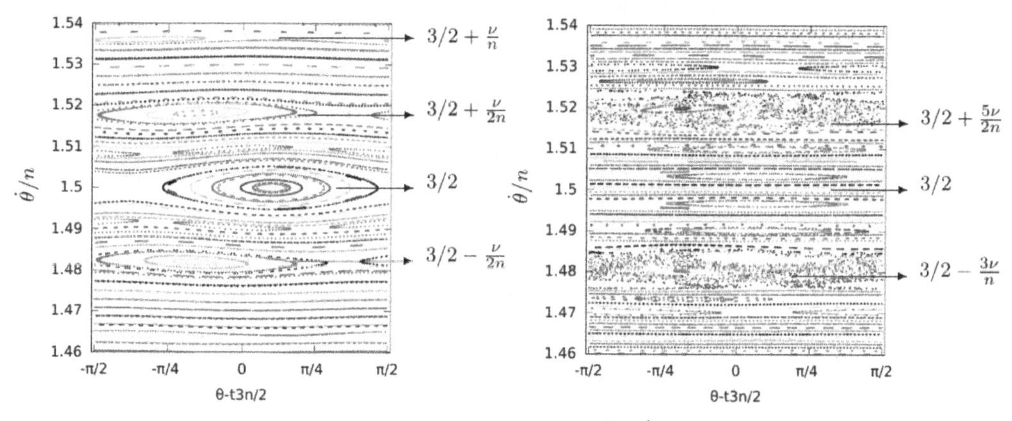

Figure 1. Poincaré surface of section in the plane $(\theta - t\frac{3n}{2}, \dot{\theta}/n)$ near the 3/2 spin-orbit eccentric resonance. (left): $\zeta_{max} - \zeta_{min} = 35°$ - tadpole configuration. (right): $\zeta_{max} - \zeta_{min} = 336°$ horseshoe configuration.

Where η_j are linear combinations with integer coefficients of the fundamental frequencies of the orbital motion (here n and ν) and ϕ_j their phases. Thus (2.1) becomes:

$$\ddot{\theta} = -\frac{\sigma^2}{2} \sum_{j \geqslant 0} \rho_j \sin\,(2\theta + \eta_j t + \phi_j). \tag{2.3}$$

For a Keplerian circular orbit, the only spin orbit resonance possible is the synchronous one, since $\rho_0 = 1$, $\eta_0 = 2n$, and $\rho_j = \eta_j = 0$ for $j > 0$. In the general Keplerian case we have the spin-orbit eccentric resonances, $\eta_j = pn$ and the ρ_j are the Hansen coefficients $X_p^{-3,2}(e)$ (see Wisdom *et al.*). For the circular coorbital case, Correia and Robutel (2013) showed that a whole family results from the splitting of the synchronous resonance of the form $\eta_j = 2n \pm k\nu$. For small amplitudes of libration around L4 or L5 (tadpole), the width of the resonant island decreases as k increases.

In the eccentric coorbital case, each eccentric spin-orbit resonance of the Keplerian case splits in resonant multiplets which are centred in $\dot{\theta} = pn/2 \pm k\nu/2$. For relatively low amplitude of libration of ζ, the width of the resonant island decreases as k increases, see Figure 1 (left). But for higher amplitude, especially for horseshoe orbit, the main resonant island may not be located at $k = 0$. In Figure 1 (right), the main islands are located at $\dot{\theta} = 3n/2 \pm 5\nu/2$ and $\dot{\theta} = 3n/2 \pm 6\nu/2$. These islands overlap, giving rise to chaotic motion for the spin, while the island located at $\dot{\theta} = 3n/2$ is much thinner.

3. Conclusion

The coorbital spin-orbit resonances populate the phase space between the eccentric resonances. Generalised chaotic rotation can be achieved when harmonics of co-orbital spin-orbit resonances overlap each other, which is a different mechanism than the one described by Wisdom *et al.* (1984), where the eccentricity harmonics overlap.

References

Correia, A. C. M. & Robutel, P. 2013, *AJ*, 779, 20
Wisdom, J., Peale, S. J., & Mignard, F. 1984, *Icarus*, 58, 137

Complex Planetary Systems
Proceedings IAU Symposium No. 310, 2014
Z. Knežević & A. Lemaitre, eds.

doi:10.1017/S1743921314008242

On the Lagrange libration points of the perturbed Earth-Moon System

Tatiana V. Salnikova[1] **and Sergey Ya. Stepanov**[2]

[1]Moscow State Lomonosov University, Moscow, Russia
email: `tatiana.salnikova@gmail.com`

[2]Dorodnitsyn Computing Center, RAS, Moscow, Russia
email: `stepsj@ya.ru`

Abstract. In this work we discuss the elusive Kordylewski clouds – dust matter in the neighborhood of the Lagrange libration points L_4, L_5 of the Earth-Moon system. On the base of restricted planar circular four body problem we get some proof for possibility of existence of four such clouds and some rule to predict the optimal moments of time for their observation.

Keywords. Earth, Moon, libration points, stable periodic solution, interplanetary dust clouds

---

1. Introduction

In 1961 polish astronomer K. Kordylewski took photos of the interplanetary dust-clouds in the neighborhood of triangular libration point L_5. Later there were many attempts to repeat this observation or to give some theoretical explanation for this phenomenon, but almost all of them were unsuccessful.

2. Overview

We consider the problem of four bodies – the Earth (E), the Moon (M), the Sun (S) and a test particle (P). As an absolute coordinate system we take the frame with origin S and axes pointed to stars. Barycenter O of the Earth-Moon system moves around S along circular orbit with the angular velocity ω, Earth and Moon rotate as a rigid body around O in the plane of orbit of O. Let us consider the motion of the particle P in the coordinate system $Oxyz$, rotating around the z-axis, which is orthogonal to the plane (SEM), with absolute angular velocity ω_1, x-axis directed along OM. Let t be the time, t_0 – the full moon time, $p = 2\pi(\omega_1 - \omega)(t - t_0)$ – the angle between radius-vector SO and the x-axis, $T = 2\pi/(\omega_1 - \omega)$ – the length of synodic month; m_0, $(1-\mu)m$, μm – masses of the Sun, the Earth and the Moon. And let $R = |SO|$, $l = |EM|$, $r = |PE|$. We choose m, l, ω_1^{-1} as units of mass, length and time. Then due to relations $\omega^2 R^3 = \gamma m_0$ and $\omega_1^2 l^3 = \gamma m$ we obtain $m = 1$, $l = 1$, $\omega_1 = 1$, $\gamma = 1$, $m_0 = \omega^2 R^3$. For other parameters we take the values $R = 389.18$, $\omega = 1/13.36$, $\mu = 0.0122$.

The Lagrange equations for the radius-vector of particle P $Z(p) = (x(p),\ y(p),\ z(p))^T$ possess a stable in Lyapunov sense 2π-periodic solution (the black curve at top-left figure) $Z^\star(p)$, surrounding the libration point L_4 with initial conditions

$$Z^\star(t_0) = (-0.4029548,\ 0.08224355,\ 0)^T, \quad dZ^\star/dt = (-0.05135043,\ 0.1739387,\ 0)^T.$$

Numbers from 0 to 7 show positions of P for $p = 0,\ \frac{1}{4}\pi,\ \ldots,\ \frac{7}{4}\pi$ with the difference $\frac{1}{4}\pi$. The sets of points, around them correspond to positions at these times of perturbed motion with initial perturbations $(\delta x(0) = -0.008,\ \delta z(0) = 0.04)$.

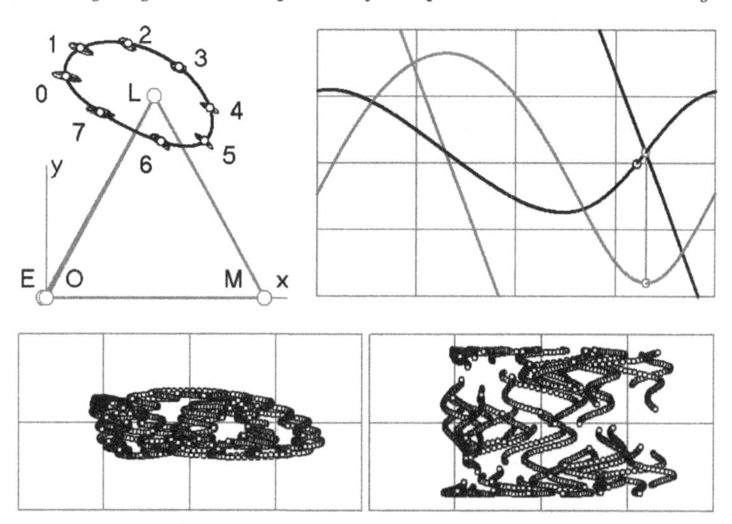

The dependence of angular distance $q = \arcsin \frac{-\sqrt{3}(x+\mu)+y}{2r}$ from L_4 to P (black curve), and the difference $(r-1)$ (gray curve) on p are shown in the rectangle $(0 < p < 2\pi;\ -0.2 < q, r < 0.2)$ at top-right figure. Straight lines show the dependence $p + q = \frac{5}{3}\pi$ (black) for full illumination and $p + q = \frac{2}{3}\pi$ (gray) for absence of illumination of the particle by the Sun. We see that three positions: passage P before L_4, shortest distance $r = 0.82$ of P from the Earth and full illumination of P – are very close to each other. The first corresponds to $t^\star = t_0 + T - \Delta t$ (Δt before full moon), $\Delta t = 5^d 20^h 24^m$. The angular velocity of the dust-cloud with respect to stars at this moment equals to $17^o 57'$ per day.

From symmetry of the equations of motion $y \to -y, p \to -p$, we get the periodic solution, capturing the point L_5. Similarly the optimal time to observe the dust-cloud in this case is the same Δt after full moon. In some approximation the equations are also invariant for $p \to p + \pi$. Hence, we get another two 2π-periodic solutions with the same graph but for $-\pi < p < \pi$ and with the role of the illumination lines interchanged. For both solutions full illumination corresponds to the furthest distance of point P from the Earth ($r = 1.164$).

The numerical experiment shows that the angular radius of the dust cloud is between 6° and 9° depending on the direction in the space. The set of positions of the mentioned above perturbed motion at the moments $t^\star + kT, k = 0, \ldots, 2000$ in rectangles $(0 < p < 2\pi,\ -0.05 < r < 0.05)$ and $(0 < p < 2\pi,\ -0.05 < \arcsin(z/r) < 0.05))$ is shown at bottom-left and bottom-right figures. The latter projection corresponds to the view from the Earth. For large perturbations of the initial conditions the motion become chaotic, and the dust density beyond the neighborhood of the periodic orbit decrease.

3. Implications

We show the possibility of existence of four dust-clouds. The optimal time for observation of two of them are at the time $\Delta t = 5^d 20^h 24^m$ before the full moon in vicinity of L_4, and at the same time after the full moon in vicinity of L_5, if it is night and the points L_4 and L_5 are up the horizon. Two other clouds are less suitable for observation from the Earth. These theoretical results agree well with the Kordylewski observations and give possibility to predict the optimal time for observations.

Reference

Kordylewski, K., 1961, *Acta Astron.*, 11, 165

Complex Planetary Systems
Proceedings IAU Symposium No. 310, 2014
Z. Knežević & A. Lemaitre, eds.

doi:10.1017/S1743921314008254

The Grand Tack model: a critical review

Sean N. Raymond[1] **and Alessandro Morbidelli**[2]

[1]Laboratoire d'Astrophysique de Bordeaux, CNRS and Université de Bordeaux, UMR 5804, F-33270 Floirac, France.
email: `rayray.sean@gmail.com`

[2]Observatoire de la Cote d'Azur, Laboratoire Lagrange, Bd. de l'Observatoire, B. P. 4229, F-06304 Nice Cedex 4, France.
email: `morby@oca.eu`

Abstract. The "Grand Tack" model proposes that the inner Solar System was sculpted by the giant planets' orbital migration in the gaseous protoplanetary disk. Jupiter first migrated inward then Jupiter and Saturn migrated back outward together. If Jupiter's turnaround or "tack" point was at $\sim$ 1.5 AU the inner disk of terrestrial building blocks would have been truncated at $\sim$ 1 AU, naturally producing the terrestrial planets' masses and spacing. During the gas giants' migration the asteroid belt is severely depleted but repopulated by distinct planetesimal reservoirs that can be associated with the present-day S and C types. The giant planets' orbits are consistent with the later evolution of the outer Solar System.

Here we confront common criticisms of the Grand Tack model. We show that some uncertainties remain regarding the Tack mechanism itself; the most critical unknown is the timing and rate of gas accretion onto Saturn and Jupiter. Current isotopic and compositional measurements of Solar System bodies – including the D/H ratios of Saturn's satellites – do not refute the model. We discuss how alternate models for the formation of the terrestrial planets each suffer from an internal inconsistency and/or place a strong and very specific requirement on the properties of the protoplanetary disk.

We conclude that the Grand Tack model remains viable and consistent with our current understanding of planet formation. Nonetheless, we encourage additional tests of the Grand Tack as well as the construction of alternate models.

Keywords. solar system: formation, minor planets, asteroids, comets: general, planetary systems: protoplanetary disks, planets and satellites: formation, methods: n-body simulations

---

1. The Grand Tack model

Until recently, models of terrestrial planet formation suffered from a debilitating problem: they could not form Mars. Rather, simulations systematically produced planets at Mars' location that were 5-10 times more massive than the real one. This issue is commonly referred to as the *small Mars* problem (Wetherill 1991; Chambers 2001; Raymond *et al.* 2009).

There have long existed solutions to the small Mars problem but none that appeared reasonable (section 3; see also recent reviews by Morbidelli *et al.* 2012 and Raymond *et al.* 2014). For instance, the terrestrial planets are easily reproduced if the initial conditions for planet formation consisted of just a narrow annulus of large planetary embryos extending from 0.7 to 1 AU (Hansen 2009; see also Morishima *et al.* 2008). Earth and Venus accreted within the annulus but Mars and Mercury were scattered beyond the edges and effectively starved. This produced a large Earth and Venus and a smaller Mercury and Mars. The problem is in justifying these initial conditions, especially the sharp outer edge of the annulus. Given that protoplanetary disks are extended objects and that additional planets exist beyond Mars' orbit, why should terrestrial embryos only exist out to 1 AU?

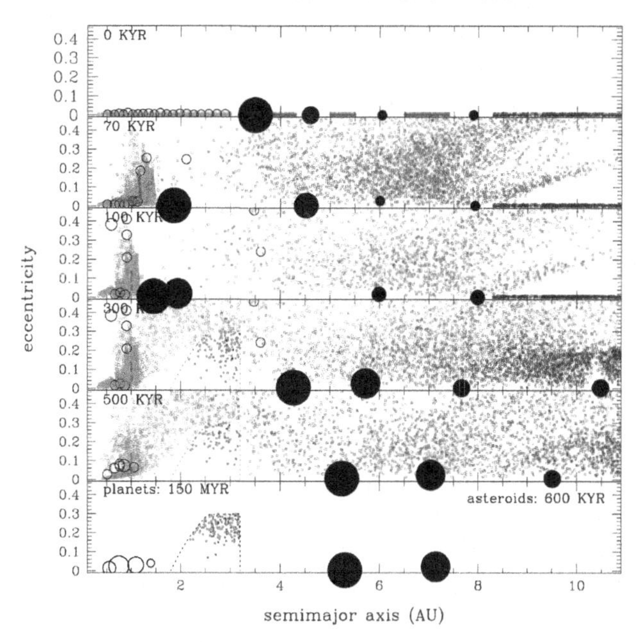

Figure 1. Snapshots in time of the evolution of the inner Solar System in the Grand Tack model. The large black dots represent the giant planets, the open circles represent terrestrial embryos, and the small symbols are planetesimals. Planetesimals originating interior to Jupiter's orbit are light grey; planetesimals initially between the giant planets are darker grey; and those beyond Neptune are even darker grey. The dashed curves represent the approximate boundaries of the present-day asteroid belt. Adapted from Walsh *et al.* (2011; this version from Raymond *et al.* 2014).

The Grand Tack model (Walsh *et al.* 2011)† proposes that the giant planets are responsible for Mars' small mass. The model proposes the following train of reasoning, illustrated in Figure 1. Jupiter was the first gas giant to form. It accreted gas onto a $\sim 10\,\mathrm{M}_\oplus$ core (e.g., Pollack *et al.* 1996), carved an annular gap in the disk and migrated inward on disk's local viscous timescale (e.g., Lin & Papaloizou 1986). Saturn grew concurrently with Jupiter but farther out and somewhat slower. Saturn accreted gas and also started to migrate inward. Saturn's migration was faster than Jupiter's (Masset & Papaloizou 2003). Saturn caught up to Jupiter and became trapped in its exterior 2:3 mean motion resonance (Masset & Snellgrove 2001; Morbidelli & Crida 2007; Pierens & Nelson 2008). This shifted the balance of disk torques acting on the planets' orbits, and Jupiter and Saturn then migrated *outward* together (Masset & Snellgrove 2001; Morbidelli & Crida 2007; Pierens & Raymond 2011; D'Angelo & Marzari 2012). Outward migration slowed and eventually stopped as the disk dissipated, stranding the gas giants on still-resonant orbits close to their current ones. This configuration is consistent with a much later instability envisioned by the Nice model (Morbidelli *et al.* 2007; Levison *et al.* 2011; Nesvorny & Morbidelli 2012).

The inner Solar System was sculpted by the giant planets' migration. As Jupiter migrated inward it shepherded most of the material interior to its orbit inward (see, e.g., Mandell *et al.* 2007), compressing the disk of embryos. Jupiter also truncated the inner disk at its closest approach to the Sun. To truncate the disk of embryos at 1 AU requires Jupiter to be located at $\sim$ 1.5 AU. This constrains the "tack point", where Jupiter's

† For a layman-level introduction to the Grand Tack, see : http://planetplanet.net/ - 2013/08/02/the-grand-tack/.

migration changed direction, which otherwise cannot be estimated *a priori*. As Jupiter and Saturn migrated outward their influence on the inner disk waned. The giant planets encountered planetesimals from a number of source regions: inner-disk planetesimals scattered outward during the giant planets' inward migration; planetesimals that originated in between the giant planets; and planetesimals that accreted beyond the giant planets, beyond roughly 10–12 AU. Jupiter and Saturn ejected the majority of these planetesimals but some survived on stable orbits in the asteroid belt.

The Grand Tack reproduces a number of aspects of the inner Solar System. The disk of embryos that was truncated at 1 AU forms Mars analogs with the correct mass (Walsh *et al.* 2011; O'Brien *et al.* 2014; Jacobson & Morbidelli 2014). The asteroid belt is strongly depleted during the giant planets' migration, consistent with the low density of the present-day belt (which contains $\lesssim 10^{-3}\,M_\oplus$). Grand Tack simulations show that the planetesimals implanted during the giant planets' migration are consistent with the observed structure of the present-day belt (Gradie & Tedesco 1982; Demeo & Carry 2014). The inner belt is dominated by particles that originated interior to Jupiter's orbit – assumed to represent S-types – and the outer belt is dominated by particles that originated between and beyond the giant planets (Walsh *et al.* 2012). Finally, the Grand Tack can also explain the origin of Earth's water. Although Earth forms mainly from the inner disk of embryos that were likely very dry, the inner regions are polluted by water-rich planetesimals. Polluting planetesimals were scattered inward by the giant planets and overshot the asteroid belt into the inner disk. Polluting particles outnumber those trapped in the asteroid belt by roughly an order of magnitude and naturally provide Earth with its current water budget (Walsh *et al.* 2011; O'Brien *et al.* 2014). These particles come from the same parent population as those implanted into the outer asteroid belt and should therefore have C-type compositions and be consistent with the chemical signature of Earth's water (Morbidelli *et al.* 2000; Marty & Yokochi 2006).

A number of questions remain for the Grand Tack model. Is Jupiter's inward-then-outward migration plausible? Is the large-scale implantation of the asteroid belt consistent with the properties known for Solar System objects? Do other, simpler models exist to explain the origin of the inner Solar System? Here we address these questions. We start by confronting specific issues with the model (section 2). We then explore the validity of alternate models (section 3). We conclude in section 4.

2. Criticisms of the Grand Tack

2.1. *Dynamics of the Tack mechanism*

The Grand Tack is built on the outward migration of Jupiter and Saturn. This is the opposite of standard type II migration, which is generally directed inward (Lin & Papaloizou 1986; Ward 1997; Kley & Nelson 2012). For the gas giants to migrate outward they must satisfy two criteria (Masset & Snellgrove 2001; Morbidelli & Crida 2007). First, the planets must orbit close enough to each other that the two planets' annular gaps in the gaseous disk overlap. The most common orbital configuration is the mutual 3:2 resonance (but other configurations – such as the 2:1 resonance – are possible; Pierens *et al.* 2014). Second, the Jupiter-to-Saturn mass ratio must be between roughly 2 and 4.

The timescale for inward migration is roughly proportional to the disk's local surface density. The timescale for accretion is also roughly proportional to the surface density. Naively, migrating a given distance should therefore increase a planet's mass by a given amount. If Jupiter and Saturn's cores formed in the same region of the Solar System, how then could Saturn have caught up with Jupiter without growing to a Jupiter-mass?

One possible solution is that the properties of planet migration change as the protoplanetary disk evolves. Type-I migration – the regime relevant for low-mass planets and giant planets' cores – is sensitive to the disk's thermodynamic properties (Masset & Casoli 2010; Paardekooper *et al.* 2011; Lega *et al.* 2014). In time, gaseous protoplanetary disks viscously evolve and dissipate (see reviews by Armitage 2011, Alexander *et al.* 2014). Bitsch *et al.* (2014) used migration maps for planets embedded in disks with a range of properties to show that rapid inward migration is more strongly inhibited in younger disks than in older ones. Jupiter's inward migration was slowed until it became massive enough to open a gap in the disk and provoke a transition to the type-II migration regime. Saturn's core, however, could migrate rapidly inward at a much lower mass, i.e. much earlier in its formation history. This may well have allowed Saturn to catch up with Jupiter and produce the configuration needed for outward migration. In simple terms, the above argument fails because migration does not scale linearly with the disk's mass.

Jupiter and Saturn may thus achieve the required configuration for outward migration, but can said configuration be maintained throughout migration? If one planet accretes much faster than the other then their mass ratio may be driven outside of the range needed for outward migration. Unfortunately, our understanding of gas accretion onto giant planet cores remains incomplete. Giant planets accrete gas first by the slow collapse of quasi-hydrostatic envelopes (Ikoma *et al.* 2000, Machida *et al.* 2010) and later by viscous transport through a circumplanetary disk (Ayliffe & Bate 2009; Uribe *et al.* 2013). Some models suggest that circumplanetary disks may have very low viscosities and thus act as bottlenecks for giant planet growth (Fujii *et al.* 2011; Rivier *et al.* 2012). Yet there remain additional sources of accretion, for example from polar inflows (Morbidelli *et al.* 2014; Szulagi *et al.* 2014). Hydrodynamical simulations of planet migration do not have the requisite resolution to realistically include gas accretion, yet these two are intimately coupled in the Grand Tack model. This is a key uncertainty for the Grand Tack: it is unclear whether long-term outward migration of Jupiter and Saturn is possible given the stringent mass ratio requirement. Yet it is a fact that even today Jupiter and Saturn have the appropriate mass ratio for outward migration.

2.2. *Compositions and isotopic ratios of small bodies*

The Grand Tack model proposes that the present-day asteroid belt was implanted from planetesimals that accreted across the Solar System (Walsh *et al.* 2011, 2012). After being scattered outward-then-inward, the S-types in the inner belt remain close to their original orbital radii. The C-types, however, were scattered inward and trapped in the main belt from between and beyond the giant planets (see Fig. 1). C-types are thought to be represented by carbonaceous chondrite meteorites. Although there is a spread in values, the D/H ratios of carbonaceous chondrites are a good match to Earth's water (Marty & Yokochi 2006; Alexander *et al.* 2012). The D/H ratios of bodies originating in the outer Solar System are more uncertain. The D/H ratios of nearly-isotropic comets thought to originate in the Oort cloud are roughly twice as high as Earth's (e.g., Bockelee-Morvan *et al.* 2012). Classically, the Oort cloud comets are expected to have formed from the giant planet region, but recently Brasser and Morbidelli (2013) argued in favor of a trans-Neptunian origin. The D/H ratio of Saturn's moon Enceladus is also roughly twice Earth's (Waite *et al.* 2009).

Alexander *et al.* (2012) used these data to argue against large-scale implantation of C-types from the outer Solar System. They proposed that Enceladus' and Oort cloud comets' elevated D/H ratios are characteristic of planetesimals formed in the giant planet region. This would mean that planetesimals formed beyond Saturn could not be precursors of C-type asteroids. However, the D/H ratio of Saturn's largest moon, Titan, has been measured to be Earth-like (Coustenis *et al.* 2008; Abbas *et al.* 2010; Nixon *et al.* 2012). Given its much larger mass, Titan's D/H – assuming it represents the bulk source of water – is much more likely than Enceladus' to be representative of the D/H of locally-grown planetesimals. In addition, the disparity in D/H between the two satellites calls into question the very notion of using moons' compositions to constrain dynamical models. Of course, there remain several uncertainties. While Enceladus' D/H ratio was measured in water, Titan's was measured in methane and acetylene. The D/H ratio of Titan's water is not certain. Strong fractionation may produce different D/H ratios for different species. In the case of comet Hale-Bopp, the D/H ratio measured in HCN was 7 times larger than that measured in water (Meier *et al.* 1998). What counts is the D/H ratio of the dominant reservoir of H. This remains unknown for Saturn's moons.

Alexander *et al.* also argued that a correlation between D/H and C/H in meteorites shows isotopic exchange between the pristine ice and organic matter within the parent bodies of carbonaceous chondrites. This would suggest that the original water reservoir had an even lower D/H than Earth, Titan or any comet, again making carbonaceous asteroids distinct from bodies formed in the giant planet region. However, such a reservoir of pristine ice has never been observed; the fact that Earth's water and other volatiles are in chondritic proportion means that carbonaceous chondrites – wherever they formed – reach their current bulk D/H ratios very quickly, before delivering volatiles to Earth. The same could have happened to comets and satellites.

Another argument for a distinction between carbonaceous asteroids and comets is that even the comets with a chondritic D/H ratio (e.g. Hartley 2; Meech *et al.*, 2011) have a non-chondritic $^{15}N/^{14}N$ ratio. Titan has a cometary $^{15}N/^{14}N$ as well (Mandt *et al.*, 2014). Here, again, a few caveats are in order. First, it is difficult to relate a satellite composition, born from a circum-planetary disk with its own thermal and chemical evolution, to the composition of bodies born at the same solar distance but on heliocentric orbits. Second, it is unclear whether any comets for which isotope ratios have been measured originate from the giant planet region, as opposed to the trans-Neptunian region (see Brasser & Morbidelli 2013).

A similarity between carbonaceous chondrites and comets has been proposed from the analysis of micro-meteorites. The isotopic ratios of most micro-meteorites – $\sim 100\mu m$ particles collected in Antarctic ice – are chondritic (with the exception of the ultra-carbonaceous particles, which constitute a small minority of micrometeorites; Duprat *et al.* 2010). However, dynamical models show that most of the dust accreted by Earth should be cometary, even taking into account the entry velocity bias (Nesvorny *et al.*, 2010; Rowan-Robinson & May 2013). For the case of the CI meteorite Orgueil, Gounelle *et al.* (2006) used an orbital reconstruction to argue for a cometary origin. These factors suggest that the rocky components of comets and carbonaceous asteroids are very similar – perhaps indistinguishable – in their bulk- and isotopic compositions.

It has been argued that if the parent bodies of carbonaceous chondrites accreted among the giant planets they should be $\sim 50\%$ water rather than the $\sim 10\%$ inferred from meteorites (Krot 2014). However, a body's original water content cannot easily be estimated from its aqueous alteration. The carbonaceous parent bodies may very well have been more water-rich than the alteration seems to imply. In addition, large main belt comets Themis (Campins *et al.* 2010; Rivkin & Emery 2010) and Ceres (Kuppers *et al.* 2014) appear to contain far more water than carbonaceous chondrites. Meteorites may simple represent rocky fragments of bodies that were far wetter/icier.

3. Alternate models

It is often assumed that giant and terrestrial planet formation can be considered separate phases. Most simulations of terrestrial accretion start from fully-formed giant planets that perturb the building blocks of terrestrial planets. Following Raymond *et al.* (2014), we refer to these as *classical* models. Of course, there are nonetheless free parameters, in particular the distribution of planetesimals and planetary embryos and the giant planets' orbits.

3.1. *Effect of the giant planets' orbits*

In order to fit in a self-consistent model of Solar System formation, the giant planets' orbits must be chosen carefully. For instance, it is inconsistent for Jupiter and Saturn to be on their current orbits during terrestrial accretion. Scattering of embryos by giant planets during accretion systematically decreases the giant planets' orbital eccentricities (e.g., Raymond 2006). If they were at their current orbital radii, Jupiter and Saturn must therefore have had higher eccentricities at early times.

The simplest version of the classical model assumes that Jupiter and Saturn's early orbits can be represented by the initial conditions for the instability presumed to have caused the late heavy bombardment (the Nice model; Gomes *et al.* 2005). A typical configuration places Jupiter and Saturn in 3:2 resonance with Jupiter at ~ 5.4 AU and low eccentricities for both planets (Morbidelli *et al.* 2007). Simulations of terrestrial accretion with Jupiter and Saturn on these orbits completely fail to reproduce the terrestrial planets (Raymond *et al.* 2006, 2009; Morishima *et al.* 2010; see also Fischer & Ciesla 2014). Mars analogs are far too large (hence the small Mars problem) and Mars-sized embryos are systematically stranded in the asteroid belt.

Some classical model simulations can reproduce Mars' small size. To date, the most successful is the *EEJS* – for 'Extra Eccentric Jupiter and Saturn' – model of Raymond *et al.* (2009) and Morishima *et al.* (2010). In this model the gas giants' primordial eccentricities were somewhat larger than their current values: 0.07-0.1 rather than ~ 0.05. The ν_6 secular resonance at 2.1 AU – which marks the inner edge of the main asteroid belt – is much stronger with this orbital configuration of the giant planets. Particles that enter the ν_6 quickly have their eccentricities pumped to high values and are lost from the system (Gladman *et al.* 1997). Given the orbital 'jostling' caused by interactions between embryos, the ν_6 acts to drain material out of Mars' feeding zone and restrict its mass.

A similar model was proposed by Nagasawa *et al.* (2005) and Thommes *et al.* (2008). In their model the ν_5 secular resonance swept inward through the asteroid belt as the protoplanetary disk dissipated, asymptotically reaching its current location at ~ 0.7 AU (Ward 1981). Strong dynamical excitation from the sweeping secular resonance efficiently cleared out the Mars region without depleting the growing Earth or Venus.

Despite the apparent success of the EEJS and secular resonance sweeping models they are hard to put in context. The sweeping secular resonance model (Thommes *et al.* 2008) requires the presence of a gas disk, yet it is not consistent with accepted theory for planet-disk interactions as it requires Jupiter and Saturn to maintain their current orbits in the presence of the disk. The disk should realistically damp out any residual eccentricity and drive the planets into a resonant configuration (e.g., Morbidelli *et al.* 2007; Kley & Nelson 2012). The EEJS model suffers from a similar problem, as no model has ever been proposed to explain such large eccentricities of the giant planets either late in the gaseous disk phase or immediately after. One might imagine that a Nice model-like instability could have occurred immediately after the dispersal of the disk; however, such an instability can only reasonably excite Jupiter and Saturn's eccentricities to their current values, not to values twice as large. In addition, there would necessarily be a prolonged phase of planetesimal scattering associated with such an instability, which would act to spread the orbits of Jupiter and Saturn and effectively change the position of the ν_6 resonance. Finally, if the EEJS model represents the correct initial orbital configuration, then no later orbital migration of the giant planets is allowed. Of course, late migration is needed to explain the orbital distribution of the Kuiper belt (Malhotra 1995; Levison *et al.* 2008; Dawson & Murray-Clay 2012) and the late heavy bombardment (Gomes *et al.* 2005; Morbidelli *et al.* 2007; Levison *et al.* 2011).

3.2. *Effect of the disk properties*

A depletion of mass in the Mars region is clearly required to solve the small Mars problem. Perhaps small bodies simply migrated inward and cleared the Mars region (Kobayashi & Dauphas 2013). Inward migration has been invoked to explain systems of super-Earths observed on close-in orbits around other stars (Terquem & Papaloizou 2007; Cossou *et al.* 2014). While no super-Earths exist in the Solar System, could a similar mechanism have been at play? Indeed, large-scale migration could explain the asteroid belt's strong mass depletion. However, if embryos migrated inward *en masse* from the asteroid belt, the eccentricities and inclinations of planetesimals that were left behind should be very low due to aerodynamic drag from the gaseous disk. This conflicts with the belt's broad eccentricity and inclination distributions. Apart from the Grand Tack, the only known mechanism capable of producing the observed distributions is scattering by embryos within the belt *after* the dissipation of the disk (Petit *et al.* 2001, Chambers & Wetherill 2001; O'Brien *et al.* 2007). Of course, invoking this mechanism invalidates our assumption that embryos migrated inward away from the belt.

Jin *et al.* (2008) proposed that the viscosity structure of the protoplanetary disk could produce the required mass depletion. In their model the inner and outer parts of the disk are MRI-active and therefore high-viscosity as they are ionized by the irradiation from the central star and external OB stars. However, the middle region has a low viscosity and is essentially a dead zone. A deficit in the disk's surface density is created at the inner edge of the dead zone, with a viscous inner disk and an inviscid outer disk. For certain disk parameters, Jin *et al.* found that this deficit can be located close to Mars' orbit. Izidoro *et al.* (2014) tested the effect of such a deficit on the accretion of the terrestrial planets. They found that in some cases a small Mars could indeed form, and water could also be delivered to Earth from more distant C-type material (as in the classical model; Morbidelli *et al.* 2000; Raymond *et al.* 2007). Their successful simulations reduced the surface density in the Mars region by a factor of 4 and placed the giant planets at their current orbital radii and with their current orbital eccentricities. As discussed above, it is not strictly self-consistent for the giant planets to be on their current orbits during this epoch. This model therefore places not one but two stringent requirements on Solar System formation. First, the disk must have a carefully-specified viscosity structure. And second, no late migration or eccentricity damping of the giant planets is allowed. As discussed above, this second requirement flies in the face of current thinking about the evolution of the outer Solar System.

4. Conclusions

This short review served to address common criticisms of the Grand Tack model. Given our current understanding the Grand Tack is self-consistent and provides a reasonable solution to the small Mars problem. There is one clear loose end related to the coupling between gas accretion onto the giant planets and outward migration (section 2.1). This is important because a Jupiter-to-Saturn mass ratio of 2-4 is required for outward migration to occur. We showed that the compositions and isotopic ratios of known Solar System bodies cannot currently be used to refute the Grand Tack model (section 2.2).

We found that each competing model that form a small Mars has strong limitations (section 3). The EEJS model (section 3.1) is inconsistent with models for the evolution of the outer Solar System. The sweeping secular resonance model (section 3.1) is not consistent with the well-developed theory of planet-disk interactions. Finally, the model proposing a mass deficit in the disk in the Mars region (section 3.2) both requires a fine-tuned disk structure and does not allow late evolution of the outer Solar System.

Nonetheless, we hope and fully expect that new models will provide alternate pathways to solving the small Mars problem.

Acknowledgments. This work was funded by the Agence Nationale pour la Recherche via grant ANR-13-BS05-0003-002 (project *MOJO*).

References

Abbas, M. M., *et al.*, 2010, *ApJ*, 708, 342
Alexander, C. M. O. '., Bowden, R., Fogel, M. L., Howard, K. T., Herd, C. D. K., & Nittler, L. R., 2012, *Science*, 337, 721
Alexander, R., Pascucci, I., Andrews, S., Armitage, P., & Cieza, L., 2013, arXiv, arXiv:1311.1819
Armitage, P. J., 2011, *ARA&A*, 49, 195
Ayliffe, B. A. & Bate, M. R., 2009, *MNRAS*, 397, 657
Bitsch, B., Morbidelli, A., Lega, E., & Crida, A., 2014, *A&A*, 564, A135
Bockelée-Morvan, D., *et al.*, 2012, *A&A*, 544, L15
Brasser, R. & Morbidelli, A., 2013, *Icarus*, 225, 40
Campins, H., *et al.*, 2010, *Nature*, 464, 1320
Chambers, J. E., 2001, *Icarus*, 152, 205
Chambers, J. E. & Wetherill, G. W., 2001, *M&PS*, 36, 381
Cossou, C., Raymond, S. N., Hersant, F., & Pierens, A., 2014, arXiv, arXiv:1407.6011
Coustenis, A., *et al.*, 2008, *Icarus*, 197, 539
D'Angelo, G. & Marzari, F., 2012, *ApJ*, 757, 50
Dawson, R. I. & Murray-Clay, R., 2012, *ApJ*, 750, 43
DeMeo, F. E. & Carry, B., 2014, Nature, 505, 629
Duprat, J., *et al.*, 2010, *Science*, 328, 742
Fujii, Y. I., Okuzumi, S., & Inutsuka, S.-i., 2011, *ApJ*, 743, 53
Fischer R. A. & Ciesla F. J., 2014, *E & PSL*, 392, 28
Gladman, B. J., *et al.*, 1997, *Science*, 277, 197
Gomes, R., Levison, H. F., Tsiganis, K., & Morbidelli, A., 2005, *Nature*, 435, 466
Gounelle, M., Spurný P., & Bland, P. A., 2006, *M&PS*, 41, 135
Gradie, J. & Tedesco, E., 1982, *Science*, 216, 1405
Hansen, B. M. S., 2009, *ApJ*, 703, 1131
Hartogh, P., *et al.*, 2011, *Nature*, 478, 218
Ikoma, M., Nakazawa, K., & Emori, H., 2000, *ApJ*, 537, 1013
Izidoro, A., Haghighipour, N., Winter, O. C., & Tsuchida, M., 2014, *ApJ*, 782, 31
Jacobson, S. A. & Morbidelli, A., 2014, arXiv, arXiv:1406.2697
Jin, L., Arnett, W. D., Sui, N., & Wang, X., 2008, *ApJ*, 674, L105
Küppers M., *et al.*, 2014, *Nature*, 505, 525
Kley, W. & Nelson, R. P., 2012, *ARA&A*, 50, 211
Kobayashi, H. & Dauphas, N., 2013, *Icarus*, 225, 122
Lega, E., Crida, A., Bitsch, B., & Morbidelli, A., 2014, *MNRAS*, 440, 683
Levison, H. F., Morbidelli, A., Tsiganis, K., Nesvorný D., & Gomes, R., 2011, *AJ*, 142, 152

Levison, H. F., Morbidelli, A., Van Laerhoven, C., Gomes, R., & Tsiganis, K., 2008, *Icarus*, 196, 258
Lin, D. N. C. & Papaloizou, J., 1986, *ApJ*, 309, 846
Lis, D. C., *et al.*, 2013, *ApJ*, 774, L3
Machida, M. N., Kokubo, E., Inutsuka, S.-I., & Matsumoto, T., 2010, *MNRAS*, 405, 1227
Malhotra, R., 1995, *AJ*, 110, 420
Mandell, A. M., Raymond, S. N., & Sigurdsson, S., 2007, *ApJ*, 660, 823
Mandt, K. E., Mousis, O., Lunine, J., & Gautier, D., 2014, *ApJ*, 788, L24
Marty, B. & Yokochi, R., 2006, *Reviews in Mineralogy and Geochemistry*, 62, 421
Masset, F. & Snellgrove, M., 2001, *MNRAS*, 320, L55
Masset, F. S. & Casoli, J., 2010, *ApJ*, 723, 1393
Masset, F. S. & Papaloizou, J. C. B., 2003, *ApJ*, 588, 494
Meech, K. J., *et al.*, 2011, *ApJ*, 734, L1
Meier, R., Owen, T. C., Matthews, H. E., Jewitt, D. C., Bockelee-Morvan, D., Biver, N., Crovisier, J., & Gautier, D., 1998, *Science*, 279, 842
Morbidelli, A., Lunine, J. I., O'Brien, D. P., Raymond, S. N., & Walsh, K. J., 2012, *AREPS*, 40, 251
Morbidelli, A., Chambers, J., Lunine, J. I., Petit, J. M., Robert, F., Valsecchi, G. B., & Cyr, K. E., 2000, *M&PS*, 35, 1309
Morbidelli, A., Szulágyi J., Crida, A., Lega, E., Bitsch, B., Tanigawa, T., & Kanagawa, K., 2014, *Icarus*, 232, 266
Morbidelli, A. & Crida, A., 2007, *Icarus*, 191, 158
Morbidelli, A., Tsiganis, K., Crida, A., Levison, H. F., & Gomes, R., 2007, *AJ*, 134, 1790
Morishima, R., Schmidt, M. W., Stadel, J., & Moore, B., 2008, *ApJ*, 685, 1247
Morishima, R., Stadel, J., & Moore, B., 2010, *Icarus*, 207, 517
Nagasawa, M., Lin, D. N. C., & Thommes, E., 2005, *ApJ*, 635, 578
Nesvorný D. & Morbidelli, A., 2012, *AJ*, 144, 117
Nesvorný D., Jenniskens, P., Levison, H. F., Bottke, W. F., Vokrouhlický D., & Gounelle, M., 2010, *ApJ*, 713, 816
Nixon, C. A., *et al.*, 2012, *ApJ*, 749, 159
O'Brien, D. P., Morbidelli, A., & Bottke, W. F., 2007, *Icarus*, 191, 434
O'Brien, D. P., Walsh, K. J., Morbidelli, A., Raymond, S. N., & Mandell, A. M., 2014, *Icarus*, 239, 74
Paardekooper, S.-J., Baruteau, C., & Kley, W., 2011, *MNRAS*, 410, 293
Petit, J.-M., Morbidelli, A., & Chambers, J., 2001, *Icarus*, 153, 338
Pierens, A. & Nelson, R. P., 2008, *A&A*, 482, 333
Pierens, A. & Raymond, S. N., 2011, *A&A*, 533, A131
Pierens, A., Raymond, S. N., Nesvorny, D. & Morbidelli, A. 2014, arXiv, arXiv:1410.0543
Pollack, J. B., Hubickyj, O., Bodenheimer, P., Lissauer, J. J., Podolak, M., & Greenzweig, Y., 1996, *Icarus*, 124, 62
Raymond, S. N., 2006, *ApJ*, 643, L131
Raymond, S. N., Kokubo, E., Morbidelli, A., Morishima, R., & Walsh, K. J., 2013, arXiv, arXiv:1312.1689
Raymond, S. N., O'Brien, D. P., Morbidelli, A., & Kaib, N. A., 2009,*Icarus*, 203, 644
Raymond, S. N., Quinn, T., & Lunine, J. I., 2007, *Astrobiology*, 7, 66
Raymond, S. N., Quinn, T., & Lunine, J. I., 2006, *Icarus*, 183, 265
Rivier, G., Crida, A., Morbidelli, A., & Brouet, Y., 2012, *A&A*, 548, A116
Rivkin, A. S. & Emery, J. P., 2010, *Nature*, 464, 1322
Rowan-Robinson, M. & May, B., 2013, *MNRAS*, 429, 2894
Szulágyi J., Morbidelli, A., Crida, A., & Masset, F., 2014, *ApJ*, 782, 65
Terquem, C. & Papaloizou, J. C. B., 2007, *ApJ*, 654, 1110
Thommes, E., Nagasawa, M., & Lin, D. N. C., 2008, *ApJ*, 676, 728
Uribe, A. L., Klahr, H., & Henning, T., 2013, *ApJ*, 769, 97
Waite, J. H., Jr., *et al.*, 2009, *Nature*, 460, 487

Walsh, K. J., Morbidelli, A., Raymond, S. N., O'Brien, D. P., & Mandell, A. M., 2012, *M&PS*, 47, 1941

Walsh, K. J., Morbidelli, A., Raymond, S. N., O'Brien, D. P., & Mandell, A. M., 2011, *Nature*, 475, 206

Ward, W. R., 1981, *Icarus*, 47, 234

Ward, W. R., 1997, *Icarus*, 126, 261

Wetherill, G. W., 1991, *LPI*, 22, 1495

Complex Planetary Systems
Proceedings IAU Symposium No. 310, 2014
Z. Knežević & A. Lemaitre, eds.

doi:10.1017/S1743921314008266

Planetesimal fragmentation and giant planet formation: the role of planet migration

O. M. Guilera[1], D. Swoboda[2], Y. Alibert[2,3], G. C. de Elía[1], P. J. Santamaría[1] and A. Brunini[1]

[1]Grupo de Ciencias Planetarias, Facultad de Ciencias Astronómicas y Geofísicas & Instituto de Astrofísica de La Plata (CONICET-UNLP), Argentina.

[2]Physics Institute and Center for Space and Habitability, University of Bern, Switzerland.

[3]Observatoire de Besançon, France.
email: oguilera@fcaglp.unlp.edu.ar

Abstract. In the standard model of core accretion, the cores of the giant planets form by the accretion of planetesimals. In this scenario, the evolution of the planetesimal population plays an important role in the formation of massive cores. Recently, we studied the role of planetesimal fragmentation in the in situ formation of a giant planet. However, the exchange of angular momentum between the planet and the gaseous disk causes the migration of the planet in the disk. In this new work, we incorporate the migration of the planet and study the role of planet migration in the formation of a massive core when the population of planetesimals evolves by planet accretion, migration, and fragmentation.

Keywords. planetary systems: formation, planetesimal fragmentation, planet migration

1. Introduction

In the standard model of core accretion, the formation of a giant planet occurs by four principal stages (Pollack *et al.* 1996, Fortier *et al.* 2009): first, a solid core is formed by the accretion of planetesimals; as this solid core grows, it binds the surrounding gas and a gaseous envelope grows in hydrostatic equilibrium; initially, the planetesimal accretion rate is higher than the gas accretion rate, so the solid core grows faster than the gaseous envelope, but when the mass of the envelope equals the mass of the core (it is often said that the mass of the core reaches a critical value) the planet triggers the gas accretion and big quantities of gas are accreted in a short period of time; finally, for some mechanism poorly understood the planet stops the accretion of gas and evolves in isolation, contracting and cooling at constant mass.

The mass of the core to start the gaseous runaway phase is found to be $\gtrsim 10\ M_\oplus$ (although, recent works showed that if the envelope's grain opacity is lower than the values of the ISM (Movshovitz *et al.* 2010) or if there is an increment of the envelope's abundance of heavy elements (Hori & Ikoma, 2011), the critical core mass could be much lower than in the classical scenario). So, the real bottleneck for giant planet formation in the core accretion model, is the growth of the critical core mass before the dissipation of the disk. In a recent work (Guilera *et al.* 2014), we found that if planetesimal fragmentation is taken into account, the formation of massive cores in a few million years is only possible starting with a population of big planetesimals (of 100 km of radius) and massive disks, and if most of the mass loss in planetesimal collisions is distributed in larger fragments. However, in this work the migration of the planet is neglected. The exchange of angular momentum between the planet and the gaseous disk forces the planet to migrate along the disk, entering in new zones of the population of planetesimals which could help in the formation of a massive core. However, type I migration (in idealized isothermal disks) predicts rapid inward migration rates, thus it is necessary to reduce the migration rates using an ad-hoc factor to reproduce observations (Ida & Lin 2008, Mordasini *et al.* 2009, Alibert *et al.* 2005 b, 2011, Miguel *et al.* 2011a, b). We incorporate type I migration in our model of giant planet formation with the aim of study the role of planet migration in the formation of a massive core when the population of planetesimals evolves by planet accretion, migration due to the nebular drag, and fragmentation due to planetesimal collisions.

2. The model

Following the work of Alibert *et al.* (2005 a), we incorporated in our model the prescription of type I migration derived by Tanaka *et al.* (2002)†, with an ad-hoc reduction factor, to calculate the velocity migration of the planet given by:

$$\frac{da_P}{dt} = -2\ f_I\ a_P \frac{\Gamma}{\mathrm{L}_P}, \tag{2.1}$$

where a_P represents the planet's semi-major axis, f_I is the reduction factor, and $\mathrm{L}_P = \mathrm{M}_P\sqrt{G\mathrm{M}_\star a_P}$ is the angular momentum of the planet. Γ is the total torque, which is given by:

$$\Gamma = (1.364 + 0.541\alpha)\left(\frac{\mathrm{M}_P\ a_P\ \Omega_P}{\mathrm{M}_\star c_{s_P}}\right)^2 \Sigma_P\ a_P^4\ \Omega_P^2, \tag{2.2}$$

where Ω_P, c_{s_P} and Σ_P are the values of the keplerian frequency, the sound speed, and the gas surface density at the position of the planet, respectively. The factor α is defined by $\alpha = d\log\Sigma/d\log R$ evaluated at $R = a_P$, with R the radial coordinate.

Eq. (2.1) is couple self-consistently to the growth of the core and the gaseous envelope. The rest of the model is the same as the one described in Guilera *et al.* (2010, 2014).

3. Results

We studied the formation of a giant planet (until the planet's core reaches the critical mass) with an initial semi-major axis of 5 au. We focused on the comparison of two cases: the in situ formation of the planet, and when the planet migrates in the disk under type I migration. We considered an initial homogeneous population of planetesimals of 100 km of radius and a disk ten times more massive than the Minimum Mass Solar Nebula (Hayashi, 1981). As in Guilera *et al.* (2014), we carried out two sets of simulations: when the population of planetesimals evolves by planet accretion and planetesimal migration (hereafter case a), and when the population of planetesimals evolves by planet accretion, planetesimal migration and planetesimal fragmentation (hereafter case b).

In Fig. 1, we plot (for the case a) the time evolution of the planet's semi-mayor axis (left panel) and the time evolution of core mass and envelope mass (right panel) for the case of in situ formation ($f_I = 0$), and for different values of the reduction factor of type I migration. We found that the planet quickly achieves the inner radius of the disk (at 0.7 au) if type I migration is not strongly reduced ($f_I = 0.01$) or not considered. Moreover, only when $f_I = 0$ and $f = 0.01$ the planet core reaches the critical mass before the dissipation of the disk (6 Myr). For these cases, when planet migration is considered in the model, the planet reaches the critical core mass in ~ 2.65 Myr, implying a reduction of $\sim 35\%$ in time respect to the case of in situ formation. This is due to an increment in the planetesimal surface density at the planet's feeding zone as a consequence of the inward migration of the planet (Fig. 2-iii, curve C).

† Paardekooper *et al.* (2010, 2011) derived new formulations for type I migration rates in non-isothermal disks. To incorporate these migration rates in our model, the evolution of the gaseous disk should be calculated in a more realistic way. This will be study in a future work.

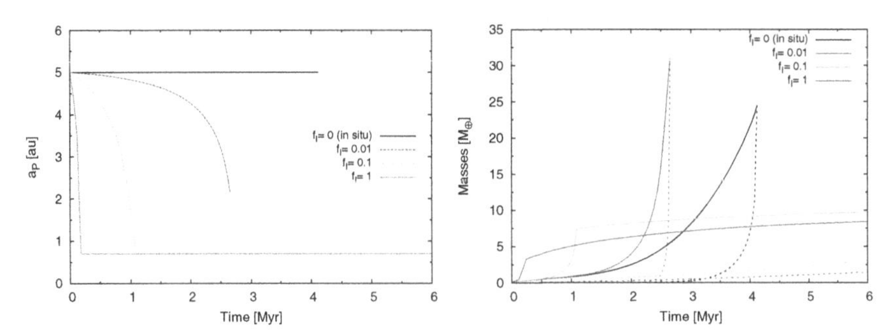

Figure 1. Left panel: time evolution of the planet's semi-mayor axis. Right panel: time evolution of the core mass (solid line) and envelope mass (dashed line). Both cases correspond to an initial embryo of 0.005 $M_\oplus$ located at 5 au, and for different values of the ad hoc reduction factor of the planet's velocity migration. Color figure only available in the electronic version.

For case b, we considered only the cases when $f_I = 0.01$ and $f_I = 0$. Fig. 2 represents the time evolution of: the planet's semi-mayor axis (i), core mass and envelope mass (ii), the mean value of the total planetesimal surface density at the planet feeding zone (iii), and the total planetesimal accretion rate (iv). When planetesimal fragmentation and planet migration are considered the planet reaches the critical core mass in ~ 2 Myr (curves D). This implies a reduction in the time of $\sim 25\%$ in comparison to the case of planet migration without planetesimal fragmentation (curves C), $\sim 45\%$ in comparison to the case of in situ formation considering planetesimal fragmentation (curves B), and $\sim 52\%$ in comparison to the case of in situ formation without planetesimal fragmentation (curves A). We note that despite the total planetesimal surface density in the planet's feeding zone for the case when planetesimal fragmentation and planet migration are considered (iii, curve D) being smaller than the case when only planet migration is considered (iii, curve C), the time at which the planet reaches the critical core mass is shorter (ii, curves D and C, respectively). This is because the accretion of small fragments (when planetesimal fragmentation is considered) causes that the total planetesimal accretion rate becomes greater (iv, curves D and C, respectively).

4. Conclusions

Our results are in concordance with those found by Alibert *et al.* (2005 b): moderate migration of the planet favors the formation of a massive core. The combination of planet migration and planetesimal fragmentation reduces the time at which the planet reaches the critical core mass more than 50% in comparison to the case of in situ formation without planetesimal fragmentation. We remark that the accretion of small fragments (products of the planetesimal fragmentation) increases the total planetesimal accretion rate of the planet even if the planetesimal surface density is smaller than the case where planetesimal fragmentation is not considered.

If type I migration is not strongly reduced the planet quickly reaches the inner radius of the disk and does not reach the critical core mass. Nelson & Papaloizou (2004) found that the inclusion of magnetic fields could reduce, or even stop, type I migration rates. Moreover, Guilet *et al.* (2013) found that outward type I migration can occur in isothermal disks if full MHD turbulence is considered. When moderate migration is considered, together with planetesimal fragmentation, the planet reaches the critical core mass in a few million years. This result could have important implications linking models that invoke the need for an inward migration of a proto Jupiter (Walsh *et al.* 2011) and models that invoke the need for a population of initial big planetesimals (Morbidelli *et al.* 2009).

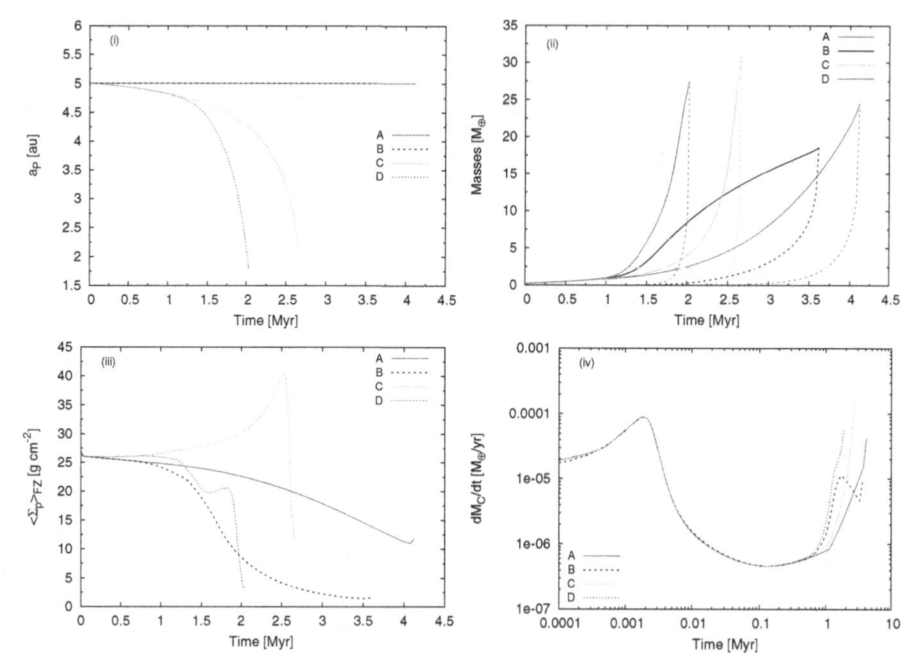

Figure 2. Time evolution of: planet's semi-major axis (i), core mass and envelope mass (ii), total planetesimal surface density at the planet's feeding zone (iii), and total planetesimal accretion rate (iv). The solid line A corresponds to the case of in situ formation when planetesimal fragmentation is not considered (case a). The large dashed line B corresponds to the case of in situ formation when planetesimal fragmentation is considered (case b). The short dashed line C corresponds to the case when planet migration is considered ($f_I = 0.01$) but planetesimal fragmentation is not considered (case a), and the dotted line D corresponds to the case when planet migration ($f_I = 0.01$) and planetesimal fragmentation are considered (case b). Color figure only available in the electronic version.

References

Alibert, Y., Mordasini, C., Benz, W., & Winisdoerffer, C. 2005 a, *A&A*, 434, 343
Alibert, Y., Mousis, O., Mordasini, C., & Benz, W. 2005 b, *ApJL*, 626, L57
Alibert, Y., Mordasini, C., & Benz, W. 2011, *A&A*, 526, A63
Fortier, A., Benvenuto, O. G., & Brunini, A. 2009, *A&A*, 500, 1249
Guilera, O. M., Brunini, A., & Benvenuto, O. G. 2010, *A&A*, 521, A50
Guilera, O. M., de Elía, G. C., Brunini, A., & Santamaría, P. J. 2014, *A&A*, 565, A96
Guilet, J., Baruteau, C., & Papaloizou, J. C. B. 2013, *MNRAS*, 430, 1764
Hayashi, C. 1981, *Progress of Theoretical Physics Supplement*, 70, 35
Hori, Y. & Ikoma, M. 2011, *MNRAS*, 416, 1419
Ida, S. & Lin, D. N. C. 2008, *ApJ*, 673, 487
Miguel, Y., Guilera, O. M., & Brunini, A. 2011 a, *MNRAS*, 412, 2113
Miguel, Y., Guilera, O. M., & Brunini, A. 2011 b, *MNRAS*, 417, 314
Morbidelli, A., Bottke, W. F., Nesvorný, D., & Levison, H. F. 2009, *Icarus*, 204, 558
Mordasini, C., Alibert, Y., & Benz, W. 2009, *A&A*, 501, 1139
Movshovitz, N., Bodenheimer, P., Podolak, M., & Lissauer, J. J. 2010, *Icarus*, 209, 616
Nelson, R. P. & Papaloizou, J. C. B. 2004, *A&A*, 350, 849
Paardekooper, S.-J., Baruteau, C., Crida, A., & Kley, W. 2010, *MNRAS*, 401, 1950
Paardekooper, S.-J., Baruteau, C., & Kley, W. 2011, *MNRAS*, 410, 293
Pollack, J. B., Hubickyj, O., Bodenheimer, P., *et al.* 1996, *Icarus*, 124, 62
Tanaka, H., Takeuchi, T., & Ward, W. R. 2002, *ApJ*, 565, 1257
Walsh, K. J., Morbidelli, A., Raymond, S. N., *et al.* 2011, *Nature*, 475, 206

Complex Planetary Systems
Proceedings IAU Symposium No. 310, 2014
Z. Knežević & A. Lemaitre, eds.

doi:10.1017/S1743921314008278

Rapid planetesimal formation in the inner protoplanetary disk

Joanna Drążkowska[1], Fredrik Windmark[1] and Satoshi Okuzumi[2]

[1]Heidelberg University, Center for Astronomy, Institute of Theoretical Astrophysics, Albert-Ueberle-Str. 2, 69120 Heidelberg, Germany
email: drazkowska@uni-heidelberg.de

[2]Tokyo Institute of Technology, Department of Earth and Planetary Sciences, Meguro-ku, Tokyo 152-8551, Japan

Abstract. Growth barriers, including the bouncing, fragmentation and radial drift problems, are still a big issue in planetesimal and thus planet formation theory. We present a new mechanism for very rapid planetesimal formation by sweep-up growth. Planetesimal formation is extremely fast in the inner protoplanetary disk where the growth rate exceeds the radial drift rate, leading to local planetesimal formation and pile-up inside of 1 AU. This scenario is very appealing particularly in the context of explaining the low mass of Mars, as well as the formation of recently discovered multi-transiting systems with tightly-packed inner planets.

Keywords. accretion, accretion disks, stars: circumstellar matter, planetary systems: protoplanetary disks, planetary systems: formation, solar system: formation, methods: numerical

1. Introduction

Planet formation takes place in disks surrounding young stars. It starts with μm-sized grains, which are already present in the interstellar medium. The journey from a μm-sized monomer to a 1000 km-sized planet covers 40 orders of magnitude in mass and comprises many intermediate steps. Tiny monomers are known to easily coagulate to mm-sized aggregates that are held together solely by material strength. On the other side of the mass range, km-sized planetesimals are held together by self-gravity. The particles between the dust aggregates and planetesimals are often called pebbles. However, a number of growth barriers have been identified that aggravates the pebble formation.

Evolution of the solid material in a protoplanetary disk is driven by its interaction with gas. Due to pressure support, the gas disk is rotating with a slightly sub-Keplerian velocity and thus the dust particles feel a constant headwind. The grains lose their angular momentum due to aerodynamic drag and drift towards the star. The velocity of the inward drift is determined by the radial pressure gradient and particle size. Particles of different sizes acquire different systematic drift speeds and thus relative velocities that drive their collisions. As the radial drift velocity in a standard disk model is as high as 30 m s^{-1}, the loss of solids and the high impact speed collisions pose major problems in growing large bodies. Panel a) of Fig. 1 shows a map of the growth barriers in a Minimum-Mass Extrasolar Nebula disk (Chiang & Laughlin 2013) in terms of the distance from the star and dust grain size. The gray region corresponds to the radial drift barrier, where the timescale of inward drift is shorter than the growth timescale. It is located in the outer part of the disk, meaning that all larger grains in this region are efficiently removed by the drift. The large grains drift until they reach inner regions (< 10 AU in this example) where the Stokes drag triggers rapid grain growth (Birnstiel *et al.* 2010, Okuzumi *et al.* 2012). However, the high velocity collisions leading to bouncing, erosion

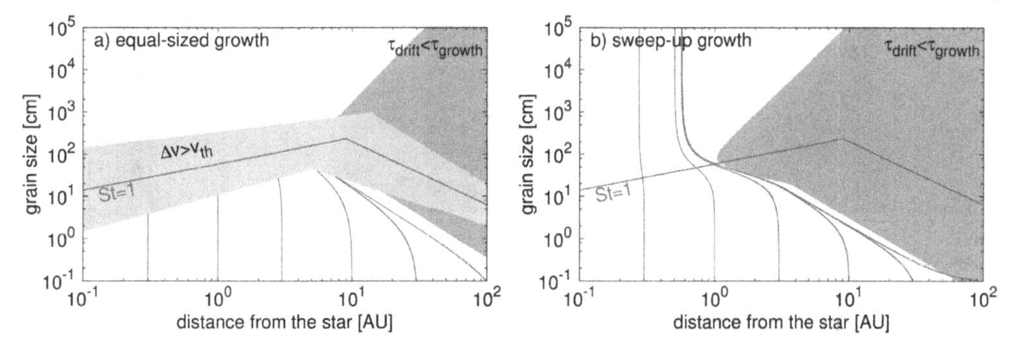

Figure 1. Maps of the growth barriers: the radial drift barrier (gray region) and the fragmentation barrier (yellow region), and evolution of test particles (gray lines): a) for the usual, equal–sized growth mode, b) for the sweep-up growth. The red line corresponds to aggregates with Stokes number of unity, which have the highest radial drift and impact velocities. The change of slope comes from the change of the drift regime form Epstein (outer disk) to Stokes (inner disk). The fragmentation barrier corresponds to impact velocities higher than $v_{\rm th} = 10$ m s^{-1}.

and fragmentation of dust aggregates do not allow formation of pebbles in the inner part of the disk. The interplay between drift and growth timescales leads to redistribution of initially homogeneous material and a solid-depleted outer disk (Birnstiel *et al.* 2012).

A number of possible solutions to the growth barriers problem have been suggested over the years. Whipple (1972) proposed that local inhomogeneities in the protoplanetary disk structure can reduce or reverse the pressure gradient (*pressure bumps*). This leads to a reduction of the inward drift and local enhancement of dust abundance as well as limiting the impact speeds, and therefore facilitates planetesimal formation (Brauer *et al.* 2008, Drążkowska *et al.* 2013). Specific kinds of pressure bumps are also required by the streaming instability scenario, where pebbles form clumps that are dense enough to gravitationally collapse to 100 km-sized planetesimals (Johansen *et al.* 2007). However, the formation mechanism and lifetimes of such pressure bumps are not yet well understood and the grain sizes required for streaming instability to trigger are hard to obtain due to bouncing and fragmentation. In this work, we focus on the idea of sweep-up growth and show that planetesimals can form in the inner disk, even without pressure bumps.

2. Sweep-up and pile-up scenario

Sweep-up growth Laboratory experiments show that even very high impact speed collisions may lead to a net growth of a target particle if the mass ratio of the colliding aggregates is high (Wurm *et al.* 2005, Teiser & Wurm 2009, Meisner *et al.* 2013). These are fragmentation with mass transfer collisions and the corresponding growth mode is called sweep-up growth (Windmark *et al.* 2012a). The growth barriers, however, hinder the formation of any larger aggregates that could benefit from this process. A solution for this problem may be the impact velocities distribution produced by turbulence present in the disk. Around 1 in 10^{30} aggregates may be "lucky" and participate only in low-velocity collisions until they grow to a size from which they can grow by the sweep-up process (Windmark *et al.* 2012b, Garaud *et al.* 2013, Drążkowska *et al.* 2014).

Pile-up While it has been previously postulated that the radial drift may lead to a pile-up of the solid material in the inner disk (Youdin & Shu 2002, Laibe *et al.* 2012), the interplay between dust drift and growth was investigated only briefly. We show that including both growth and drift leads to the conclusion that planetesimals can only form and survive in the inner part of the disk, where the growth rate exceeds the drift rate.

Panel b) of Fig. 1 shows the evolution of test particles undergoing the radial drift and

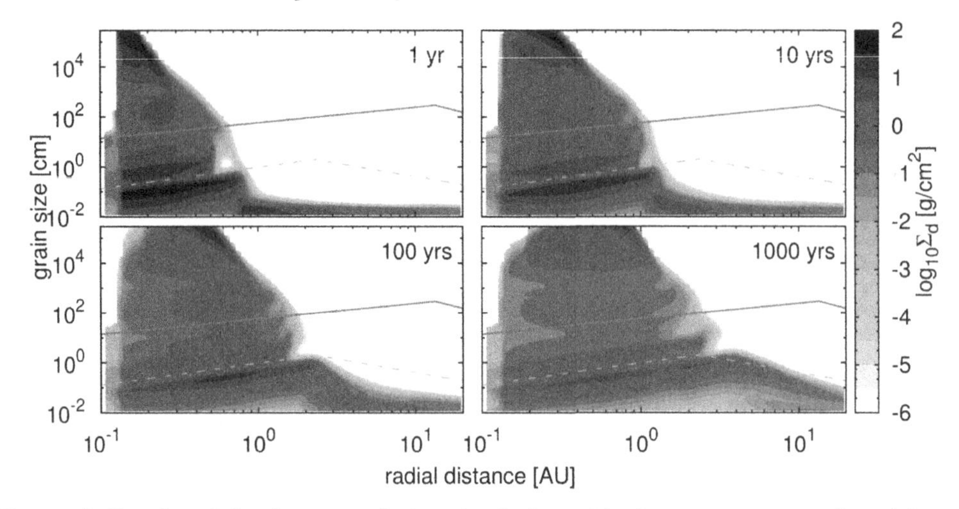

Figure 2. Results of the dust coagulation simulation with the sweep-up growth and impact velocity distribution enabled. The panels show color-coded surface density of dust aggregates. Planetesimals are formed very efficiently inside of 1 AU. The original fragmentation barrier, marked with the orange dashed line, is still impacting the outcome: only some fraction of dust aggregates is able to overcome it. The solid red line corresponds to the Stokes number of unity.

growing via the sweep-up of μm-sized monomers. The drift barrier is more pronounced than in the equal-sized growth case (panel a) because the sweep-up growth rate is lower. However, the region that is free of the drift barrier still exists inside 1 AU and this is where our growing test particles pile-up, indicating that planetesimals should be able to both form and remain there.

We check the predictions of the toy model, which comprises only a simplified prescription for drift and growth of test particles, by direct numerical simulations using a dust coagulation code based on the code presented by Birnstiel *et al.* (2010). We implement a Maxwellian impact velocity distribution that allows us to overcome the growth barriers (Windmark *et al.* 2012b). Results obtained with this model are presented in Fig. 2. A population of km-sized planetesimals is formed very quickly inside of 1 AU, but a large population of small aggregates is also present due to the original fragmentation barrier. Redistribution of the solid material by the radial drift leads simultaneously to significant pile-up in the inner disk, reaching a few times the initial dust-to-gas ratio, and to depletion of solids beyond 1 AU. The results presented in this contribution will be described in more detail in Windmark *et al.* (2014).

3. Summary and possible applications

We present a new scenario of rapid planetesimal formation in the inner part of the protoplanetary disk. We explain this scenario using a simple toy model and confirm it with direct numerical simulations. Including both the radial drift and dust growth is crucial for this scenario, as it emerges from the interplay between the two processes. This interplay leads to redistribution of solids, causing a significant pile-up in the inner disk. At the same time, larger bodies are growing very efficiently inside of 1 AU and a relatively narrow planetesimal rim is formed. Comparing to other planetesimal formation scenarios, the scenario we suggest is particularly appealing because it does not require any additional conditions beyond a standard disk model. It is particularly interesting in the context of the following issues:

Low mass of Mars Raymond *et al.* (2009) showed that attempting to reproduce the final assembly of the inner Solar System starting from an initially uniform distribution of planetesimals leads to mass for Mars significantly higher than observed. Hansen (2009) found that the mass of Mars can be reproduced if all the planetesimals are initially packed in a narrow annulus between 0.7 AU and 1 AU. Standard explanation of this setup involves the Grand Tack scenario, where Jupiter migrates inwards and the gravitational interactions truncate the planetesimal disk at roughly 1 AU. Subsequent planetesimal accretion gives a mass for Mars consistent with observations (Walsh *et al.* 2011). Most recently, Izidoro *et al.* (2014) showed that the low mass of Mars can be reproduced using a disk with an initial ad-hoc depletion of solids between 1 AU and 2 AU. Our scenario may naturally produce planetesimal distribution required to reproduce the masses of inner Solar System planets and we are going to investigate this in our future work.

Systems with tightly-packed inner planets The *Kepler* mission has found hundreds of new exoplanets. Many of them are in multiple systems, with 3 to 5 planets with orbital periods of less than 100 days and very low inclinations (Fang & Margot 2012). These multi-transiting systems with tightly-packed inner planets could form by migration of planets from the outer disk (Raymond & Cossou 2014), but this would lead to mean-motion resonances, which are only observed in some of the systems. The planetesimal formation scenario we present suggests that in-situ formation of these systems is a natural outcome of solid material evolution in a gas-rich protoplanetary disks.

References

Birnstiel, T., Dullemond, C. P., & Brauer, F. 2010, *A&A*, 513, A79
Birnstiel, T., Klahr, H., & Ercolano, B. 2012, *A&A*, 539, A148
Brauer, F., Henning, T., & Dullemond, C. P. 2008, *A&A*, 487, L1
Chiang, E. & Laughlin, G. 2013, *MNRAS*, 431, 3444
Drążkowska, J., Windmark, F., & Dullemond, C. P. 2013, *A&A*, 556, A37
Drążkowska, J., Windmark, F., & Dullemond, C. P. 2014, *A&A*, 567, A38
Fang, J. & Margot, J.-L. 2012, *ApJ*, 761, 92
Garaud, P., Meru, F., Galvagni, M., & Olczak, C. 2013, *ApJ*, 764, 146
Hansen, B. M. S.. 2009, *ApJ*, 703, 1131
Izidoro, A., Haghighipour, N., Winter, O. C., & Tsuchida, M. 2014, *ApJ*, 782, 31
Johansen, A., Oishi, J. S., Mac Low, M.-M., Klahr, H., Henning, T., & Youdin, A. 2007, *Nature*, 448, 1022
Laibe, G., Gonzalez, J.-F., & Maddison, S. T. 2012, *A&A*, 537, A61
Meisner, T., Wurm, G., Teiser, J., & Schywek, M. 2013, *A&A*, 559, A123
Okuzumi, S., Tanaka, H., Kobayashi, H., & Wada, K. 2012, *ApJ*, 752, 106
Raymond, S. N. & Cossou, C. 2014, *MNRAS*, 440, L11
Raymond, S. N., O'Brien, D. P., Morbidelli, A., & Kaib, N. A. 2009, *Icarus*, 203, 644
Teiser, J. & Wurm, G. 2009, *MNRAS*, 393, 1584
Walsh, K. J., Morbidelli, A., Raymond, S. N.,O'Brien, D. P., & Mandell, A. M. 2011, *Nature*, 475, 206
Windmark, F., Birnstiel, T., Güttler, C., Blum, J., Dullemond, C. P., & Henning, T. 2012a, *A&A*, 540, A73
Windmark, F., Birnstiel, T., Ormel, C. W., & Dullemond, C. P. 2012b, *A&A*, 544, L16
Windmark, F., & Okuzumi, S., Drążkowska, J. 2014, in prep.
Whipple, F. L. 1972, in: Elvius, A. (ed.), *From Plasma to Planet* (New York: Wiley Interscience Division), p. 211
Wurm, G., Paraskov, G., & Krauss, O. 2005, *Icarus*, 178, 253
Youdin, A. N. & Shu, F. H. 2002, *ApJ*, 580, 494

Complex Planetary Systems
Proceedings IAU Symposium No. 310, 2014
Z. Knežević & A. Lemaitre, eds.

doi:10.1017/S174392131400828X

Effects of planetary mass accretion on the planets migration: the disk structures

Ramiro Álvarez and Erick Nagel

Departamento de Astronomía, Universidad de Guanajuato,
Guanajuato, Gto 36240, Mexico
email: ramiro@astro.ugto.mx

Abstract. We develop numerical tools to detect and quantify the mass and area for all structures of a protoplanetary disk with different optical depths. We consider two planets of 0.1 M_J. They are initially at 12.5 and 20 AU embedded in a viscous disk, where planetary mass accretion and migration are allowed. We find a gap region with optically thick streams, this region is characterized by an optically thin filling factor near to 1. Additionally, we generate the Spectral Energy Distribution of each structure of the disk considering silicate dust. Our tools are useful to identify the locations of structures and sub-structures as possible sites where the material is accumulated.

Keywords. planets and satellites: formation, planetary systems: protoplanetary disk, structure formation, pre-transitional disk, transitional disk

---

1. Introduction

The transitional disks show optically thick emission of hot dust close to the star (Calvet *et al.* 2005), while pre-transitional disks show optically thin emission (Espaillat *et al.* 2007). They have (1) mass accretion rate onto the star of $\sim 10^{-10} - 10^{-8} M_{\odot}/yr$, (2) regions with holes or cavities with sizes around tens of AU. Despite the lack of dust, the observed stellar accretion rates are inconsistent to dust-to-gas standard ratio.

A feasible way to explain this kind of disks is the planetary mass accretion. This problem was previously studied by Zhu *et al.* (2011), They considered accretion rates of mass onto the planets, which first are able to form a gap and then an internal hole. Furthermore, Dodson-Robinson & Salyk (2011) suggested that the planetary system induces a gap and structures formation into the hole. These structures consist of currents (like filaments) and an optically thick residual inner disk, and they can be optically thick or optically thin. Particularly, the optically thick region can hide disk material.

In this work, we perform hydrodynamical simulations with FARGO code (Masset 2000). Additionally, we develop numerical tools to quantify the physical properties of disk structures. Finally, considering silicate dust, we make the SEDs of each disk structure using RADMC3D (Dullemond *et al.* in prep).

2. Initial conditions

For the simulations we use the initial conditions similar to Zhu *et al.* (2011): 1) the disk density is $\Sigma = 178(r/\mathrm{AU})^{-1}(0.01/\alpha)$ g cm^{-2}, with $\alpha = 0.1$, (Shakura & Sunyaev 1973), in steady state with accretion stellar rate of 10^{-8} $M_{\odot}$, the central stellar mass is 1 $M_{\odot}$; 2) the disk has 0.75-200 AU in size, resulting in 0.025 $M_{\odot}$; 3) the temperature radial profile (isothermal vertically) is $T = 221(r/\mathrm{AU})^{-1/2}$ K; 4) the disk aspect ratio is $h/r = 0.029(r/\mathrm{AU})^{0.25}$, where h is the height scale; 5) the planetary accretion is

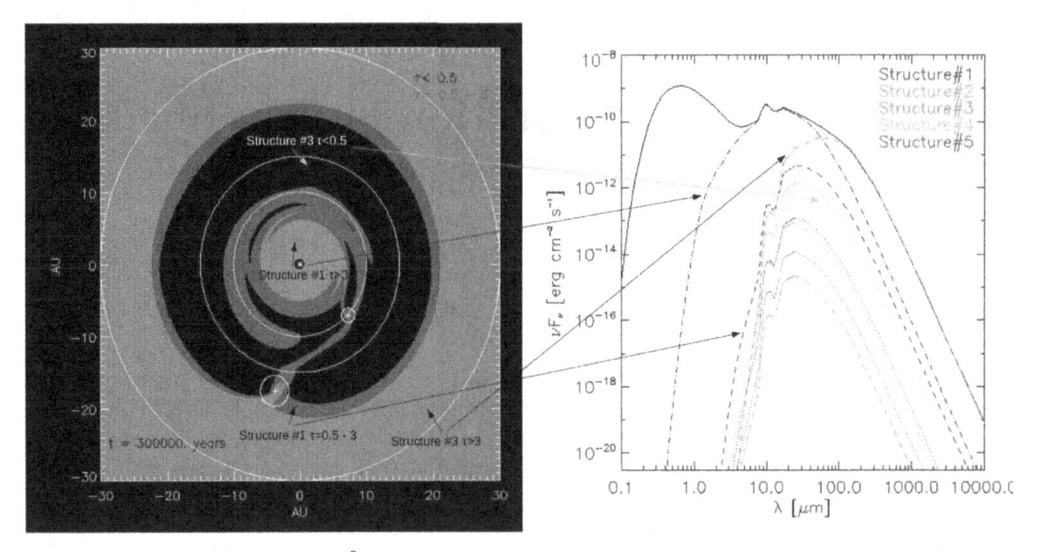

Figure 1. Snapshots at 3×10^5 years. Connection between the structures on ranges of optical depths and their SEDs. **Left:** Great circles shows radii at 10, 15 and 30 AU. The small ones shows the Hill Radii of the planets. The number of structure with their respective range on optical depth is showed by arrows. **Right:** SEDs at 140 pc. The black line shows the total emission. The colors (see electronic version) correspond to individual structure. The lines: doted, dashed and dash dot, correspond to structures optically thin, intermediate and thick, respectively.

parametrized by the α parameter inside of the Hill Radii of the planets $\alpha_\mathrm{p} = 0.4$, the case showed here only consider initially two planets with 0.1 M_J.

3. Results and conclusions

In order to characterize the structures, we consider the following physical characteristics of the dust: 1) ratio 1:100 of dust-to-gas; 2) Silicates, Organics and Troilite with $R_\mathrm{min} = 0.005$ μm, $R_\mathrm{max} = 0.25$ μm, size distribution of $R^{-3.5}$, and opacity at 10 μm of 9.6 cm^2 g^{-1}. We use iteration of the optical depth, as a method to find structures on different opacity ranges. This means that neighbour cells with the same range of optical depth become a member of the same structure.

The figure 1 (left), shows a snapshot at 3×10^5 years with structures and ranges on optically depth: $\tau < 0.5$ (thin), $0.5 < \tau < 3$ (intermediate), $3 < \tau$ (thick). Every found structure has a corresponding SED in figure 1 (Right).

The tools shown here allow us to constrain all the disk by structures or substructures, and to explore physical quantities such as the mass, area, temperature and spectrum. Specially, the region of the spiral arm between planets is well characterized.

References

Calvet, N., D'Alessio, P., Watson, D. M., Franco-Hernández, R., Furlan, E., Green, J., Sutter, P. M., Forrest, W. J., Hartmann, L., Uchida, K. I., Keller, L. D., Sargent, B., Najita, J., Herter, T. L., Barry, D. J., & Hall, P. 2005, *ApJ*, 630, 185

Dodson-Robinson, S. E. & Salyk, C. 2011, *ApJ*, 738, 131

Espaillat, C., Dalvet, N., D'Alessio, P., Hernández, J., Qi, C., Hartmann, L., Furlan, E., & Watson, D. M. 2007, *ApJ*, 670, 135

Masset, F. 2000, *A&A*, 141, 165

Zhu, Z., Nelson, R. P., Hartmann, L., Espaillat, C., & Calvet, N. 2011, *ApJ*, 729, 47

Complex Planetary Systems
Proceedings IAU Symposium No. 310, 2014
Z. Knežević & A. Lemaitre, eds.

doi:10.1017/S1743921314008291

Reversing Type I Migration in Gap Shadows

Hannah Jang-Condell

Dept. of Physics & Astronomy, University of Wyoming
1000 E University, Dept 3905
Laramie, WY 82071, USA
email: hjangcon@uwyo.edu

Abstract. Type I migration refers to the radial drift of a sub-Jupiter mass planet resulting from tidal interactions with a protoplanetary disk. It results in the rapid inward migration of small planets and planet cores through the disk. Type I migration is so rapid compared to disk dissipation time scales that explaining distant planets, such as the HR 8799 planets and Jupiter itself, is problematic because their growing cores should have been lost. Here, we present a scenario for solving the Type I migration problem. As a planet grows in mass, its Type I migration rate should increase, assuming that disk properties are not significantly altered by the forming planet. But a growing planet clears some, but not all, material from its orbital path, creating a partial gap in the disk. The trough of a partially cleared gap such as this is shadowed from stellar illumination while the far side of the gap is illuminated. Since stellar irradiation is the primary heat source of passively accreting protoplanetary disks, gap self-shadowing can significantly change the local temperature profile. This change to the local temperature gradient can significantly slow, or even reverse Type I migration.

Keywords. planetary systems: formation, Type I migration, planetary dynamics

---

Type I migration occurs because of exchange of angular momentum between a planet and a viscous gas disk. The planet loses angular momentum to the outer disk and gains angular momentum from the inner disk, with the direction of migration dependent on the balance of the torques. Ward (1997) showed that the migration is inward in typical protoplanetary disks because the pressure support shifts the position of the Lindblad resonances systematically inward. So long as the pressure gradient is negative (pressure decreasing with radius) inward migration results. Here, we show that shadowing in gaps can sufficiently reverse the temperature gradient so as to reverse the balance of torques.

Jang-Condell & Sasselov (2005) showed that shadowing near local planet perturbations in disks change the temperature gradient to slow Type I migration near planets by up to a factor of 2, but did not include the effects of gap clearing by the planet. Planets of about a Jupiter mass or larger can halt Type I migration by completely clearing their resonances of disk material. Then, the planet undergoes Type II migration, moving in lock step with the accretion of the disk onto the star. At intermediate masses ($\gtrsim 30\,M_\oplus$), the planet does not fully clear a gap, but clears a partial gap, where the density is non-zero but less than what it would be in an unperturbed disk. Shadowing and heating on this gap is more substantial than that produced by the local perturbations modeled in Jang-Condell & Sasselov (2005), and Type I migration rates should be more affected.

In Jang-Condell & Turner (2012), we modeled the tempertures in gaps produced by planets, accounting for shadowing and illumination effects and self-consistently calculating the density and temperature structure of the gap. The resulting temperatures for varying planet mass and position are shown in Figure 1. As the depth of the gap increases, so does the magnitude of the midplane temperature perturbation. At 1 AU, the temperature decreases in the shadow caused by the gap. At farther distances, the

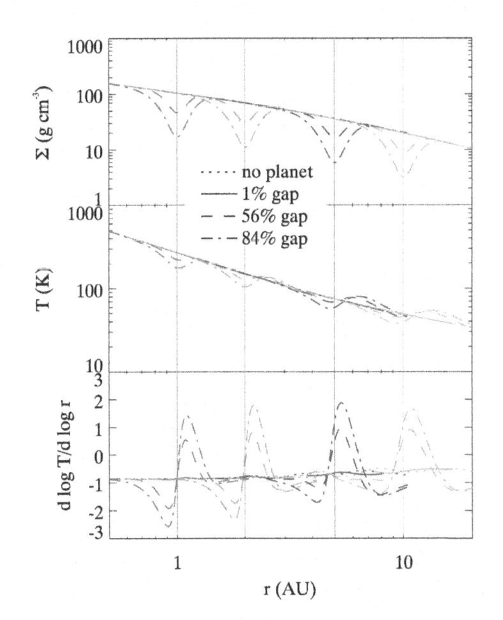

Figure 1. Surface density and midplane temperatures in the vicinity of partial gaps produced by planets at 1 (red), 2 (green), 5 (blue), and 10 (cyan) AU. The vertical lines show the position of the planet.

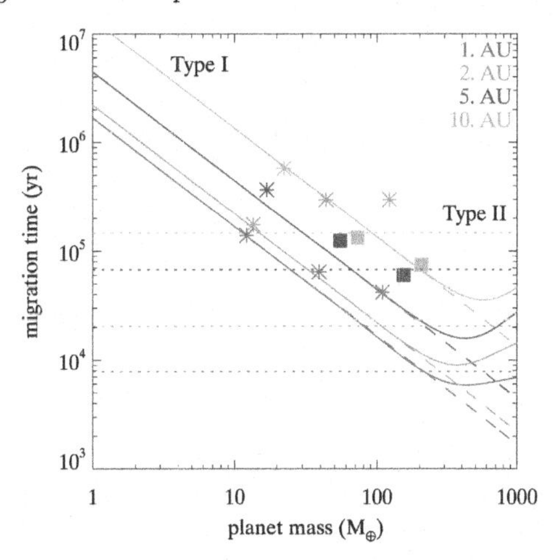

Figure 2. Type I migration time scales calculated for planets at 1 (red), 2 (green), 5 (blue), and 10 (cyan) AU. Type I and Type II migration rates in the absences of gap shadowing are shown by dashed and dotted lines, respectively, and the solid line shows the transition between the two. The points indicate migration rates calculated accounting for temperature perturbations caused by shadowing in the gaps, with asterisks indicating inward migration and squares indicating outward migration.

temperature decrease at or interior to the planet's distance is paired with a temperature increase outside the planet's orbit. The temperature increase is caused by the illumination of the exposed shoulder of the far side of the gap. The effect is smaller at smaller distances from the star because viscous heating sets a lower bound for the temperature, leading to a smaller shadow to begin with. The paired cooling and heating leads to a local inversion to the temperature profile, as shown in the plot of $d\log T/d\log r$ in the bottom panel of Figure 1. As the temperature gradient becomes increasingly positive, the more the balance of torques will lead to outward rather than inward migration.

In Figure 2, we show the final migration time scales calculated with the inclusion of cooling in the shadows of the gaps carved by planets. Asterisks indicate inward migration while squares indicate outward migration. At 1 AU (red), as the gap size increases, the migration rate increases (time scale decreases) with planet mass, but not as rapidly as compared to the predictions of Type I migration. At 2 AU (green), the migration rate actually decreases with increasing planet mass. At 5 and 10 AU, the temeprature gradient is sufficiently reversed to lead to outward migration on rapid time scales.

In conclusion, we have found a way to halt or even reverse Type I migration as planets grow in mass. The timing of the growth of planets may determine which way they migrate. Planets that grow quickly can reverse migration early on and end up very far from the star, whereas planets that grow slowly will migrate inward.

References

Jang-Condell, H. & Sasselov, D. D. 2005, *ApJ*, 619, 1123
Jang-Condell, H. & Turner, N. J. 2012, *ApJ*, 749, 153
Ward, W. R. 1997, *Icarus*, 126, 261

Complex Planetary Systems
Proceedings IAU Symposium No. 310, 2014
Z. Knežević & A. Lemaitre, eds.

doi:10.1017/S1743921314008308

Planet Formation in Close Binaries

Hannah Jang-Condell

Dept. of Physics & Astronomy, University of Wyoming
1000 E University, Dept 3905, Laramie, WY 82071, USA
email: hjangcon@uwyo.edu

Abstract. Several exoplanets have been discovered in close binaries ($a < 30$ AU) to date. The fact that planets can form in these dynamically challenging environments says that planet formation must be a robust process. Disks in these systems should be tidally truncated to within a few AU, so if they form in situ, the efficiency of planet formation must be high. While the dynamical capture of planets is also a possibility, the probability of these interactions is low, so in situ formation is the more plausible explanation. I examine the truncation of protoplanetary disks in close binary stars, studying how the disk mass is affected as it evolves from higher accretion rates to lower rates. In the gamma Cephei system, a protoplanetary disk around the primary star should be truncated to within a few AU, but enough mass still remains for planets to form. However, if the semimajor axis of the binary is too small or its eccentricity is too high, such as in HD 188753, the disk will have too little mass for planet formation to occur. I present a way to characterize the feasibility of planet formation based on binary orbital parameters such as stellar mass, companion mass, eccentricity and semi-major axis. Using this measure, we can quantify the robustness of planet formation in close binaries and better understand the overall efficiency of planet formation in general.

Keywords. planets and satellites: formation, planetary dynamics, stars: binaries

---

We calculate the feasibility of planet formation in close binaries by considering the truncation radius of the initial protoplanetary disk and determine whether the disk could have supported planet formation despite its truncation. We follow the procedure carried out for HD 188753 (Jang-Condell 2007) and γ Cep (Jang-Condell *et al.* 2008) for the binary systems tabulated in Table 1.

Since the initial properties of the disk are unconstrained, we calculate a suite of disk models with $\alpha \in \{0.001, 0.01, 0.1\}$ and $\dot{M} \in \{10^{-9}, 10^{-8}, 10^{-7}, 10^{-6}, 10^{-5}, 10^{-4}\}\, M_{\odot}\,\mathrm{yr}^{-1}$, for a total of 18 disk models for each binary pair. The truncation radius for each disk is calculated following Artymowicz & Lubow (1994). We then enumerate how many of these 18 disks have at least 10 M_J of mass (N_M), how many contain at least 10 $M_{\oplus}$ of solids (N_{CA}), and how many have $Q < 1$ (N_{DI}).

In Table 1, we show these statistics for the binaries under study in this paper. In each binary in which a substellar companion exists, $N_{\mathrm{DI}} = 1$ and $N_{\mathrm{CA}} \geqslant 4$. The one disk model in which disk instability is possible has the most extreme accretion rate, which suggests that core accretion is the preferred planet formation mechanism in close binaries.

We then explore a wide range of binary parameters (M_1, μ, a, e), and tabulate values of N_{CA} and N_{DI} for each set of values. We then fit a polynomial to those values,

$$N = \sum_{i,j,k,l} c_{ijkl} \left(\frac{a}{1\ \mathrm{AU}}\right)^i e^j \mu^k \left(\frac{M_*}{M_{\odot}}\right)^l. \tag{0.1}$$

The coefficients may be found in Jang-Condell (2014). Based on the values for N_{CA} tabulated in Table 1, we can set $N_{\mathrm{CA}} = 4$ or possibly $N_{\mathrm{CA}} = 6$ as the limiting value for which planet formation by core accretion can occur. These contours are show in Figure 1.

Table 1. Binary systems with planets examined in this paper

	HD 188753A	γ Cep	HD 41004A	HD 41004B	HD 196885A	α Cen B
M_1	1.06 $M_\odot$	1.4 $M_\odot$	0.7 $M_\odot$	0.4 $M_\odot$	1.3 $M_\odot$	0.93 $M_\odot$
M_2	1.63 $M_\odot$	0.41 $M_\odot$	0.4 $M_\odot$	0.7 $M_\odot$	0.45 $M_\odot$	1.1 $M_\odot$
μ	0.39	0.78	0.64	0.36	0.74	0.46
a	12.3 AU	20 AU	20 AU	20 AU	21 AU	23.5 AU
e	0.5	0.41	0.4	0.4	0.42	0.52
M_p	—	$\geqslant$1.85 M_J	$\geqslant$2.5 M_J	$\geqslant$18.4 M_J	$\geqslant$2.96 M_J	$\geqslant$1.13 $M_\oplus$
a_p	—	2.1 AU	6×10^{-3} AU	7.4×10^{-4} AU	2.6 AU	0.04 AU
refs	1,2	3,4	5	5,6	7	8
N_M	6	8	7	6	8	6
$N_{\rm CA}$	0	6	6	4	6	4
$N_{\rm DI}$	0	1	1	1	1	1

References: (1) Konacki (2005), (2) Eggenberger *et al.* (2007), (3) Hatzes *et al.* (2003), (4) Endl *et al.* (2011), (5) Zucker *et al.* (2004), (6) Zucker *et al.* (2003), (7) Correia *et al.* (2008), (8) Dumusque *et al.* (2012).

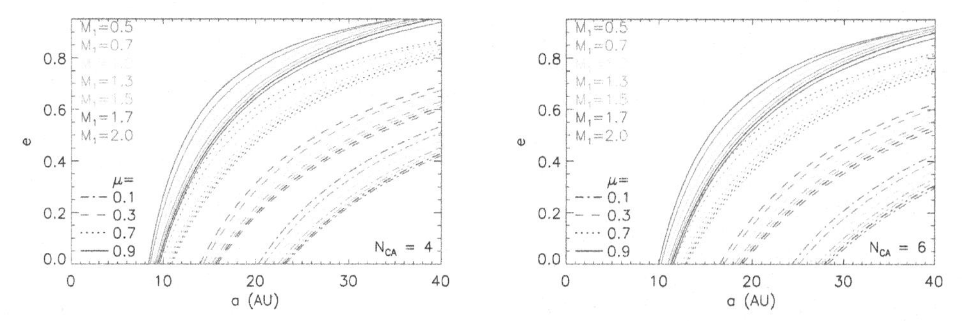

Figure 1. Binary parameters allowing giant planet core formation, using $N_{\rm CA} = 4$ (left) and $N_{\rm CA} = 6$ (right) as limits.

Then, for a given stellar mass and mass ratio ($\mu = M_1/(M_1 + M_2)$), the selected curve shows in what region of $a - e$ space planet formation by core accretion can occur.

References

Artymowicz, P. & Lubow, S. H. 1994, *ApJ*, 421, 651

Correia, A. C. M., Udry, S., Mayor, M., Eggenberger, A., Naef, D., Beuzit, J.-L., Perrier, C., Queloz, D., Sivan, J.-P., Pepe, F., Santos, N. C., & Ségransan, D. 2008, *A&A*, 479, 271

Dumusque, X., Pepe, F., Lovis, C., Ségransan, D., Sahlmann, J., Benz, W., Bouchy, F., Mayor, M., Queloz, D., Santos, N., & Udry, S. 2012, *Nature*, 491, 207

Eggenberger, A., Udry, S., Mazeh, T., Segal, Y., & Mayor, M. 2007, *A&A*, 466, 1179

Endl, M., Cochran, W. D., Hatzes, A. P., & Wittenmyer, R. A. 2011, in *American Institute of Physics Conference Series*, Vol. 1331, American Institute of Physics Conference Series, ed. S. Schuh, H. Drechsel, & U. Heber, 88–94

Hatzes, A. P., Cochran, W. D., Endl, M., McArthur, B., Paulson, D. B., Walker, G. A. H., Campbell, B., & Yang, S. 2003, *ApJ*, 599, 1383

Jang-Condell, H. 2007, *ApJ*, 654, 641

Jang-Condell, H. 2014, *ApJ*, submitted.

Jang-Condell, H., Mugrauer, M., & Schmidt, T. 2008, *ApJ* (Letters), 683, L191

Konacki, M. 2005, *Nature*, 436, 230

Zucker, S., Mazeh, T., Santos, N. C., Udry, S., & Mayor, M. 2003, *A&A*, 404, 775

Zucker, S., Mazeh, T., Santos, N. C., Udry, S., & Mayor, M. 2004, *A&A*, 426, 695

Complex Planetary Systems
Proceedings IAU Symposium No. 310, 2014
Z. Knežević & A. Lemaitre, eds.

doi:10.1017/S174392131400831X

Terrestrial planet formation in low-mass disks: dependence with initial conditions

M. P. Ronco, G. C. de Elía and O. M. Guilera

Grupo de Ciencias Planetarias, Facultad de Ciencias Astronómicas y Geofísicas & Instituto de Astrofísica de La Plata (CONICET - UNLP), Argentina.
email: mpronco@fcaglp.unlp.edu.ar

Abstract. In general, most of the studies of terrestrial-type planet formation typically use ad hoc initial conditions. In this work we improved the initial conditions described in Ronco & de Elía (2014) starting with a semi-analytical model wich simulates the evolution of the protoplanetary disk during the gas phase. The results of the semi-analytical model are then used as initial conditions for the N-body simulations. We show that the planetary systems considered are not sensitive to the particular initial distribution of embryos and planetesimals and thus, the results are globally similar to those found in the previous work.

Keywords. planetary systems; formation, terrestrial planets, methods: n-body simulations

---

1. Introduction

Many observational and theoretical works suggest that planetary systems with only rocky planets are the most common in the Universe. In particular, Miguel *et al.* (2011) indicated that a planetary system with only small rocky planets is the most common outcome obtained from a low-mass disk ($\lesssim 0.03\ M_{\odot}$) for different surface density profiles. In general, most of the studies of terrestrial-type planet formation typically use ad hoc initial conditions (Kokubo & Ida, 1998; Raymond *et al.* 2005). Here, we complement the results of *N-body high-resolution simulations* performed by Ronco & de Elía (2014) starting from a semi-analytical model developed by Brunini & Benvenuto (2008) and Guilera *et al.* (2010) wich simulates the evolution of the protoplanetary disk during the gas phase. We analyze the formation of terrestrial planets and water delivery without gas giants and compare the results with the previous work.

The semi-analytical model is used to calculate the formation of several embryos between 0.5 AU and 5 AU (as in de Elía *et al.* 2013). These embryos were separated by 10 mutual Hill radii and their initial masses correspond to the transition mass between runaway and oligarchic growth (Ida & Makino, 1993). In our previous work we adopted three different values for the exponent γ that characterize the slope of the surface density ($\gamma = 0.5$, 1 and 1.5). The planetary systems formed with $\gamma = 1.5$ were the most distinctive ones from an astrobiological point of view. Thus, in this new work we developed three N-body simulations with new initial conditions given by the semi-analytical model, only for $\gamma = 1.5$. For this profile we found the same proportion for both populations: half the mass in embryos and half the mass in planetesimals after the gas is completely dissipated in 3 Myr.

2. Results

These new planetary systems are globally similar to those found in our previous work. Each simulation formed between 1 and 2 planets in the habitable zone (HZ) (between

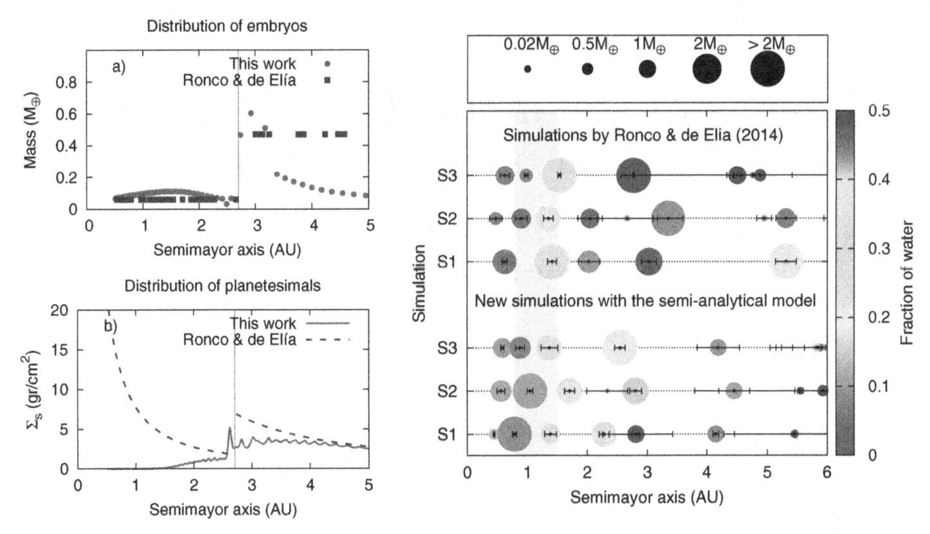

Figure 1. Left: a) Distributions of embryos used to start the N-body simulations. The squares represent the distribution of embryos used by Ronco & de Elía (2014) and the circles represent the final results obtained with the semi-analytical model. b) Surface density profiles used to distribute 1000 planetesimals to start the N-body simulations. The dashed line represents the surface density used in Ronco & de Elía (2014) and the solid line represents the final results obtained with the semi-analytical model. Right: Final configuration of the simulations obtained in Ronco & de Elía (2014) and the new ones obtained with the semi-analytical model. The color scale represent the water content and the shaded region, the HZ. The excentricity of each planet is shown over it, by its radial movement over an orbit. Color figure only available in the electronic version.

0.8 AU and 1.5 AU) after 200 Myr of evolution (Fig. 1). Their masses range between $1.18M_\oplus$ and $2M_\oplus$ and their water content between 7.5% and 24.3% by mass, which represent between 427 and 1671 Earth oceans (1 Earth ocean $= 2.8 \times 10^{-4} M_\oplus$). We also found planets with masses from $1M_\oplus$ to $2.36M_\oplus$ near the snow line (located at 2.7 AU), which can be discovered by the microlensing technique. The masses of the planets in the HZ are large enough to retain an atmosphere and to sustain plate tectonics, and as we also formed in the previous simulations, this profile formed *water worlds* that come from beyond the snow line. Thus, the planets that remain in the HZ present the characteristics to be potencially habitable.

We therefore conclude that the results are globally similar to those found by Ronco & de Elía (2014). These planetary systems do not seem to be sensitive to the particular initial distribution of embryos and planetesimals and we suggest that the strong dependence on the final results would go with the initial mass proportion used in both populations. However, these more realistic initial conditions allowed us to find more reliable results concerning the water delivery and the global dinamics of the planetary systems.

References

Brunini, A. & Benvenuto, O. G. 2008, *Icarus*, 194, 800
de Elía, G. C., Guilera, O. M., & Brunini, A. 2013, *A&A*, 557, A42
Guilera, O. M., Brunini, A., & Benvenuto, O. G. 2010, *A&A*, 521, A50
Ida, S. & Makino, J. 1993, *Icarus*, 106, 210
Kokubo, E. & Ida, S. 1998, *Icarus*, 131, 171
Miguel, Y., Guilera, O. M., & Brunini, A. 2011, *Icarus*, 417, 314
Raymond, S. N., Quinn,T. & Lunine, J. I. 2005, *ApJ*, 632, 670
Ronco, M. P. & de Elía, G. C. 2014, *A&A*, 567, A54

Complex Planetary Systems
Proceedings IAU Symposium No. 310, 2014
Z. Knežević & A. Lemaitre, eds.

doi:10.1017/S1743921314008321

Influence of the inclination damping on the formation of planetary systems

Sotiris Sotiriadis[1], Anne-Sophie Libert[1] and Kleomenis Tsiganis[2]

[1]naXys, Department of Mathematics, University of Namur, Belgium
email: sotiris.sotiriadis@unamur.be, anne-sophie.libert@unamur.be

[2]Department of Physics, Aristotle University of Thessaloniki, Greece
email: tsiganis@auth.gr

Abstract. Highly non-coplanar extrasolar systems (e.g. Upsilon Andromedae) and unexpected spin-orbit misalignment of some exoplanets have been discovered. In Libert and Tsiganis (2011), a significant increase of the mutual inclination of some multi-planet systems has been observed during the type II migration, as a result of planet-planet scattering and/or resonant interactions between the planets. Here we investigate the effect of the inclination damping due to planet-disk interactions on the previous results, for a variety of planetary systems with different initial configurations and mass ratios. Using the damping formulae for eccentricity and inclination provided by the numerical hydrodynamical simulations of Bitsch *et al.* (2013), we examine their impact on the possible multiple resonances between the planets and how the growth in eccentricity and inclination is affected.

Keywords. planetary systems: formation, planet-disc interactions, resonance capture, inclination

1. Introduction

Planet-planet scattering and resonant planet-planet interactions during migration of giant planets in the protoplanetary disk have been invoked to explain the inclined orbits of extrasolar systems (Libert and Tsiganis (2011), Thommes and Lissauer (2003)). During the migration of the planets, the disk affects their eccentricity and inclination. While eccentricity damping is generally considered in the simulations, this is not the case for inclination damping. Bitsch *et al.* (2013) performed three-dimensional numerical hydrodynamical simulations of protoplanetary disks with embedded high-mass planets on fixed orbits and provided damping formulae for eccentricity and inclination that fit the numerical data. Using these formulae, we investigate here, for a variety of planetary systems with different initial configurations and mass ratios, the effect of the inclination damping on the possible multiple resonances between the planets and how the growth in eccentricity and inclination is affected.

2. Numerical set-up

We consider here two models for the protoplanetary disk. In the first case the mass of the disk is constant and there is no migration or eccentricity and inclination damping for the inner planet. The outer planets migrate with a constant rate depending on their initial semi-major axis (Ward (1997))

$$\tau_{II}(yr) = \frac{1}{9.4}\frac{1}{\alpha}\left(\frac{a}{H}\right)^2\left(\frac{a}{AU}\right)^{3/2}. \tag{2.1}$$

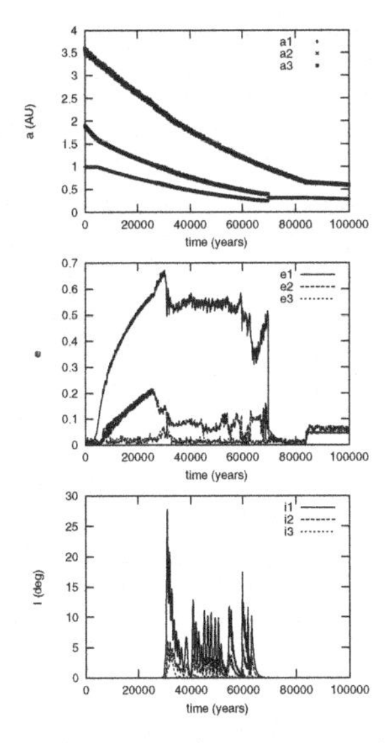

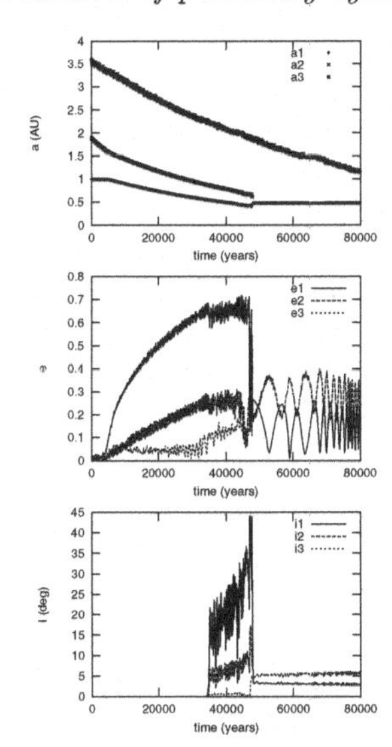

Figure 1. Left panel: A typical evolution of a three-planet system when considering inclination damping (constant mass of the disk). Right panel: Evolution of the same system without eccentricity and inclination damping, as shown in Libert and Tsiganis (2011).

In the second set of simulations, the mass of the disk is decreasing exponentially $M_{disk} = M_0 e^{-t/10^6}$ and the lifetime of the disk is $\sim 10^7$ yr. All the planets are initially affected by the protoplanetary disk and there is an inner cavity in the disk starting at 0.05 AU.

We have a total of 60 systems for each set of simulations. For both sets the initial parameters of the inner planet are $a_1 = 1$ AU and $m_1 = 1.5\ M_J$. For the outer planets, we consider the parameters: $a_2 = 1.9$ AU, $m_2 = 3, 1.5, 0.75\ M_J$, $a_3 = 2.8, 3, 1, 3.6$ AU and $m_3 = 6, 3, 1.5, 0.75, 0.375\ M_J$.

3. Results

The main observation of this preliminary work is that inclination damping has a strong impact on the evolution of the planets evolving in the gas disk, since non-coplanar systems are not a usual outcome any more. An example is given in Figure 1, where although the systems evolve in a Laplace three-body resonance, inclinations of the planets eventually tend to zero when considering the eccentricity and inclination damping of Bitsch *et al.* (2013).

In the first set of simulations, with a constant mass of the disk, the systems end-up with small mean values in eccentricity and inclination ($\bar{e} \sim 0.066, e_{max} = 0.2265, \bar{I} \sim 0.001°$), so planet-planet scattering and resonant interactions between the planets seem to be less efficient than in previous works.

However, in the second set of simulations, eccentricity and inclination excitations are observed due to the dissipation of the mass of the protoplanetary disk. Indeed, while in the Laplace resonance, mergings between the planets and ejections of one (or two) companions are usual outcomes and the final distribution of the systems is quite similar to the one of Libert and Tsiganis (2011). Final system configurations of a bigger set of simulations have to be further analyzed.

Acknowledgement

This work is part of the F.R.S.-FNRS "ExtraOrDynHa" research project.

References

Bitsch, B., Crida, A., Libert, A.-S., & Lega, E. 2013, *A&A*, 555
Libert, A.-S. & Tsiganis, K. 2011, *Celes Mech. Dyn. Astr.*, 111
Thommes, E. W. & Lissauer, J. J. 2003 *ApJ*, 597
Ward, W. R. 1997, *Icarus*, 126

Complex Planetary Systems
Proceedings IAU Symposium No. 310, 2014
Z. Knežević & A. Lemaitre, eds.

doi:10.1017/S1743921314008333

Rapid clump formation in discs of young stellar objects class O-I

Olga P. Stoyanovskaya[1,2] **and Valeriy N. Snytnikov**[1,2]

[1]Boreskov Institute of Catalysis SB RAS
Pr. Academika Lavrentieva, 5, 630090, Novosibirsk, Russia
email: stop@catalysis.ru, snyt@catalysis.ru
[2]Novosibirsk State University,
Str. Pirogova, 2, 630090, Novosibirsk, Russia

Abstract. The formation of planetesimals and large bodies in circumstellar discs can be referred to the stage of massive unstable disc emerging due to the gravitational collapse in the molecular cloud. We have simulated the dynamics of gas and boulder clumps in such systems. The features of instability development in massive disc is discussed. Unlike medium-massive discs the massive disc can provide conditions for simultaneous formation of several clumps as a result of overdensity ring fragmentation. We found regimes where 3, 2 or 1 overdensity ring appear in the disc and then fragment into clumps collecting about a half of the disc mass.

Keywords. circumstellar matter, Young Stellar Object (YSO) class O-I, gravitational instability, self-gravitating clump formation.

Circumstellar discs of young stellar objects of class O-I can reach the threshold of gravitational instability development leading to the appearance of solitary overdensity areas or clumps. Such clumps may play a critical role in planetary system formation, being embryos of gas giant planets or brown dwarfs, collecting solids for planetesimal and large body formation. Numerical modeling of gravitational instability development demonstrates that depending on disc and central body parameters the clumps can be formed in radial fragmentation of global overdensity rings or in spiral sleeves of different nature.

In low- and medium-mass discs with intense cooling, clumps are formed in spiral arms; the first clump first appears and then affects the gas flow in the disc, which may provide the possibility of clump formation at a smaller radius Helled *et al.* (2014). For a cold massive disc, the isolated high-density regions are formed by fragmentation of a dense ring emerging due to development of the global Toomre instability. Unlike the case of medium- and low-mass disc, the formation of several rings in a massive disc allows the fragmentation to start not at a maximum radius (see Fig.1). Fig.1 shows the logarithm of the gas surface density during an orbital period.

In this study we investigate the scenario where rings are initial stage of gravitational instability development. Computational experiments were carried out within a quasi-3D model of the disc described in Snytnikov & Stoyanovskaya (2013). The hybrid model of massive disc includes Eulear equations for gas dynamics with enthropy formulation of energy equation, collisionless Boltzmann equation for 1 up to 10 m in size primary bodies dynamics, and Poisson equation for self-consistent gravitational field. The numerical model is based on SPH and PIC methods combined with grid method for Poisson equation solving.

In our simulation of massive disc dynamics we found regimes where 3, 2 or 1 overdensity rings fragment into clumps. The total mass of clumps can reach up to half mass of the

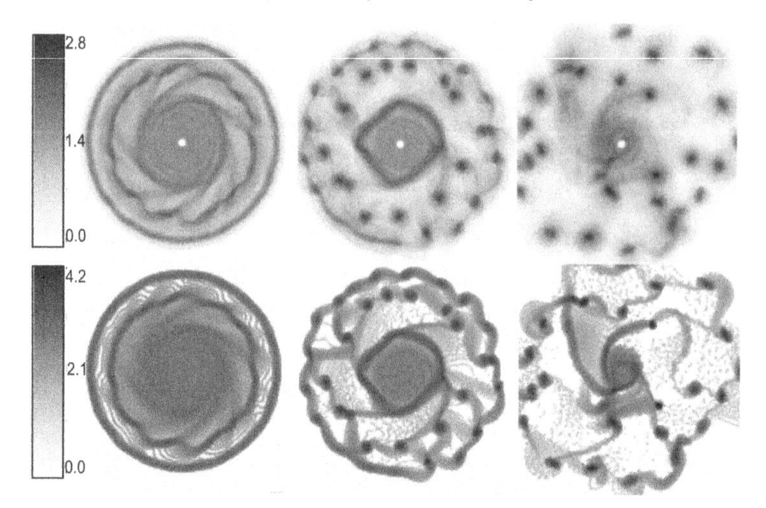

Figure 1. The surface density logarithm of the gas (bottom) and the primary solids subdisc (top) for time points T = 7.5, 10, and 12.5. The orbital time of the outer part of the disc is 15. The disc has a radius of 20 AU, the central body had mass $M_c = 0.45M_0$, and the disc was represented by a gas component with mass $M_{gas} = 0.52M_0$ and a primary solids subdisc with $M_{par} = 0.03M_0$. The effective adiabatic exponent was $\gamma = 5/3$, initial gas temperature at $R = 10$ au was 235 K, initial velocity dispersion of solids was 95 m s^{-1}. The numerical resolution of the calculation was $16 * 10^4$ SPH particles, 10^7 PIC particles, $\frac{h_r}{R} = 0.01$.

disc. The minimum found value of ratio of fragmenting ring radius to the disc radius was 0.25 for polynomial density distribution and 0.1 for exponential density distribution. Varying the central body mass we found its effect on fragmentation order of the rings: the ring of smaller radius becomes more stabilized against fragmentation by the increasing of the central mass. During the rings fragmentation simultaneous formation of several high-density areas of gas and solids at almost the same radius takes place. In a massive disc, the emerging clumps are closely surrounded by neighbour clumps, and they can move from the centre to the periphery over short time periods under the action of the gravitational fields of other regions.

We have demonstrated that the gravitational field of the isolated high-density areas could capture bodies from 1 up to 10 m in size (without regard to the frictional force between the gas and such bodies) to form epicyclic trajectories both in the initial stages of collapse and before collapse. Triggering a gravitational collapse in such objects (high-density areas) can transform them into planet embryos or into regions of intense growth of solids. A key problem here is to estimate the typical lifetime of such objects and to determine whether they can exist for a time period sufficient for launching a mechanism that moves the disc to the next stage of planetary formation.

This work was supported by RFBR project 14-01-31516, the RAS Presidium programs 'Biosphere origin and evolution' and 'Origin, structure and evolution of objects in the Universe', as well as by the SB RAS Integration Project N130.

References

Snytnikov, V. N. & Stoyanovskaya, O. P., 2013, *MNRAS*, 428, 2

Helled, R., Bodenheimer, P., Podolak, M., Boley, A., Meru, F., Nayakshin, S., Fortney, J. J., Mayer, L., Alibert, Y., & Boss, A. P. *Giant Planet Formation, Evolution, and Internal Structure*, in Protostar and Planets VI, arXiv:1311.1142v1

Complex Planetary Systems
Proceedings IAU Symposium No. 310, 2014
Z. Knežević & A. Lemaitre, eds.

doi:10.1017/S1743921314008357

Can we expect the massive discs around young stellar objects class O-I to be a birthplace of planetesimal?

Valeriy N. Snytnikov[1,2], **Olga P. Stoyanovskaya**[1,2] **and Olga A. Stadnichenko**[1,2]

[1]Boreskov Institute of Catalysis SB RAS
Pr. Academika Lavrentieva, 5, 630090, Novosibirsk, Russia
email: snyt@catalysis.ru, stop@catalysis.ru, zasypoa@catalysis.ru

[2]Novosibirsk State University,
Str. Pirogova, 2, 630090, Novosibirsk, Russia

Abstract. Possibility of large bodies formation in massive discs of young stellar objects (YSO) class O-I was investigated. On the stage of YSO O-I the whole of factors: chemical composition of gas and solids, chemical catalytic reactions, the disc self-gravitation, the increased ratio of solids to gas surface density, adiabatic gas cooling provides favorable conditions for gravitational instabilities development. We simulated 3D dynamics of gas and dust under self-consistent gravitational field and reproduced the formation and evolution of the disc around the protostar. We found that for stars of Solar mass there are regimes when the disc of variable mass is unstable for the development of fast gravitational instabilities.

Keywords. Young Stellar Object (YSO) class O-I, gravitational instability, self-gravitating clump formation.

Star and planetary system formation goes via fragmentation of a molecular cloud with gravitational collapse of a solitary clump into protostar with a circumstellar accretion disc, transformation of protostar into star, gas loss and appearance of planets and other bodies in the disc. This evolutionary sequence is distinguished as young stellar object (YSO) of class 0, I, II, III. The early phase of young stellar objects (class 0 and I) is the most difficult for observation. In such objects the protostar is surrounded by a disc of comparable mass. On this stage fast luminosity changes are recorded for some objects, which indicated that they are unstable. Gravitational instability of such objects can be responsible for planetary formation processes as well as the protostar mass increasing.

In this paper we list condition (1)-(4) for massive circumstellar disc of YSO of classes 0 and I that provides the development of gravitational instability over timescale of several orbital periods. Mass and temperature distribution for accretion disc is determined by the infall of opposing gas-dust streams and their collision. Then the gas moves along the rotation axis away from the equatorial plane. The gas expansion decreases its temperature and leaves the coagulating dust in the plane. Over the time 0.5 million year (1) the mass of the disc reaches the mass of the star (Snytnikov *et al.* (2013)), and then the disc mass decreases by the million year to 0.1 of star mass. In the circumstellar disc the the subdisc made of solid bodies form. (2) In the subdisc the ratio of solid to gas concentration is increased with respect to the molecular cloud. The composition of the solid is defined by the cosmic abundance of elements, according to which the fraction of water and organic compounds is higher than the fraction of inorganic substance (Herbst & van Dishoeck (2009)). On the stage of YSO of classes 0 and I coagulation and aggregation of such solids (when the evaporation and decomposition of organic has not led already to prevalence of inorganics) may lead to (3) the formation of metre-sized bodies. Gas drag decreases the velocity dispersion of such solids and causes their infall onto the protostar over a timescale of 0.1 Myr. Chemical reaction such as, for instance, synthesis of molecular from atomic hydrogen on the surface of solids, can affect (4) the effective adiabatic exponent.

To simulate three-dimensional dynamics of a gas cloud, we used a numerical code developed by the authors (Snytnikov & Stadnichenko (2011)). The code is based on the FLIC method and employs a uniform mesh in Cartesian coordinates. The method provides the first order approximation over space and time. The gas dynamics code solves the initial-boundary value problem for non-viscous gas in a closed region with the initial conditions coinciding with the Bonnor Ebert solution. The boundary conditions ensure the absence of shocks on the boundary at nonzero mass flow. The calculations demonstrate the development of physical gravitational instability in the formed gaseous disc.

We carry out numerical simulation of multiphase massive disc dynamics with the development of global gravitational instability. We found that local gravitational collapse which can form big planetesimal or gas giant can be an outcome of the instability development over a timescale of several orbital period (Snytnikov & Stoyanovskaya (2013)).

This work was supported by RFBR project 14-01-31516, the RAS Presidium programs 'Biosphere origin and evolution' and 'Origin, structure and evolution of objects in the Universe', the SB RAS Integration Project N130. The work was also supported by the Ministry of Education and Science of the Russian Federation.

References

Snytnikov, V. N. & Stoyanovskaya O. P., 2013, *MNRAS*, 428, 2

Snytnikov, V. N., Stoyanovskaya, O. P., & Stadnichenko, O. A., 2013, *EPJ Web of Conferences*, 46, 07004

Snytnikov, V. N. & Stadnichenko, O. A., 2011, *Astron. Rep.*, 55, 214.

Herbst, E. & van Dishoeck, E. F., 2009, *Ann. Rev. Astron. and Astrophys.*, 47, 427

Author Index

Printed in the United States
by Baker & Taylor Publisher Services